EPA-600/8-78-017
December 1978

MICROBIOLOGICAL METHODS FOR MONITORING THE ENVIRONMENT

Water and Wastes

Edited by
Robert Bordner and John Winter
Environmental Monitoring and Support Laboratory-Cincinnati
Cincinnati, Ohio 45268
and Pasquale Scarpino, University of Cincinnati
Cincinnati, Ohio 45219

Prepared in part under EPA Contract No. 68-03-0431

Project Officer

John Winter
Environmental Monitoring and Support Laboratory
Cincinnati, Ohio 45268

ENVIRONMENTAL MONITORING AND SUPPORT LABORATORY
OFFICE OF RESEARCH AND DEVELOPMENT
U.S. ENVIRONMENTAL PROTECTION AGENCY
CINCINNATI, OHIO 45268

DISCLAIMER

This report has been reviewed by the Environmental Monitoring and Support Laboratory-Cincinnati, U.S. Environmental Protection Agency and approved for publication. Mention of trade names or commercial products does not constitute endorsement or recommendation for use.

FOREWORD

Environmental measurements are required to determine the quality of ambient waters and the character of waste effluents. The Environmental Monitoring and Support Laboratory (EMSL)—Cincinnati conducts research to:

- Develop and evaluate techniques to measure the presence and concentration of physical, chemical, and radiological pollutants in water, wastewater, bottom sediments, and solid waste.
- Investigate methods for the concentration, recovery, and identification of viruses, bacteria and other microorganisms in water.
- Conduct studies to determine the responses of aquatic organisms to water quality.
- Conduct an Agency-wide quality assurance program to assure standardization and quality control of systems for monitoring water and wastewater.

This publication of EMSL-Cincinnati, contains the methods selected by consensus of EPA senior microbiologists for parameters of interest to the Agency. Federal agencies, states, municipalities, universities, private laboratories, and industry should find this manual of assistance in monitoring and controlling microbiological pollution in the environment.

Dwight G. Ballinger
Director, EMSL-Cincinnati

PREFACE

The Federal Water Pollution Control Act Amendments of 1972, the Marine Protection, Research, and Sanctuaries Act of 1972, and the Safe Drinking Water Act of 1974, require that EPA develop and select methods for environmental monitoring and research on public and private water supplies, rivers, lakes, ground waters, wastewaters and the marine environment for the purposes of setting and enforcing environmental standards and ultimately enhancing the quality of the environment. This manual of methodology supports these needs.

Under the direction of a Steering Committee formed for the development of an Agency microbiology manual, a seminar was held among representative Agency microbiologists in San Francisco, January, 1973. Assignments were made to committee members for the preparation of first draft material. The basic design, format and content of the manual were established and the first drafts presented and reviewed at the second meeting of the Committee in January, 1974 at Cincinnati.

The drafts submitted by the Steering Committee members were formatted and developed into the initial version under EPA Contract No. 68–03–0431 by Dr. Pasquale Scarpino, Professor of Environmental Engineering, Department of Civil and Environmental Engineering, University of Cincinnati, working with the two EPA editors: Robert Bordner, Chief Microbiology Section, Biological Methods Branch and John Winter, Chief, Quality Assurance Branch, both of EMSL-Cincinnati. Subsquently, these editors added technical detail and the necessary information reflecting Agency policies. Valuable source documents for This Manual were Current Practices in Water Microbiology, National Training and Operational Technology Center and Handbook for Evaluating Water Bacteriological Laboratories, Municipal Environmental Research Center, both of U.S. EPA, Cincinnati, Ohio. The refined product is presented here.

Comments or questions concerning the manual should be directed to:

Robert Bordner or John Winter
U.S. Environmental Protection Agency
EMSL-Cincinnati
Cincinnati, OH 45268

TABLE OF CONTENTS

APPENDICES

FIGURES

TABLES

THE MICROBIOLOGY METHODS STEERING COMMITTEE OF EPA

Cochairpersons: Robert Bordner and John Winter
Environmental Monitoring and Support Laboratory-Cincinnati

Members:

William Stang
Edwin Geldreich
Kathleen Shimmin
Harold Jeter (retired)
Francis Brezenski

CONTRIBUTORS BY SECTION

Sampling
William Stang
NEIC-Denver

General Laboratory Equipment/Media
Robert Bordner and John Winter
EMSL-Cincinnati

Pasquale Scarpino
UC Dept. of Environ. Engineering

Isolation and Enumeration of Bacteria
Robert Bordner and John Winter
EMSL-Cincinnati

Pasquale Scarpino
UC Dept. of Environ. Engineering

Selection of Analytical Methods
Robert Bordner and John Winter
EMSL-Cincinnati

Standard Plate Count
Raymond Taylor
MERL-Cincinnati

Total Coliforms
Harold Jeter (retired)
National Training Center
ERC-Cincinnati

Fecal Coliforms
Edwin Geldreich
MERL-Cincinnati

Fecal Streptococci
Francis Brezenski
Region II

Salmonella
Kathleen Shimmin
Alameda Laboratory
Region IX

Donald Spino
MERL-Cincinnati

Actinomycetes
Robert Safferman
EMSL-Cincinnati

Quality Control
Robert Bordner
EMSL-Cincinnati

Development of a Quality Control Program
John Winter
EMSL-Cincinnati

Manpower and Analytical Costs
Robert Bordner and John Winter
EMSL-Cincinnati

Legal Considerations
Dave Shedroff
Office of Enforcement
Washington, DC

Carroll Wills
NEIC-Denver

Safety
Robert Bordner and John Winter
EMSL-Cincinnati

Pasquale Scarpino
UC Dept of Environ. Engineering

ACKNOWLEDGEMENTS

The Committee wishes to acknowledge the many EPA microbiologists and others who participated in the development or review of the manual. These include, in regional and program order:

Region I
Howard Davis and Edward Gritsavage
Regional Laboratory
Needham Heights, MA

Victor Cabelli, Alfred Dufour and Morris Levin
Environmental Research Laboratory
Narragansett, RI

Region II
Isidore Seidenberg (retired)
Edison Water Laboratory
Edison, NJ

Region III
Leonard Guarraia
Office of Water & Hazardous Materials
Washington, DC

Don Lear
Annapolis Field Station
Annapolis, MD

Region IV
Bobby Joe Carroll and Ralph Gentry
S & A Division, SERL
Athens, GA

Al Bourquin
Pensacola Station
Pensacola, FL

Region V
James Adams
Central Regional Laboratory
Chicago, IL

Region VI
Harold Cumiford
Houston Facility
Houston, TX

Region VII
Carl Bailey
Regional Laboratory
Kansas City, MO

Region VIII
John Manhart
Regional Laboratory, Denver Federal Center
Denver, CO

Region IX
Harold Scotten
Alameda Laboratory
Alameda, CA

Region X
George J. Vasconcelos and Richard Bauer
Regional Laboratory
Seattle, WA

Martin Knittel
Environmental Research Laboratory
Corvallis, OR

Ronald Gordon
Alaska Water Laboratory
College, AK

Cincinnati Environmental Research Center
Joseph Santner and Rocco Russomanno
National Training Center

Martin Allen, Harry Nash and Don Reasoner
Municipal Environmental Research Laboratory

Louis Resi
Division of Technical Support

Bernard Kenner (retired)
Municipal Environmental Reserach Laboratory (AWTRL)

Paul Britton, Terry Covert and Herbert Manning
Environmental Monitoring and Support Laboratory

Elmer Akin and Walter Jakubowski
Health Effects Research Laboratory

PREPARATION OF THIS VOLUME

The editors acknowledge gratefully the excellent technical skills of organization, proofreading, typing and computerized text editing performed by M. Mary Sullivan. Her contribution of hard work and sacrifice of personal time to this manual cannot be overstated.

PART I. INTRODUCTION

As the only direct measures of pollution by man and other warm-blooded animals, microbiological parameters contribute unique information on water and wastewater quality and public health risk from waterborne disease. Microbiological analyses are conducted to:

Monitor ambient water quality for recreational, industrial, agricultural and water supply uses,

Assure the safety of potable water,

Monitor municipal and industrial discharges,

Identify the sources of bacterial pollutants,

and evaluate water resources.

Role of the Aquatic Microbiologist

Although their primary role is to produce valid data for management decisions, microbiologists should also participate in survey planning and evaluation, develop new microbial parameters and methodology, consult on microbiological problems, establish and monitor criteria and standards, testify in administrative hearings and court cases, train laboratory staffs and research special problems. Microbiologists should also go beyond sanitary microbiology to solve taste and odor problems, to study microbiological transformations, and to apply other measurements to the aquatic ecosystem.

Scope of the Microbiology Manual Series

This EPA manual provides uniform laboratory and field methods for microbiological analyses of the environment. The analytical methods are standardized procedures recommended for use in enforcement, monitoring and research. However, they are not intended to inhibit or prevent methods research and development. Exploratory and developmental methods are compiled separately for evaluation but are part of the EPA Microbiology Manual Series.

The environmental areas covered will include:

- All waters — fresh, estuarine, marine, shellfish-growing, agricultural, ground, surface, finished, recreational and industrial processing.

- All wastewaters of microbiological concern – domestic waste effluents, industrial wastes such as food, dairy, meat, tanning, sugar, textile, pulp and paper, shellfish processing and agricultural wastes such as feedlot and irrigation runoff.

- Other areas of the environment – air, sediments, soils, sludges, oils, leachates, vegetation, etc.

Coverage of the First Edition of the Manual

Although the scope of the Manual Series is broad and inclusive of many parameters and sample types, the first edition describes primarily the analytical methods that meet the immediate needs of the Agency. These are the key parameters that are accepted and used for water quality, compliance monitoring and enforcement under Federal Water Pollution Control Act, PL 92–500, Marine Protection, Research, and Sanctuaries Act, PL 92–532 and the Safe Drinking Water Act, PL 93–523. The necessary supportive sections include: sample collection, equipment and techniques, cultural media, glassware preparation, quality control, data handling, safety, legal considerations and selection of analytical methods.

Focus of the Manual

This Manual is intended for use by the supervisor or analyst who may be a professional microbiologist, a technician, chemist, engineer or plant operator. Regardless of other skills, the supervisor and analyst should have received at least two weeks training in each parameter from a federal or state agency or from a university.

To assist the new analyst, Part II has been prepared as a basic discussion on laboratory operations and for general guidance to permit use of the manual by those required to do microbiological analyses. The trained analyst will be familiar and knowledgeable of most of these techniques. The analytical procedures in Part III are written in a stepwise manner so that the manual can be used both at bench level and as a reference book. Part IV emphasizes the important, but often neglected need for quality control in microbiological analyses, while Part V describes general considerations for laboratory management.

Objectives

The objectives of This Manual are to:

- Select the best method currently available for use in the environmental monitoring, compliance monitoring, enforcement and research activities of the Agency.

- Establish uniform application of microbiological methods so that only the best methods are used and perpetuated, data from different laboratories or surveys can be fairly compared and/or results can be stored in a common data bank, e.g., STORET, for later use.

- Provide guidance on the use of these methods, their advantages, limitations and application to various types of water and wastes.

- Establish recognized procedures for method selection and evaluation that will form the baseline against which other tests for the same or new parameters can be measured.

- Emphasize the analytical quality control and management practices that should be performed in the laboratory to assure valid data.

Criteria

The first edition of This Manual describes the parameters of health and sanitary significance. In the future, the criteria for addition of a method to the Manual are:

- The method is required to satisfy new or changing needs of the Agency.
- The method is practical for field and laboratory use. Equipment, supplies and media are available and the procedure provides results within reasonable time limits.
- The method offers significant advantages over current methods.
- The method has been validated by the developer or by others according to the criteria for Comparative Testing of Methodology and Method Characterization. (See IV-C-1).
- The method criteria and characterization have been reviewed and accepted by the EPA Steering Committee for Microbiology.

PART II. GENERAL OPERATIONS

This Part describes the general procedures which are applicable to the methods of analysis for all parameters. The Sections provide the basic background information that must be understood when the analytical procedures are carried out. The procedures are divided here into broad areas of function:

Section A	Sample Collection, Preservation and Storage
Section B	General Laboratory Equipment, Techniques and Media
Section C	Isolation and Enumeration of Bacteria
Section D	Selection of Analytical Methodology

PART II. GENERAL OPERATIONS

Section A Sample Collection, Preservation and Storage

Collection, preservation and storage of water samples are critical to the results of water quality analyses. The data are only as valid as the water sample.

A sampling program must be planned to satisfy the objectives of the study yet remain within the limitations of available manpower, time and money. The survey should use the minimum number of samples that adequately represent the effluent or body of water from which they are taken. The number of samples and location of sampling sites should be determined prior to the survey and must satisfy the requirements needed to establish water quality standard or effluent permit violations.

The microbiologist should participate in the planning which specifies the microbiological tests needed, the number of analyses to be performed, and the equipment required. Consideration should be given to the weather and other local conditions prior to the formulation of a final plan. For example, seasonal variations in water temperature and flows would be important factors in deciding when to study the effects of thermal pollution on bacteria. Sample collectors must know the exact location of the sampling sites and be fully trained in the aseptic technique of sample collection as well as the use of any specialized sampling equipment. The sample collector is responsible for the recording of all pertinent information about the sample that might be significant in the evaluation and interpretation of the laboratory data or that might be necessary in potential enforcement action.

This Section is organized as follows:

1. Sample Containers

1.1 Sample Bottles: Bottles must be resistant to sterilizing conditions and the solvent action of water. Wide-mouth borosilicate glass bottles with screw-cap or ground-glass stopper or heat-resistant plastic bottles may be used if they can be sterilized without producing toxic materials (see suggested sample containers in Figure II-A-1). Screw-caps must not produce bacteriostatic or nutritive compounds upon sterilization.

1.2 Selection and Cleansing of Bottles: Sample bottles should be at least 125 ml volume for adequate sampling and for good mixing. Bottles of 250 ml, 500 ml and 1000 ml volume are often used for multiple analyses. Discard bottles which have chips, cracks, and etched surfaces. Bottle closures must be water-tight. Before use, thoroughly cleanse bottles and closures with detergent and hot water, followed by a hot water rinse to remove all trace of detergent. Then rinse them three times with laboratory-pure water (II-B,6). A test for the biological examination of glassware where bacteriostatic or inhibitory residues may be present, is described in Part IV-A, 5.1.

1.3 Dechlorinating Agent: The agent must be placed in the bottle when water and wastewater samples containing residual chlorine are anticipated. Add sodium thiosulfate to the bottle before sterilization at a concentration of 0.1 ml of a 10 percent solution for each 125 ml (4 oz.) sample volume (1). This concentration will neutralize approximately 15 mg/l of residue chlorine.

1.4 Chelating Agent: A chelating agent should be added to sample bottles used to collect samples suspected of containing >0.01 mg/liter concentrations of heavy metals such as copper, nickel or zinc, etc. Add 0.3 ml of a 15 percent solution of ethylenediaminetetraacetic acid (EDTA) tetrasodium salt, for each 125 ml (4 oz.) sample volume prior to sterilization (2, 3).

1.5 Wrapping Bottles: Protect the tops and necks of glass stoppered bottles from contamination by covering them before sterilization with aluminum foil or kraft paper.

1.6 Sterilization of Bottles: Autoclave glass or heat-resistant plastic bottles at 121 C for 15 minutes. Alternatively, dry glassware may be sterilized in a hot air oven at 170 C for not less than two hours. Ethylene oxide gas sterilization is acceptable for plastic containers that are not heat-resistant. Sample bottles sterilized by gas should be stored overnight before being used to allow the last traces of gas to dissipate. See Part II-B, 3 for sterilization procedures.

1.7 Plastic Bags: The commercially-available bags (Whirl-pak) are a practical substitute for plastic or glass sample bottles in sampling soil or sediment. See Figure II-A-1. The bags are sealed in manufacture and opened only at time of sampling. The manufacturer states that such bags are sterilized.

2. Sampling Techniques

Samples are collected by hand or with a sampling device if (1) depth samples are required or (2) the sampling site has difficult access such as a manhole, dock, bridge or bank adjacent to a surface water.

2.1 Chlorinated Samples: When samples, such as treated waters, chlorinated wastewaters or recreational waters, are collected, the sample bottle must contain a dechlorinating agent (see this Section, 1.3).

2.2 Composite Sampling: In no case should a composite sample be collected for bacteriological examination. Data from individual samples show a range of values. A composite sample will not display this range. Individual results will give information about industrial process variations in flow and compo-

FIGURE II-A-1. Suggested Sample Containers

A	Screw-cap Glass or Plastic Bottle.
B	Plastic Bag (Whirl-pak).
C	Glass Stoppered Bottle.

sition. Also, one or more portions that make up a composite sample may contain toxic or nutritive materials and cause erroneous results.

2.3 Surface Sampling by Hand: A grab sample is obtained using a sample bottle prepared as described in 1. above. Identify the sampling site on a chain of custody tag if required, or on the bottle label and on a field log sheet (see 3). Remove the bottle covering and closure and protect from contamination. Grasp the bottle at the base with one hand and plunge the bottle mouth down into the water to avoid introducing surface scum. Position the mouth of the bottle into the current away from the hand of the collector and away from the side of the sampling platform or boat (see Figure II-A-2). The sampling depth should be 15 to 30 cm (6 to 12 inches) below the water surface. If the water body is static, an artificial current can be created, by moving the bottle horizontally in the direction it is pointed and away from the sampler. Tip the bottle slightly upwards to allow air to exit and the bottle to fill. After removal of the bottle from the stream, pour out a small portion of the sample to allow an air space of 2.5 to 5 cm (1 to 2 inches) above each sample for proper mixing of the sample before analyses. Tightly stopper and label the bottle.

2.4 Surface Sampling by Weighted Bottle Frame: When sampling from a bridge or other structure above a stream or body of water, the sample collector places the bottle in a weighted frame (see Figure II-A-3) that holds the bottle securely. Remove cover and lower the device to the water. It is preferable to use nylon rope which does not absorb water and will not rot. Face the bottle mouth upstream by swinging the sampling device first downstream, and then allow it to drop into the water, without slack in the rope. Pull the sample device rapidly upstream and out of the water, thus simulating the scooping motion of grab sampling described in 2.3. Take care not to dislodge dirt or other material that might fall into the open bottle from the sampling platform.

2.5 Depth Sampling: Several additional devices are needed for collection of depth samples from lakes, reservoirs, estuaries and the oceans. These depth samplers require lowering the sampling device and/or container to the desired depth, then opening, filling, and closing the container and returning the device to the surface. Although depth measurements are best made with a pre-marked steel cable, the sample depths can be determined by premeasuring and marking the nylon rope at intervals with a non-smearing ink, paint, or fingernail polish. The following list of depth samplers is not inclusive but can serve as a guide:

2.5.1 ZoBell J-Z Sampler: This sampler described by ZoBell in 1941 (4) was designed for deep sea sampling but is also used in fresh waters. Figure II-A-4 shows its general appearance. It has a metal frame (A), a heavy metal messenger (B), a sealed glass tube (C) attached to a rubber tube (D), and a sterile 350 ml glass bottle (E) or a collapsible neoprene rubber bulb for shallow waters. The messenger (B) is released at the surface when the sampler reaches the desired depth, and breaks the glass tubing (C) at a file mark. The bent rubber tubing (D) then straightens out and the water is drawn in several inches from the sampler. A partial vacuum created by autoclaving of the sealed unit draws the water into the bottle.

2.5.2 Niskin Sampler: This is sometimes called a sterile-bag or "Book" sampler (see Figure II-A-5) (5). A messenger triggers the opening of two plates (A) in V-fashion by spring power, and causes the sterile plastic bag (B) to inflate. At the same time a plastic filler tube (C) leading to the plastic container is cut by a guillotine knife (D) and the bag fills with water. The bag is then automatically sealed with a clamp (E) and the apparatus is brought to the surface. Samplers are available that will hold 1, 2, 3, or 5 liters of water.

2.5.3 New York Dept. of Health Depth Sampler: This device (see Figure II-A-6) depends upon a vane (A) and lever (B) mechanism to lift the glass stopper (C) as water inertia is applied by a sharp upward tug on the line (D) attached to the apparatus. As the stopper is lifted, the bottle fills before the detachment of the stopper from the vane occurs and closes the sample bottle (6).

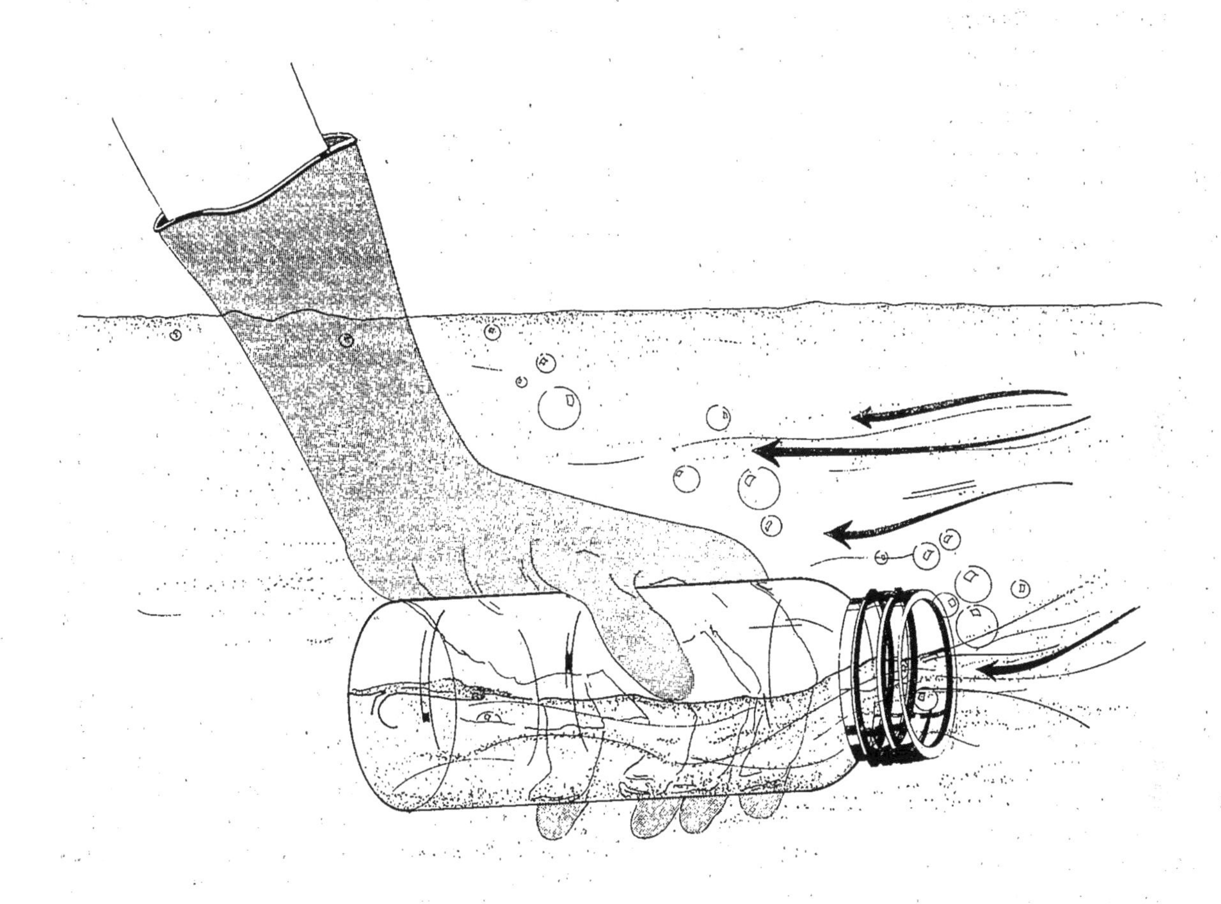

FIGURE II-A-2. Demonstration of Technique Used in Grab Sampling of Surface Waters.

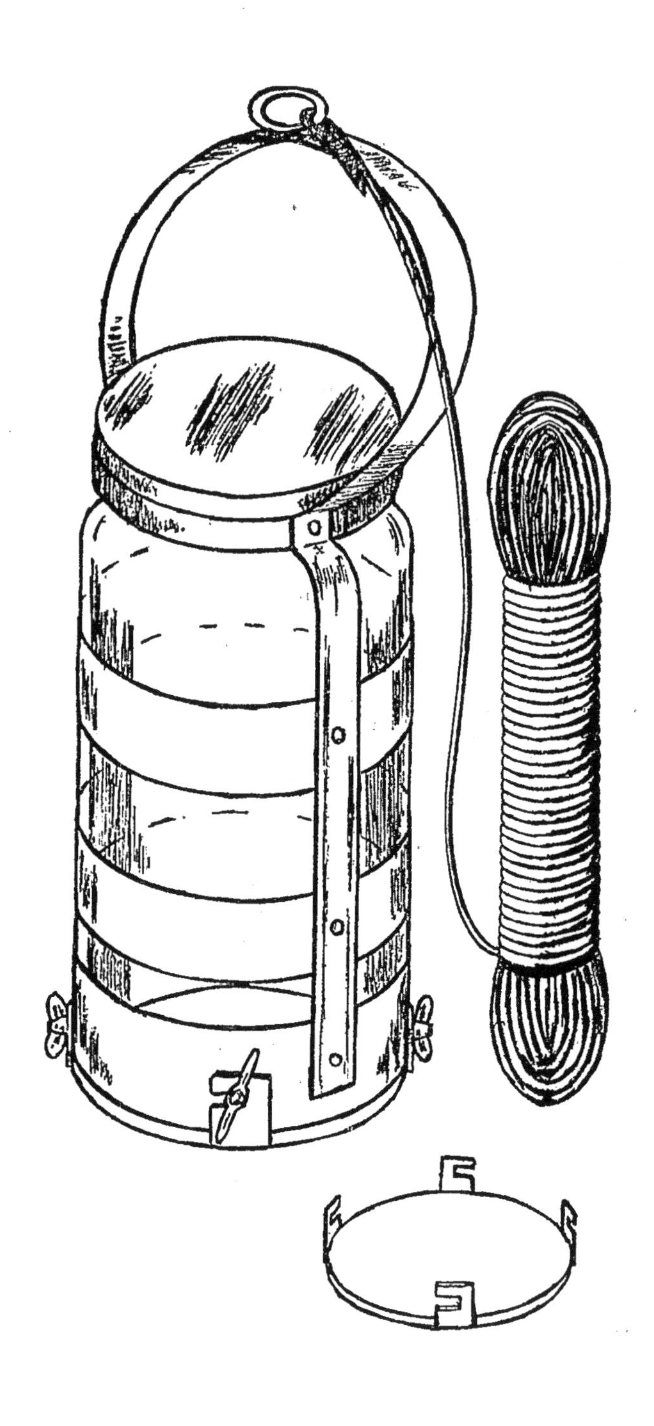

FIGURE II-A-3. Weighted Bottle Frame and Sample Bottle for Grab Sampling.

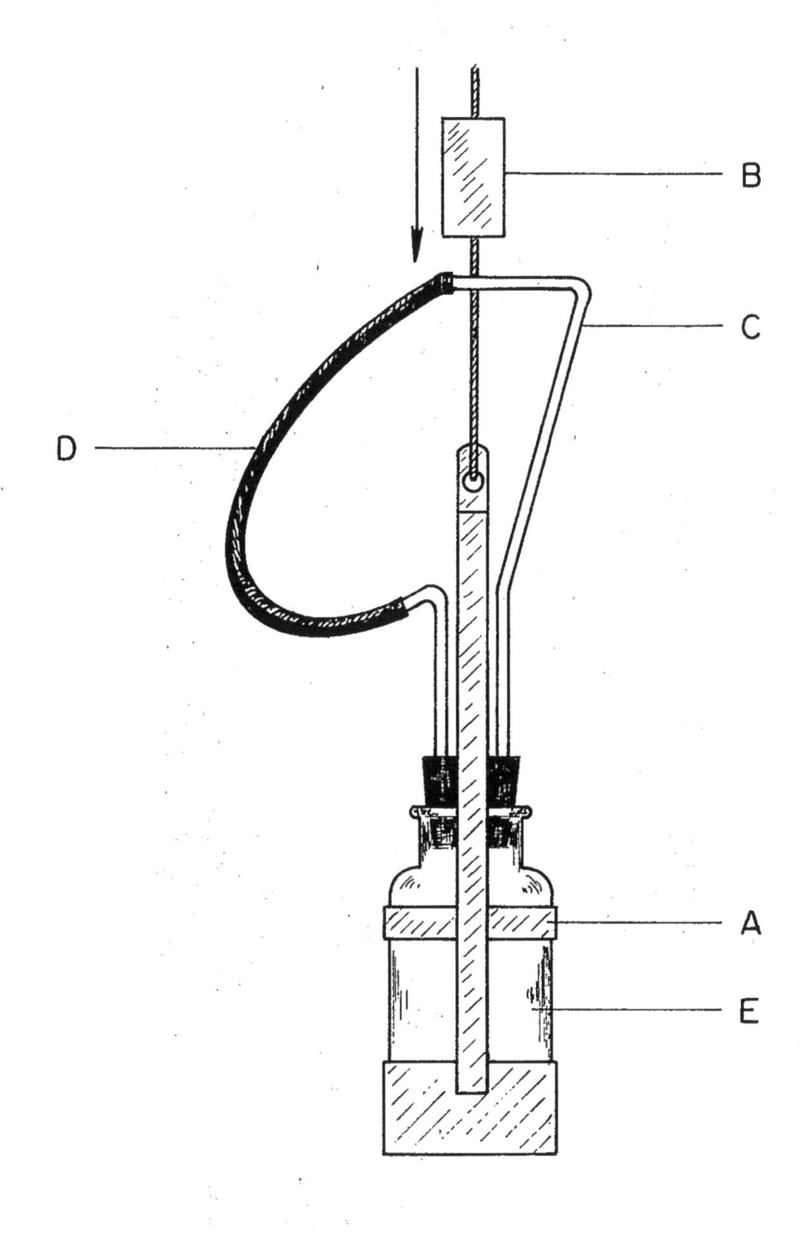

FIGURE II-A-4. Zobell J-Z Sampler. (A) metal frame, (B) messenger, (C) glass tube, (D) rubber tube and (E) sterile sample bottle.

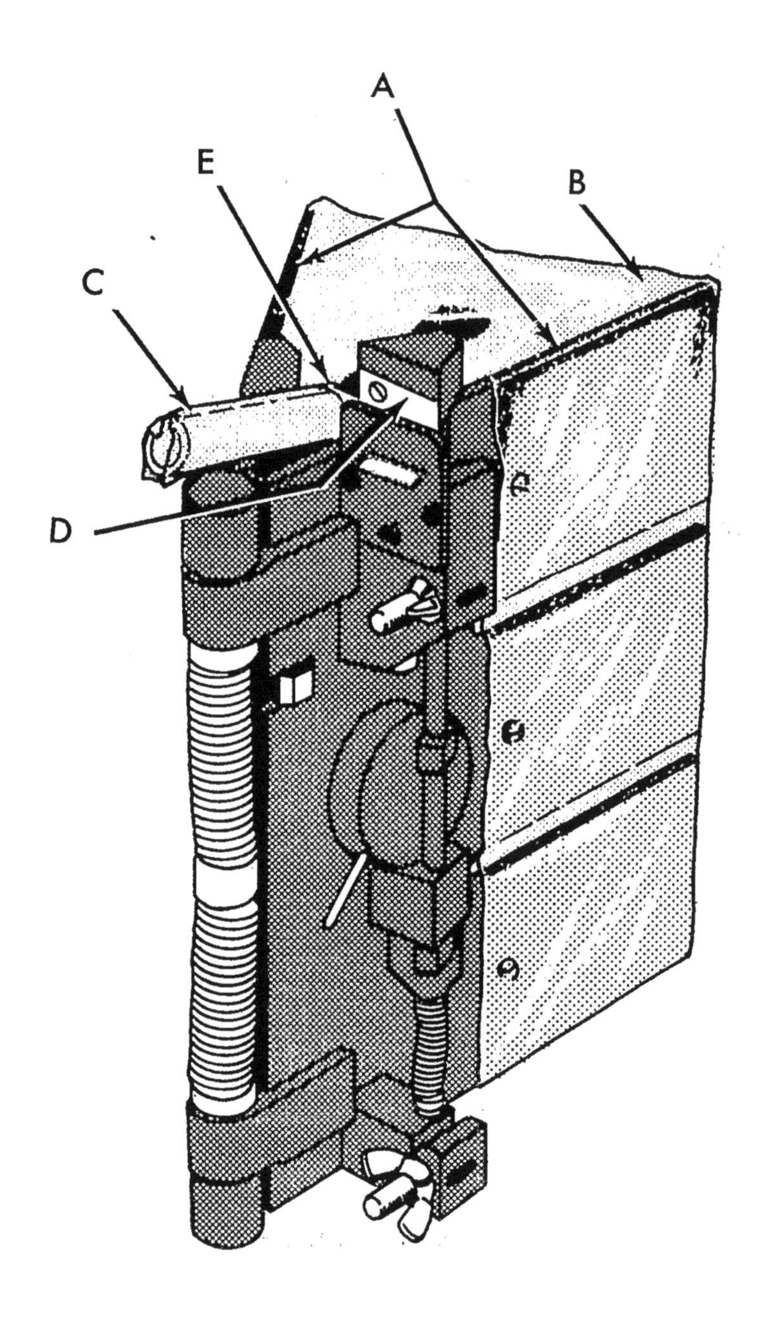

FIGURE II-A-5. Niskin Depth Sampler. (A) hinged plates, (B) plastic bag, (C) plastic filler tube in sheath, (D) guillotine knife and (E) closure clamp.

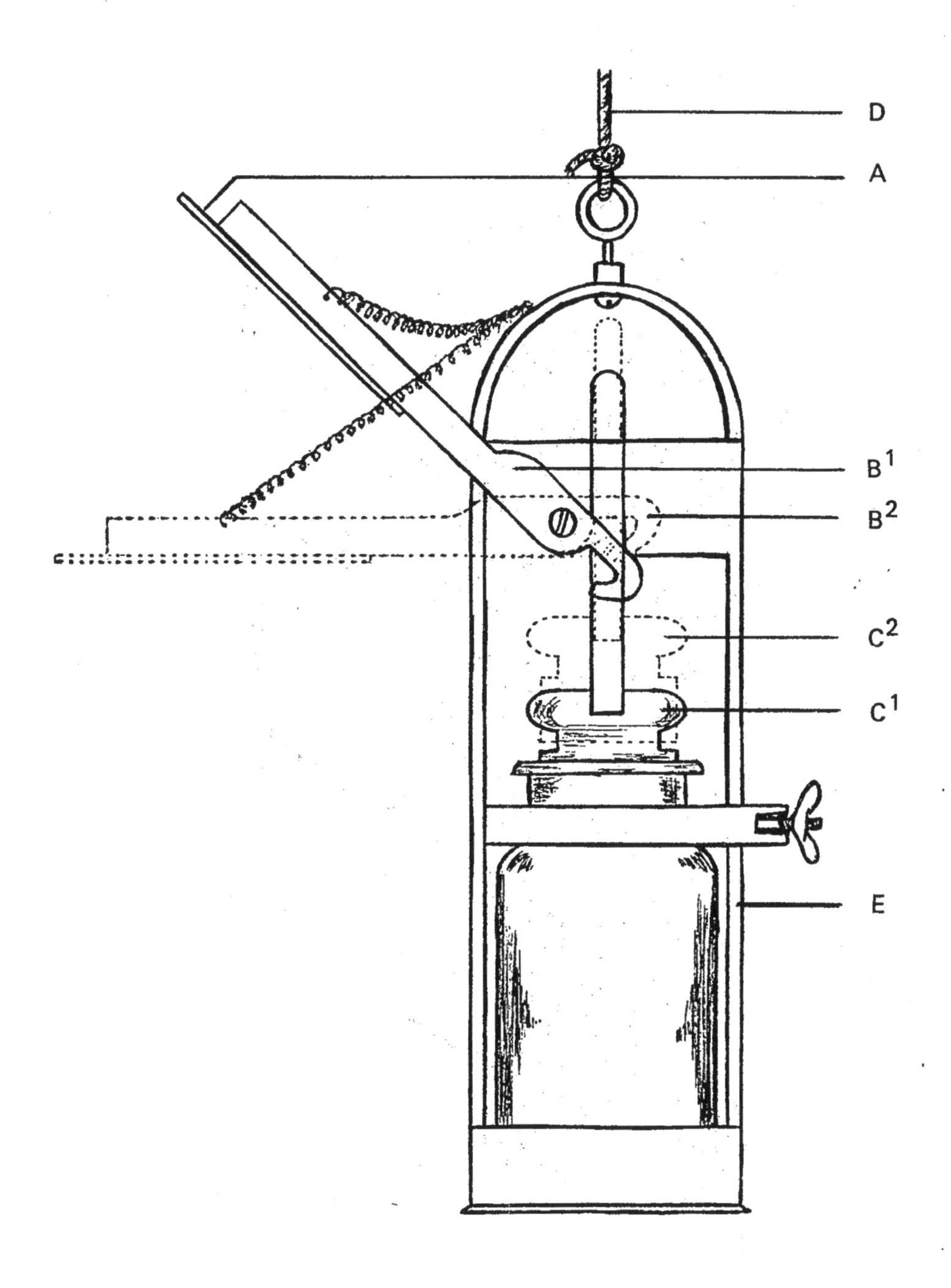

FIGURE II-A-6. New York State Dept. of Health Depth Sampler. (A) vane, (B[1]) lever in closed position, (B[2]) lever in open position, (C[1]) glass stopper in closed position, (C[2]) glass stopper in open position, (D) suspension line, and (E) metal frame.

2.5.4 Kemmerer Sampler (7): This depth sampler (see Figure II-A-7) has been used without sterilization to collect bacteriological water samples in high pollution areas. The sampler consists of a cylindrical brass or plastic tube (F) that contains a rubber stopper or valve at either end (D and G). The valves are connected to a rod (E) that passes through the center of the cylinder. The device is lowered into the water in the open position, and a water sample is trapped in the cylinder when the valves are closed by a dropped messenger (B). The Kemmerer sampler should not be used for collecting bacteriological samples without obtaining data that support its use without sterilization.

2.6 Sediment Sampling with Van Donsel-Geldreich Sampler (8): This device (see Figure II-A-8) collects sediment or mud in sterile "Whirl-Pak" plastic bags (A) down to 60 foot depth. The bag mouth is wrapped over a nosepiece (B), and the bag is kept closed during descent to the bottom by a bag clamp bar (H). As the mud plate (D) contacts the bottom, the nosepiece (B) is driven into the sediment by the weight (C) of the sampler. As the nosepiece (B) moves downward, the bag (A) slides through the bag clamp bar (H), opens, and fills with sediment. The bag is sealed when the double noose (F) tied to the bottom of the bag is pulled, before the apparatus is returned to the surface.

2.7 Water Tap Sampling: Make certain that samples are not collected from spigots that leak around their stems, or from spigots that contain aeration devices or screens within the faucet. For samples taken from direct water main connections, the spigot should be flushed for 2-3 minutes to clear the service line. For wells equipped with hand or mechanical pumps, pump the water to waste for five minutes before the sample is collected. Remove the cap aseptically from the sample bottle. Hold the sample bottle upright near the base while it is being filled. Avoid splashing. Do not rinse the bottle with the sample; fill it directly to within 2.5 cm (1 inch) from the top. Replace bottle closure and hood covering. Caution must be used to prevent contaminating the sample with finger, gloves or other materials. If the well does not have pumping machinery, collect the sample using a weighted sterilized sample bottle, such as described in 2.4 above, and shown in Figure II-A-3. Care must be taken to avoid contaminating the sample with the surface scum from the water surface.

2.8 Soil Sampling

2.8.1 Selection of the sampling site is based on knowledge of the area and the purposes of the analyses, i.e., surface sampling for natural background, surface contamination, or below surface sampling to monitor treatment effect such as irrigation, or stormwater runoff.

The actual sites for sampling and the number of points to be sampled must be predetermined by the survey objectives. Soil sampling has the advantage of permitting the survey planners to lay out a stable grid network for sampling and resampling over a given time period.

2.8.2 If a surface sample is desired, scrape the top one inch of soil from a square foot area using a sterile scoop or spoon.

If a subsurface sample is desired, use a sterile scoop or spatula to remove the top surface of one inch or more from a one foot square area. Use a second sterile scoop or spoon to take the sample.

Place samplings in a sterile one quart screw-cap bottle until it is full. Depending on the amount of moisture, a one quart bottle holds 300–800 grams of soil. Label and tag the bottle carefully and store at 4 C until analyzed.

3. Sample Identification and Handling

3.1 Specific details on sample identification are entered on a permanent label. Take care in transcribing sampling information to the label, because the enforcement action may depend upon evidence of primary labeling. See 4. in This Section. Labels must be clean,

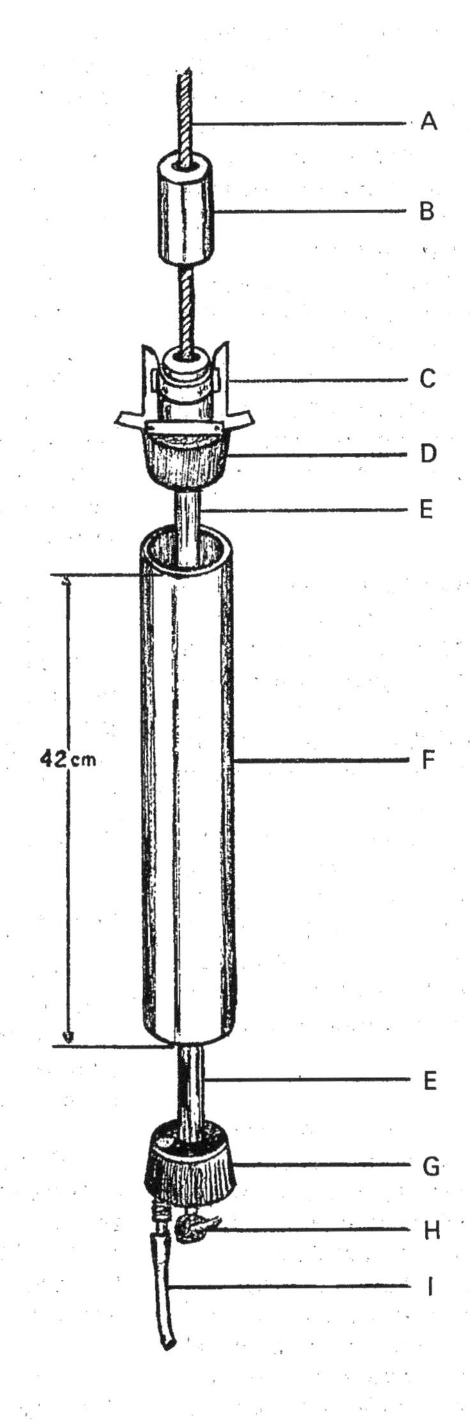

FIGURE II-A-7. Kemmerer Depth Sampler. (A) nylon line, (B) messenger, (C) catch set so that the sampler is open, (D) top rubber valve, (E) connecting rod between the valves, (F) tube body, (G) bottom rubber valve, (H) knot at the bottom of the suspension line and (I) rubber tubing attached to the spring loaded check valve.

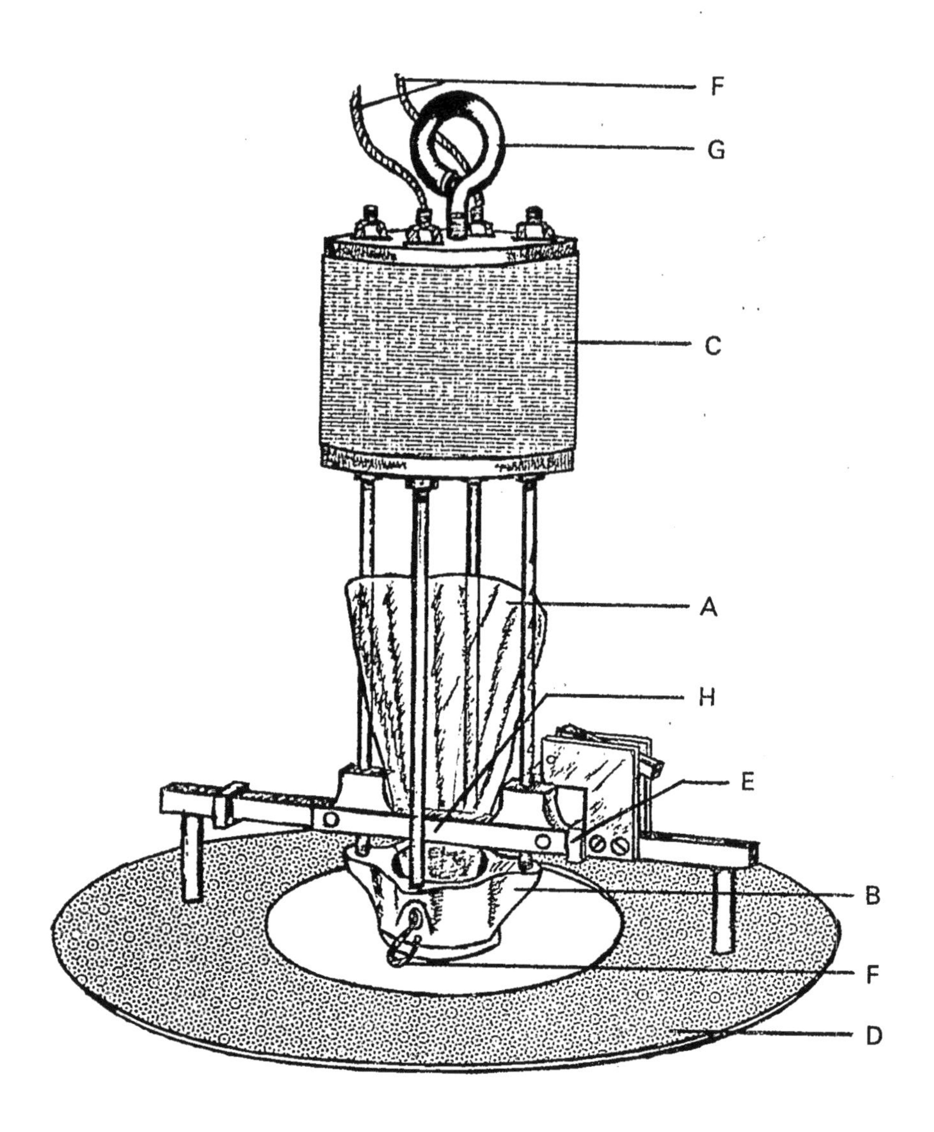

FIGURE II-A-8. Van Donsel-Geldreich Sediment Sampler. (A) sterile "Whirl-Pak" plastic bag, (B) nose piece, (C) weight, (D) mud plate, (E) slide bar, (F) part of the double noose, (G) attachment for the suspension line and (H) bag clamp bar.

waterproof, non-smearing and of sufficient size for the necessary information. Label must be securely attached to the sample bottle, but removable when necessary. Do not accept insufficiently or improperly labeled samples for examination. A sample label showing the minimum information required is pictured in Figure II-A-9.

3.2 Field Data Record: A field record should be completed on each sample to record the full details on sampling and other pertinent remarks such as flooding, rain or extreme temperature which are relevant to interpretation of results. This record also provides a back-up record of sample identification. One example is shown in Figure II-A-10.

3.3 Marking Device: A marking pen or other device must be non-smearing if wetted, and maintain a permanent legible mark.

3.4 Transport Container: Insulated ice containers in which the sample can be held, are recommended.

3.5 Storage of Samples: A refrigerator is necessary for storage of samples at the laboratory. The temperature range of the refrigerator is 1-4 C.

4. Chain of Custody Procedures

4.1 General: An agency must demonstrate the reliability of its evidence in pollution cases by proving the chain of possession and custody of samples which are offered for evidence or which form the basis of analytical results introduced into evidence. It is imperative that the office and laboratory prepare written procedures to be followed whenever evidence samples are collected, transferred, stored, analyzed, or destroyed.

4.1.1 The primary objective of these procedures is to create an accurate written record which can be used to trace the possession of the sample from the moment of its collection through its introduction into evidence. A sample is in custody if it is:

(a) in actual physical possession, or

(b) in view after being in physical possession, or

(c) in physical possession and locked up so that no one could tamper with it.

4.1.2 Personnel should receive copies of study plans and know the contents prior to the study. A pre-study briefing shall be held to appraise participants of the objectives, sample locations and chain of custody procedures. After chain of custody samples are collected, a de-briefing is held in the field to determine adherence to chain of custody procedures and whether additional samples are required.

4.2 Rules for Sample Collection

4.2.1 Handle the samples as little as possible.

4.2.2 Obtain stream and effluent samples using standard microbiological sampling techniques.

4.2.3 Attach sample tag or label (Figure II-A-9) to the sample container. The tag or label should contain as a minimum: serial number of label, location, date and time taken, type of sample, sequence number (first sample of the day – sequence No. 1, second sample, sequence No. 2, etc), analyses required and sample collector. The tags must be filled out legibly in waterproof ink.

4.2.4 Use a bound notebook to record field measurements and other pertinent information necessary to refresh the sampler's memory if the person later becomes a witness in an enforcment proceeding. A separate set of field notebooks should be maintained for each study and stored in a safe place where it can be protected and accounted for. A sample log sheet with a standard format should be established to minimize field entries and include the date, time, survey, type of samples, volume of each sample, type of analyses, label and sample numbers, sample location, field measurements such as temperature, conductivity, DO, pH, and any other pertinent

EPA, NATIONAL ENFORCEMENT INVESTIGATIONS CENTER

Station No.	Date	Time	Sequence No.

Station Location ______ Grab ______ Comp.

______ BOD ______ Metals
______ Solids ______ Oil and Grease
______ COD ______ D.O.
______ Nutrients ______ Bact.
______ Other

Remarks/Preservative:

Samplers:

SERIAL TAG #

FIGURE II-A-9. Example of a Sample Label.

SERIAL SHEET NO. ______

Samplers: ______

FIELD DATA RECORD

STATION	SAMPLE NUMBER	DATE OF COLLECTION	TIME (HRS)		pH	TEMP °C	OTHER	REMARKS
			SAMPLE TAKEN	SAMPLE RECEIVED BY LAB				

US EPA, NEIC-Denver

FIGURE II-A-10. Field Data Record.

information or observation (Figure II-A-11). The entries should be signed by the sample collector. The responsibility for preparing and storing sample notebooks during and after a study should be assigned to a study coordinator, or his designated representative.

4.2.5 A field collector is responsible for the samples collected until properly dispatched to the receiving laboratory or turned over to an assigned custodian. He must assure that each container is in his physical possession or in his view at all times, or stored in a locked place where no one can tamper with it.

4.2.6 Color slides or photographs should be taken of the sample location and any visible water pollution. The signature of the photographer, time, date, and site location must be written on the back of the photo. Such photographs should be handled according to the established chain of custody procedures to prevent alteration.

4.3 Transfer of Custody and Shipment

In transfer of custody procedures, each custodian of samples must sign, record and date the transfer. Most regulatory agencies develop chain of custody procedures tailored to their needs. These procedures may vary in format and language but contain the same essential elements. Historically, sample transfer under chain of custody has been on a sample by sample basis which is awkward and time-consuming. However, EPA's National Enforcement Investigation Center (NEIC), Denver has set a precedent with its bulk transfer of samples. Bulk transfer is speedier, reduces paperwork and the number of sample custodians. The following description of chain of custody is essentially that of NEIC-Denver (9).

4.3.1 Samples must be accompanied by a Chain of Custody Record which includes the name of the study, collector's signature, station number, station location, date, time, type of sample, sequence number, number of containers and analyses required (Figure II-A-12). When turning over the possession of samples, the transferor and transferee sign, date and note time on the sheet. This record sheet allows transfer of custody of a group of samples in the field, to the mobile laboratory or to the NEIC-Denver laboratory. When a custodian transfers a portion of the samples identified on the sheet to the field mobile laboratory, the individual samples must be noted in the column with the signature of the person relinquishing the samples. The field laboratory person receiving the samples acknowledges receipt by signing in the appropriate column.

4.3.2 If a custodian has not been assigned, the field custodian or field sampler has the responsibility for packaging and dispatching samples to the laboratory for analysis. The "Dispatch" portion of the Chain of Custody Record must be filled out, dated, and signed.

4.3.3 Samples must be carefully packed in shipment containers such as ice chests, to avoid breakage. The shipping containers are padlocked for shipment to the receiving laboratory.

4.3.4 Packages must be accompanied by the Chain of Custody Record showing identification of the contents. The original must accompany the shipment. A copy is retained by the survey coordinator.

4.3.5 If samples are delivered to the laboratory when appropriate personnel are not there to receive them, the samples must be locked in a designated area within the laboratory so that no one can tamper with them. This same person must return to the laboratory, unlock the samples and deliver custody to the appropriate custodian.

4.4 Laboratory Custody Procedures

4.4.1 The laboratory shall designate a "sample custodian" and an alternate to act in his absence. In addition, the laboratory shall set aside as a "sample storage security area", an isolated room with sufficient refrigerator space, which can be locked or just a locked refrigerator in smaller laboratories.

4.4.2 Samples should be handled by the minimum possible number of persons.

SERIAL SHEET NO. ______

FOR ______ SURVEY, PHASE ______, DATE ______

TYPE OF SAMPLE ______ SAMPLERS: *(Signature)* ______

STATION NUMBER	STATION DESCRIPTION	TOTAL VOLUME	TYPE CONTAINER	PRESERVATIVE	NUTRIENTS	BOD	COD	TOC	TOTAL SOLIDS	SUSPENDED SOLIDS	ALKALINITY	DO	pH*	CONDUCTIVITY*	TEMPERATURE*	TOTAL COLIFORM	FECAL COLIFORM	TURBIDITY	OIL AND GREASE	METALS	BACTI	PESTICIDES	HERB	TRACE ORGANICS	PHENOL	CYANIDE

(Columns NUTRIENTS through CYANIDE: A N A L Y S E S   R E Q U I R E D)

REMARKS

*Field Measurements

US EPA, NEIC-Denver

FIGURE II-A-11. Sample Log Sheet.

SERIAL SHEET NO. ________

ENVIRONMENTAL PROTECTION AGENCY
Office Of Enforcement
NATIONAL ENFORCEMENT INVESTIGATIONS CENTER
Building 53, Box 25227, Denver Federal Center
Denver, Colorado 80225

CHAIN OF CUSTODY RECORD

SURVEY	SAMPLERS: *(Signature)*

STATION NUMBER	STATION LOCATION	DATE	TIME	SAMPLE TYPE			SEQ. NO.	NO. OF CONTAINERS	ANALYSIS REQUIRED
				Water		Air			
				Comp.	Grab.				

Relinquished by: *(Signature)*	Received by: *(Signature)*	Date/Time
Relinquished by: *(Signature)*	Received by: *(Signature)*	Date/Time
Relinquished by: *(Signature)*	Received by: *(Signature)*	Date/Time
Relinquished by: *(Signature)*	Received by Mobile Laboratory for field analysis: *(Signature)*	Date/Time

Dispatched by: *(Signature)*	Date/Time	Received for Laboratory by:	Date/Time
Method of Shipment:			

Distribution: Orig.—Accompany Shipment
1 Copy—Survey Coordinator Field Files

GPO 854-809

FIGURE II-A-12. Chain of Custody Record.

4.4.3 Incoming samples shall be received only by the custodian, who will indicate receipt by signing the Chain of Custody Record Sheet accompanying the samples and retaining the sheet as a permanent record. Couriers picking up samples at the airport, post office, etc. shall sign jointly with the laboratory custodian.

4.4.4 Immediately upon receipt, the custodian places samples in the sample room, which will be locked at all times except when samples are removed or replaced by the custodian. To the maximum extent possible, only the custodian should be permitted in the sample room.

4.4.5 The custodian shall ensure that microbiological samples are properly stored and maintained at 1-4 C.

4.4.6 Only the custodian will distribute samples to personnel who are to perform tests.

4.4.7 The analyst records information in his laboratory notebook or analytical worksheet, describing the sample, the procedures performed and the results of the testing. The notes shall be dated and indicate who performed the tests. The notes shall be retained as a permanent record in the laboratory and should include any abnormalities which occurred during the testing procedure. In the event that the person who performed the tests is not available as a witness at time of trial, the government may be able to introduce the notes in evidence under the Federal Business Records Act.

4.4.8 Standard methods of laboratory analyses shall be used as described in the "Guidelines Establishing Test Procedures for Analysis of Pollutants," (10) and amendments. If laboratory personnel deviate from standard procedures, they should be prepared to justify their decision during cross-examination.

4.4.9 Laboratory personnel are responsible for the care and custody of a sample once it is handed over to them and should be prepared to testify that the sample was in their possession and view or secured in the laboratory at all times from the moment it was received from the custodian until the tests were run.

4.4.10 Once the sample testing is completed, microbiological samples can be discarded but identifying tags and laboratory record should be returned to the custodian. Other documentation of work will also be given to the custodian.

4.4.11 Tags and laboratory records of tests may be destroyed only upon the order of the Laboratory Director, who will first confer with the Chief, Enforcement Specialist Office, to make certain that the information is no longer required.

5. Selection of Sampling Sites and Frequency

These will be described for potable and recreational waters, streams, lakes, reservoirs, estuarine, and marine waters as well as domestic and industrial wastewaters.

5.1 Potable Water Supplies

An expanded program to maintain the sanitary quality of potable water supplies has been recently established by the National Interim Primary Drinking Water Regulations (11). The sampling program includes examination of water as it enters and flows throughout the distribution system. For application of the EPA Drinking Water Standards, the frequency of sampling and the location of sampling points are established jointly by the utility, the Reporting Agency, and the Certifying Authority. Additionally, the laboratory, the methods of analyses, and the technical competence of personnel must be inspected and approved by the Reporting Agency and the Certifying Authority.

5.1.1 Sampling Water Supplies: Figure II-A-13 shows how reservoirs and lakes used as water supplies are sampled: (A) at inlets, (B) at other possible sources of pollution, (C) at the draw-off point, (D) at quarter point intervals around the draw-off point at about the same depth and (E) at the reservoir outlet.

FIGURE II-A-13. Sampling a Water Supply Reservoir. (A) influent stream, (B) possible agricultural contamination, (C) water plant intake, (D) multi-point sampling around intake and (E) reservoir outlet.

5.1.2 Sampling Treatment Systems: Sampling should be representative of the distribution system and include sites such as municipal buildings, public schools, airports and parks, hydrants, restaurants, theaters, gas stations, industrial plants and private residences. A systematic coverage of such points in the distribution system should insure the detection of contamination from breaks in waterlines, loss of pressure or crossconnections. The sampling program should also include special sampling locations such as dead-end distribution lines that are sources of bacterial contamination (12).

5.1.3 Sample Frequency: The minimum number of samples which must be collected and examined each month is based upon the population density served by the distribution system (Table II-A-1). Samples should be collected at evenly spaced time intervals throughout the month. In the event of an unsatisfactory sample, repetitive samples must be collected until two consecutive samples yield satisfactory quality water. Repetitive samples from any single point or special purpose samples must not be counted in the overall total of monthly samples.

5.1.4 Standard Sample: The standards for microbiological quality are based upon the number of organisms allowable in a standard sample. A standard sample for the membrane filter technique is at least 100 ml. For the MPN test, a standard sample consists of five standard portions of either 10 ml or 100 ml.

5.2 Lakes and Impoundments

Figure II-A-14 shows the range of sampling points in a recreational impoundment or lake: (A) inlets, (B) source of pollution, (C) grid or transect across the long axis of the water body, (D) bathing beach and (E) outlet.

5.3 Stream Sampling

The objectives of the initial survey dictate the location, frequency and number of samples to be collected.

5.3.1 Selection of Sampling Sites: A typical stream sampling program includes sampling locations upstream of the area of concern, upstream and downstream of waste discharges, upstream and downstream from tributary entrances to the river and upstream of the mouth of the tributary. For more complex situations, where several waste discharges are involved, sampling includes sites upstream and downstream from the combined discharge area and samples taken directly from each industrial or municipal waste discharge. Using available bacteriological, chemical and discharge rate data, the contribution of each pollution source can be determined. See Figure II-A-14 and II-A-15.

5.3.2 Small Streams: Small streams should be sampled at background stations upstream of the pollution sources and at stations downstream from pollution sources. Additional sampling sites should be located downstream tc delineate the zones of pollution. Avoid sampling areas where stagnation may occur (backwater of a tributary) and areas located near the inside bank of a curve in the stream which may not be representative of the main channel.

5.3.3 Large Streams and Rivers: Large streams are usually not well mixed laterally for long distances downstream from the pollution sources. Sampling sites below point source pollution should be established to provide desired downstream travel time and dispersal as determined by flow rate measurements. Particular care must be taken to establish the proper sampling points as shown in Figure II-A-15: Sampling point (A) is the upper reach control station, (B) monitors a non-point source of pollution, (C) samples the waste discharge as it enters the stream, (D) shows quarter-point sampling below the pollution to detect channeling, (D) also serves as an upstream monitor on the tributary measured as (E), and (F) monitors the downstream effect of the tributary after mixing. Occasionally, depth samples are necessary to determine vertical mixing patterns.

TABLE II-A-1

Sampling Frequency for Drinking Waters Based on Population

Population served:	Minimum number of samples per month	Population served:	Minimum number of samples per month
25 to 1,000	1	90,001 to 96,000	95
1,001 to 2,500	2	96,001 to 111,000	100
2,501 to 3,300	3	111,001 to 130,000	110
3,301 to 4,100	4	130,001 to 160,000	120
4,101 to 4,900	5	160,001 to 190,000	130
4,901 to 5,800	6	190,001 to 220,000	140
5,801 to 6,700	7	220,001 to 250,000	150
6,701 to 7,600	8	250,001 to 290,000	160
7,601 to 8,500	9	290,001 to 320,000	170
8,501 to 9,400	10	320,001 to 360,000	180
9,401 to 10,300	11	360,001 to 410,000	190
10,301 to 11,100	12	410,001 to 450,000	200
11,101 to 12,000	13	450,001 to 500,000	210
12,001 to 12,900	14	500,001 to 550,000	220
12,901 to 13,700	15	550,001 to 600,000	230
13,701 to 14,600	16	600,001 to 660,000	240
14,601 to 15,500	17	660,001 to 720,000	250
15,501 to 16,300	18	720,001 to 780,000	260
16,301 to 17,200	19	780,001 to 840,000	270
17,201 to 18,100	20	840,001 to 910,000	280
18,101 to 18,900	21	910,001 to 970,000	290
18,901 to 19,800	22	970,001 to 1,050,000	300
19,801 to 20,700	23	1,050,001 to 1,140,000	310
20,701 to 21,500	24	1,140,001 to 1,230,000	320
21,501 to 22,300	25	1,230,001 to 1,320,000	330
22,301 to 23,200	26	1,320,001 to 1,420,000	340
23,201 to 24,000	27	1,420,001 to 1,520,000	350
24,001 to 24,900	28	1,520,001 to 1,630,000	360
24,901 to 25,000	29	1,630,001 to 1,730,000	370
25,001 to 28,000	30	1,730,001 to 1,850,000	380
28,001 to 33,000	35	1,850,001 to 1,970,000	390
33,001 to 37,000	40	1,970,001 to 2,060,000	400
37,000 to 41,000	45	2,060,001 to 2,270,000	410
41,001 to 46,000	50	2,270,001 to 2,510,000	420
46,001 to 50,000	55	2,510,001 to 2,750,000	430
50,001 to 54,000	60	2,750,001 to 3,020,000	440
54,001 to 59,000	65	3,020,001 to 3,320,000	450
59,001 to 64,000	70	3,320,001 to 3,620,000	460
64,001 to 70,000	75	3,620,001 to 3,960,000	470
70,001 to 76,000	80	3,960,001 to 4,310,000	480
76,001 to 83,000	85	4,310,001 to 4,690,000	490
83,001 to 90,000	90	4,690,001 or more	500

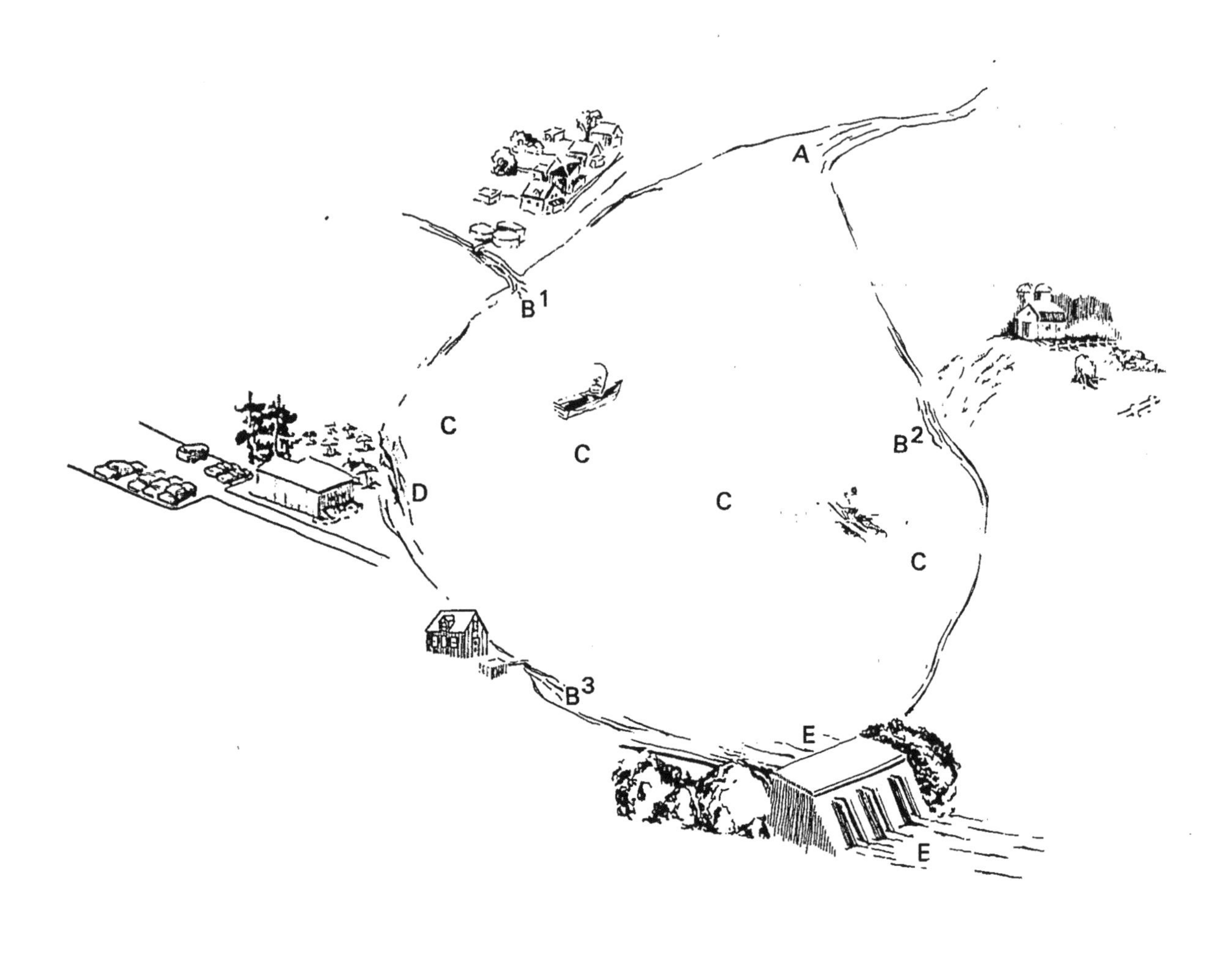

FIGURE II-A-14. Sampling a Lake or Impoundment. (A) inlets, (B) potential source of pollution, (B^1) village, (B^2) agricultural run-off, (B^3) home septic tank, (C) multi-point transect, (D) bathing beach and (E) outlet above and below dam.

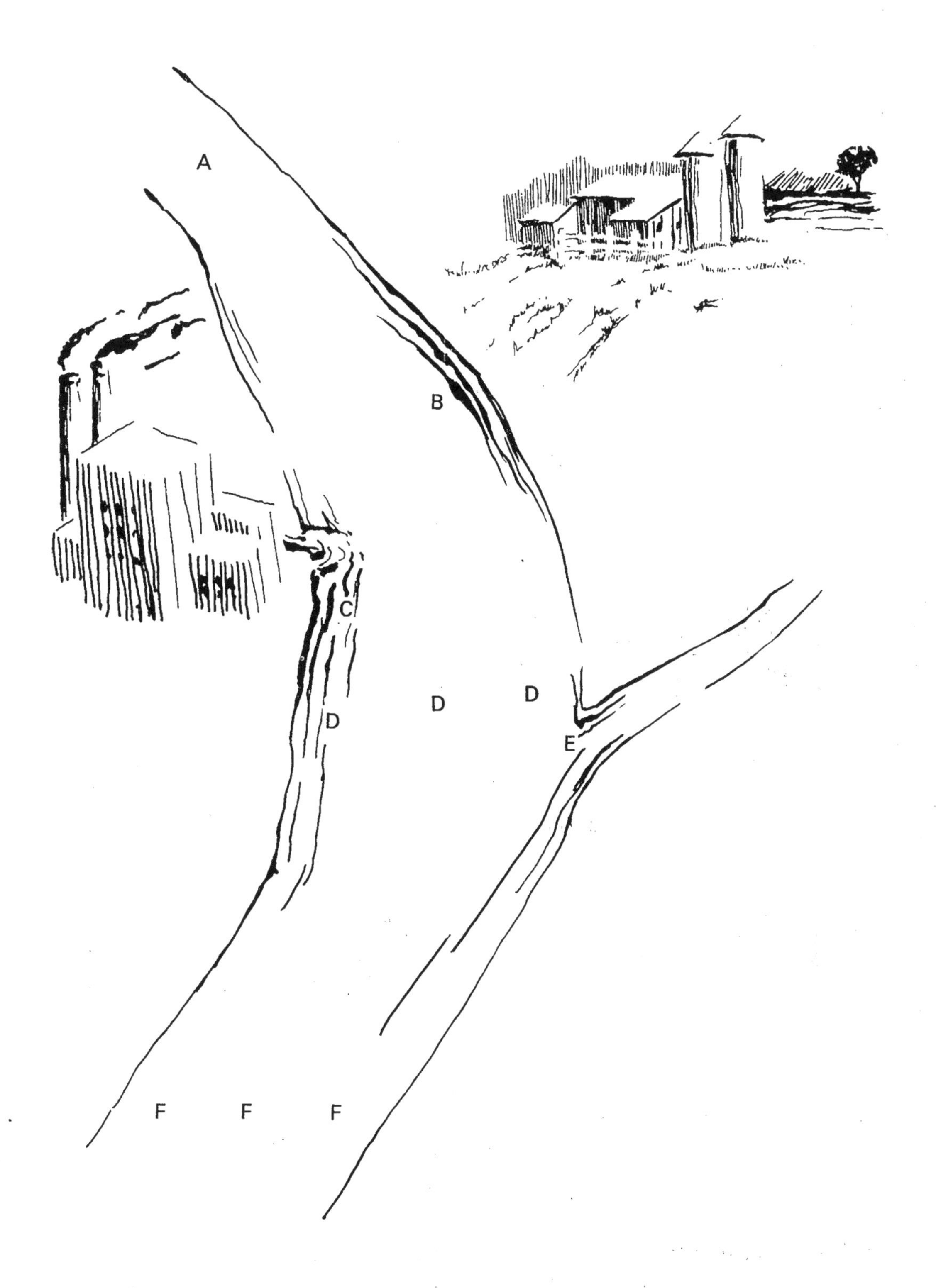

FIGURE II-A-15. Sampling a Large Stream. (A) control station, (B) agricultural pollution, (C) industrial discharge, (D) quarterpoint transect (E) tributary, and (F) downstream monitoring.

5.4 Marine and Estuarine Sampling

Sampling marine and estuarine waters requires the consideration of other factors in addition to those usually recognized in fresh water sampling. They include tidal cycles, current patterns, bottom currents and countercurrents, stratification, seasonal fluctuations, dispersion of discharges and multi-depth samplings.

The frequency of sampling varies with the objectives. When a sampling program is started, it may be necessary to sample every hour around the clock to establish pollutional loads and dispersion patterns. The sewage discharges may occur continuously or intermittently.

When the sampling strategy for a survey is planned, data may be available from previous hydrological studies done by Coast Guard, Corps of Engineers, National Oceanic and Atmospheric Administration (NOAA), U.S. Geological Survey, or university and private research investigations. In a survey, float studies and dye studies are often carried out to determine surface and undercurrents. Initially depth samples are taken on the bottom and at five feet increments between surface and bottom. A random grid pattern for selecting sampling sites is established statistically.

5.4.1 Marine Sampling: In ocean studies, the environmental conditions are most diverse along the coast where shore, atmosphere and the surf are strong influences. The shallow coastal waters are particularly susceptible to daily fluctuations in temperature and seasonal changes.

Sampling during the entire tidal cycle or during a half cycle may be required. Many ocean studies such as sampling over the continental shelf involve huge areas and no two areas of water are the same.

Selection of sampling sites and depths are most critical in marine waters. In winter, cooling of coastal waters can result in water layers which approach 0 C. In summer, the shallow waters warm much faster than the deeper waters. Despite the higher temperature, oxygen concentrations are higher in shallow than in deeper waters due to greater water movement, surf action and photosynthetic activity from macrophytes and the plankton.

Moving from the shallow waters to the intermediate depths, one observes a moderation of these shallow water characteristics. In the deeper waters, there is a marked stablization of conditions. Water temperatures are lower and more stable. There is limited turbulence, little penetration of light, sparse vegetation and the ocean floor is covered with a layer of silts and sediments.

5.4.2 Estuarine Sampling: When a survey is made on an estuary, samples are often taken from a boat, usually making an end to end traverse of the estuary. Another method involves taking samples throughout a tidal cycle, every hour or two hours from a bridge or from an anchored boat at a number of fixed points.

In a large bay or estuary where many square miles of area are involved, a grid or series of stations may be necessary. Two sets of samples are usually taken from an area on a given day, one at ebb or flood slack water, and the other three hours earlier, or later, at the half tidal interval. Sampling is scheduled so that the mid-sampling time of each run coincides with the calculated occurrence of the tidal condition.

In locating sampling sites, one must consider points at which tributary waters enter the main stream or estuary, location of shellfish beds and bathing beaches. The sampling stations can be adjusted as data accumulate. For example, if a series of stations half mile apart consistently show similar values, some of these stations may be dropped and other stations added in areas where data shows more variability.

Considerable stratification can occur between the salt water from the sea and the fresh water supplied by a river. It is essential when starting a survey of an unknown estuary to find out whether there is any marked stratification. This can be done by chloride determinations at different locations and depths. It is possible for stratification to occur in one part of an estuary and not in another.

On a flood tide, the more dense salt water pushing up into the less dense fresh river water will cause an overlapping with the fresh water flowing on top. A phenomenon called a salt water wedge can form. As a result, stratification occurs. If the discharge of pollution is in the salt water layer, the contamination will be concentrated near the bottom at the flood tide. The flow or velocity of the fresh water will influence the degree of stratification which occurs. If one is sampling only at the surface, it is possible that the data will not show the polluted underflowing water which was contaminated at a point below the fresh water river. Therefore, where stratification is suspected, samples at different depths will be needed to measure vertical distribution.

5.5 Domestic and Industrial Waste Discharges

It is often necessary to sample secondary and tertiary wastes from municipal waste treatment plants and various industrial waste treatment operations. In situations where the plant treatment efficiency varies considerably, grab samples are collected around the clock at selected intervals for a three to five day period. If it is known that the process displays little variation, fewer samples are needed. In no case should a composite sample be collected for bacteriological examination. The NPDES has established wastewater treatment plant effluent limits for all dischargers. These are often based on maximum and mean values. A sufficient number of samples must be collected to satisfy the permit and/or to provide statistically sound data and give a fair representation of the bacteriological quality of the discharge.

5.6 Recreational Waters

5.6.1 Bathing Beaches: Sampling sites at bathing beaches or other recreational areas should include upstream or peripheral areas and locations adjacent to natural drains that would discharge stormwater, or run-off areas draining septic wastes from restaurants, boat marinas, or garbage collection areas (12). Samples of bathing beach water should be collected at locations and times of heaviest use. Daily sampling, preferably in the afternoon, is the optimum frequency during the season. Weekends and holidays which are periods of highest use must be included in the sampling program. Samples of estuarine bathing waters should be obtained at high tide, ebb tide and low tide in order to determine the cyclic water quality and deterioration that must be monitored during the swimming season.

5.6.2 Swimming Pools: Swimming pool water should be monitored at least daily during maximum use periods, preferably at the overflow. It is important to test swimming pool samples for neutralization of residual chlorine at pool side to assure that the dechlorinating agent was effective.

5.7 Shellfish-Harvesting Waters

Water overlying shellfish-harvesting areas should be sampled during periods of most unfavorable hydrographic conditions, usually at low tide after heavy precipitation. However, shellfish beds are sometimes exposed during low tide and must be sampled during other tidal conditions. Procedures for sampling of shellfish and water in shellfish growing areas are governed by the National Shellfish Sanitation Program's Manual of Operations (13).

5.8 Frequency of Sampling

The frequency of sampling depends upon the type of pollution that is to be measured. Cyclic pollution and its duration are measured as frequently as practical immediately downstream from the source. Uniform pollution loads are measured at greater distances downstream from the source and at less frequent time intervals than cyclic pollution. Climatic and tidal conditions must be considered. in marine and estuarine sampling. A common approach for short-term studies is to collect samples from each site daily and advance the sampling intervals one hour during each 24-hour period to obtain data for a 7–10 day study.

Often the numbers of samples to be collected are specified by NPDES permits, drinking water regulations, or by state requirements. Some standards require a minimum number of samples to be collected each month. Other standards are less explicit and simply indicate that the geometric mean coliform density shall not exceed a certain level each month, with no more than 10%, 20%, etc. of samples exceeding a certain value. Where the number of samples required is undetermined, a sufficient number should be collected to measure the variations in stream conditions.

6. Preservation and Transit of Samples

The adherence to sample preservation and holding time limits is critical to the production of valid data. Samples exceeding these limits should not be analyzed. The following rules must be observed.

6.1 Storage Temperature and Handling Conditions

Bacteriological samples should be iced or refrigerated at a temperature of 1-4 C during transit to the laboratory. Insulated containers are preferable to assure proper maintenance of storage temperature. Care should be taken that sample bottles are not totally immersed in water during transit or storage.

6.2 Holding Time Limitations

Although samples should be examined as soon as possible after collection, they should not be held longer than six hours between collection and initiation of analyses (14). This limit is applied to fresh waters, seawaters and shellfish-bed waters. The exception is water supply samples mailed in from water treatment systems. Current regulations permit these samples to be held up to 30 hours.

6.2.1 Despite the establishment of a six hour limit, sewage samples, organically-rich wastes and marine waters are particularly susceptible to rapid increases or die-away and hence should be held for the shortest time possible to minimize change.

6.2.2 Temporary Field Laboratories: In situations where it is impossible to meet the six hour maximum holding time between collection and processing of samples, the use of temporary field laboratories located near the collection site should be considered.

6.2.3 Delayed Incubation Procedure: If sampling and transit conditions require more than six hours, and the use of field laboratories is impossible, the delayed incubation procedure for total and fecal coliforms and fecal streptococci should be considered.

6.2.4 Public Transportation: Occasionally, commercial forms of transit such as airlines, buslines or couriers can be used to transport samples contained in ice chests to the laboratory. These should be considered only when storage time and temperature requirements and the proper disposition of the samples can be assured.

REFERENCES

1. Public Health Laboratory Service Water Subcommittee, 1953. The effect of sodium thiosulfate on the coliform and *Bacterium coli* counts of non-chlorinated water samples. J. Hyg. 51:572.

2. Shipe, E. L. and A. Fields, 1954. Comparison of the molecular filter technique with agar plate counts for enumeration of *Escherichia coli* in various aqueous concentrations of zinc and copper sulfate. Appl. Microbiol. 2:382.

3. Shipe, E. L. and A. Fields, 1956. Chelation as a method for maintaining the coliform index in water supplies. Public Health Rep. 71:974.

4. Zobell, C. E., 1941. Apparatus for collecting water samples from different depths for bacteriological analysis. J. Marine Research 4:173.

5. Niskin, S., 1962. Water sampler for microbiological study. Deep Sea Research 9:501.

6. Fuhs, G. Wolfgang, 1977. Personal Communication. Director of the Environmental Health Center, Division of Lab and Research, New York State Health Department, Albany, New York.

7. Welch, P.S. 1948. Limnological Methods. Blakiston Company, Philadelphia, PA.

8. Van Donsel, D. J. and E. E. Geldreich, 1972. Relationships of Salmonella to fecal coliforms in bottom sediments. Water Research 5:1079.

9. Wills, Carroll, 1975 (June). Chain of Custody Procedures, National Enforcement Investigation Center-Denver, Colorado, U.S. EPA.

10. Guidelines Establishing Test Procedures for the Analysis of Pollutants. 40 CFR Part 136, 52780, as amended, December 1, 1976.

11. National Interim Primary Drinking Water Regulations, 40 Code of Federal Regulations, Amendments to Part 141, December 24, 1975.

12. Geldreich, E. E., 1975. Handbook for Evaluating Bacteriological Water Laboratories, (2nd ed.) U.S. Environmental Protection Agency, Municipal Environmental Research Laboratory, Cincinnati, Ohio. EPA-670/9–75–006.

13. Hauser, L. S. (ed.), 1965. National Shellfish Sanitation Program. Manual of Operations, Part I: Sanitation of shellfish growing areas. U.S. Public Health Service, Washington, D.C.

14. Public Health Laboratory Service Water Subcommittee, 1953. The effect of storage on the coliform and *Bacterium coli* counts of water samples. Storage for six hours at room and refrigerator temperatures. J. Hyg. 51:559.

PART II. GENERAL OPERATIONS

Section B Laboratory Equipment, Techniques and Media

Most equipment and supplies described in This Manual are available in well-equipped bacteriology laboratories. Other items specific for the membrane filter or multiple-tube dilution methods are used only in water laboratories. This Section describes the required equipment, media and preparation techniques used in the laboratory or in the field. The contents include:

1. Equipment and Supplies
2. Cleaning Glassware
3. Sterilization
4. Preparation and Use of Culture Media
5. Media Composition

   General Use Media
   MF Media for Coliforms
   MPN Media for Coliforms
   Media for Fecal Streptococci
   Media for *Salmonella* and Other Enterics
   Medium for Actinomycetes
6. Laboratory Pure Water
7. Dilution Water

1. Equipment and Supplies

1.1 Inoculating Needles and Loops: Needles and loops of nichrome, platinum or platinum-iridium wire are used to transfer microbes aseptically from one growth medium to another. A 24–26 gauge nichrome wire is recommended. The loop diameter should be at least 3 mm. They are sterilized by heating to redness in a gas flame or in an electric incinerator. Resterilization is not required for replicate transfers of the same bacteria or bacteria-containing materials to a sterile medium. Sterile, disposable hardwood applicator sticks or plastic loops can be used to inoculate fermentation tubes.

1.2 Plastic petri dishes (50 × 12mm) with tight-fitting lids are preferred for MF procedures because they retain humidity and are more practical for use in plastic bags submersed in a water bath incubator. Petri dishes (60 × 15 mm) with loose-fitting lids can be used in incubators with controlled humidity or in plastic boxes with tight covers, containing moist towels.

1.3 Incubators: Incubators are constant temperature air chambers or water baths which provide controlled temperature environment for microbiological tests. The incubator must control temperature within specified tolerance limits of the tests. The air type incubator can be water-jacketed, a dry air unit or aluminum block equipped with thermostat-controlled heating units that maintain 35 C ± 0.5 C. Overcrowding of incubators with plates and tubes must be avoided, because this will interfere with the constancy of the desired temperature. Water bath or aluminum block type incubators used in fecal coliform tests must control the incubation temperature of 44.5 C ± 0.2 C. Covers are required. An accurate thermometer in the water bath monitors temperature. Recording thermometers are recommended for use in the water baths.

1.4 Colony Counter: A standard counting device equivalent to the Quebec Colony Counter that provides good visibility and magnification of at least 1.5 diameters on a non-glare ruled guide plate is recommended.

1.5 pH Equipment: Electrometric pH meter must be accurate to at least 0.1 pH unit.

1.6 Balances: For routine weighing of media and reagents, use a single pan top loader balance having a sensitivity of 0.1 gram at a load of 150 grams. For weighings of less than 2 grams, use a four place analytical balance having a sensitivity of 1 mg at a load of 10 grams.

1.7 Media Preparation Utensils: Use suitable non-corrosive utensils of plastic, glass, stainless steel, or non-chipped enamel. Utensils must be chemically clean before use to prevent contamination of media with chemicals such as chlorine, copper, zinc, chromium and detergents.

1.8 Pipets and Graduated Cylinders

1.8.1 Pipets: Because transfer pipets are an important element in any microbiological method, they must be properly used. If mouth-pipetting is practiced for non-polluted waters, pipets should be cotton-plugged for safety. Blow-out pipets are not to be used.

The major types of transfer pipets (non-volumetric) used in the laboratory are based on the method of draining:

(a) To Contain (TC) Pipets: Pipets calibrated to hold or contain the exact amount specified by the calibration. Pipet must be completely emptied to provide the stated volume. Examples: Transfer Micro and Dual Purpose Pipets.

(b) To Deliver (TD) Pipets: Pipets designed to release the exact calibrated amount when the pipet tip is held vertically against the receiving vessel wall until draining stops. Examples: bacteriological, Mohr, serological, and volumetric pipets.

To Contain and To Deliver pipets are indicated by the letters TC or TD marked respectively on the neck with other calibration information.

(c) Blow-Out Pipets: These pipets are intended for rapid use in serology and are emptied by forceful blow-out. Because the blow-out action always produces an aerosol such pipets should never be used in the microbiology laboratory. The pipets do not deliver the calibrated amount until they are completely emptied.

Blow-Out pipets are marked with a double band etching or fired-in marking on the neck.

(d) Dual Purpose Pipets: A new pipet which combines three calibrations. The pipet has an upper graduation mark which is the To Deliver and Blow-Out line, a lower graduation which is the To Contain line, and carries the double band for Blow-Out.

For microbiological analyses, only bacteriological, serological or Mohr pipets are recommended for use in the To Deliver (TD) mode. Dual purpose or serological pipets which must be used as blow-out pipets, are not recommended for microbiology because of the danger of infection through aerosol and the possible mix-up of TD, TC and blow-out type pipets with subsequent misuse and improper delivery.

(e) Pipet Standards/Specifications: There are several tolerance specifications used to characterize measuring pipets. The analyst should be aware of these limits. When accuracy of measurements is critical, use only pipets within the following Class A limits:

NBS Specification for Mohr Measuring Pipets, Class A Volume Tolerances Circular 602 and Federal Specification NNN-P-350

Capacity/Grad (ml)	Tolerance (ml)
1/10 in 1/100	±0.0025
2/10 in 1/100	±0.004
1 in 1/100	±0.01
1 in 1/10	±0.01
2 in 1/10	±0.01
5 in 1/10	±0.02
10 in 1/10	±0.03
25 in 1/10	±0.05

For routine use, the following tolerances are acceptable:

APHA Specification, Bacteriological Pipets

Capacity/Grad (ml)	Tolerance (ml)
1 in 1	±0.025
1.1 in 1.1, 1.0	±0.025
2.2 in 2.2, 2.1 2.0, 1.0	±0.040
11 in 11	±0.10

USPHS Specification, Bacteriological Pipets for Water Analyses

Capacity/Grad (ml)	Tolerance (ml)
11 in 11, 10, 1	±0.06

1.8.2 Pipetting Devices: Mechanical pipetting devices are recommended for all purposes. Although several devices have recently become available, some are not practical for water analyses because they are too slow and cumbersome. A finger-mounted safety pipettor which can be fabricated in the laboratory (1, 2) is recommended as the most efficient. It can be fitted with a mouthpiece, hand-held bulb or attached to a vacuum pump. Figure II-B-1 shows the finger-held model and Figure II-B-2 shows an enlargement of the plastic ring suction device.

1.8.3 Graduates: For normal laboratory operations, graduates are used for measuring volumes greater than 10 ml. When extreme accuracy is required above 10 ml, volumetric pipets up to 200 ml and volumetric flasks from 10 ml to 6 liters are available. The tolerances given below are acceptable for graduates.

1.9 Pipet Container: Only aluminum, stainless steel, pyrex glass or other non-corrosive heat-resistant containers, either cylindrical or rectangular in shape, should be used. Pipets may be also wrapped in kraft paper. Copper or copper alloy containers must not be used.

1.10 Dilution (Milk Dilution) Bottles or Tubes: Borosilicate or other non-corrosive glass bottles with screw-caps and inert liners.

1.11 Fermentation Tubes and Vials: The fermentation tubes should be large enough to contain the media and inocula in no more than half of the tube depth. There should be sufficient media in the outer fermentation tube to fill the enclosed inverted vial after sterilization and partially submerge it.

NBS Specification, Tolerance of Graduates, in ml

Volume	Demarcations	NBS Class A	NNN-C-940 Type 1 Style 1
2000	20		±10.0
1000	10		±5.0
500	5	±1.3	±2.6
250	2	±0.8	±1.4
100	1	±0.4	±0.6
50	1	±0.26	

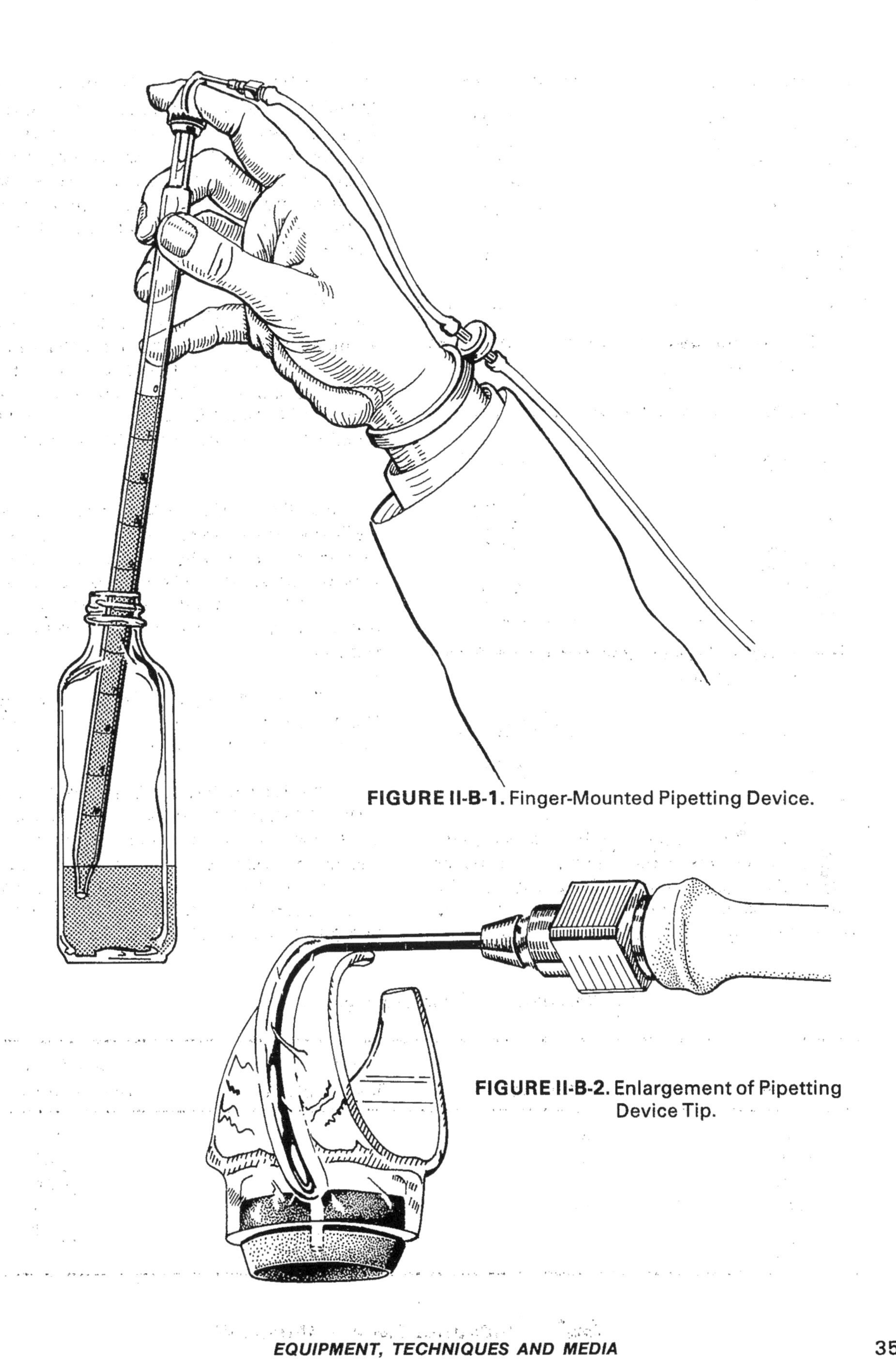

FIGURE II-B-1. Finger-Mounted Pipetting Device.

FIGURE II-B-2. Enlargement of Pipetting Device Tip.

1.12 Sample Bottles: Sample bottles should be borosilicate glass or plastic, resistant to sterilizing conditions and the solvent action of water and which do not produce toxic substances upon sterilization. Wide-mouth ground-glass stopper or screw-cap bottles are acceptable. Bottles equipped with screw-caps are acceptable provided that bacteriostatic or nutritive compounds are not produced from caps or liners. Bottles should be at least 125 ml volume.

2. Cleaning Glassware

In microbiology, clean glassware is crucial to insure valid results. Previously used or new glassware must be thoroughly cleaned with a phosphate-free laboratory detergent and hot water, then rinsed repeatedly with hot water, followed by at least three rinses with laboratory pure water. To determine whether the detergent used contains inhibitory residues, the test procedure for detergent suitability should be performed each time a new type detergent is purchased. See Part IV-A, 5.1.

3. Sterilization

Sterilization is the process that eliminates living organisms from treated substances or objects. Disinfection is the destruction or removal of the infectious agents by chemical or physical means. Usually chemical agents are used as disinfectants (germicide and bactericide are synonymous with the term, disinfectant). Pasteurization is a form of disinfection used for materials which may be altered or damaged by excessive heat. Low heat is applied once or repeatedly to sensitive liquids to destroy vegetative cells. Sterilization can be accomplished by moist heat, dry heat, incineration, filtration, radiation or by the use of the chemical agent, ethylene oxide. Bottles should have loosened caps for penetration of steam or gas.

3.1 Moist Heat: The autoclave is used in the laboratory for moist heat sterilization. It is normally operated at 15 lbs. per sq. in. steam pressure for 15 minutes, producing a temperature inside the autoclave of 121.6 C (250F) at sea level. Steam under pressure provides effective sterilization since it has good penetration power and coagulates microbial protoplasm. The temperature in the autoclave should be monitored. The relationship between the pressure of steam and the temperature is shown in Table II-B-1. The autoclave is used to sterilize solid and liquid media, contaminated materials, discarded cultures, glassware of all types, filtering units, etc. Pressure cookers and vertical autoclaves are not recommended.

3.2 Dry Heat: The hot-air oven is used to sterilize glassware such as petri dishes, pipets, sample bottles and flasks, hardwood applicator sticks, and other articles, but not liquids or materials which will evaporate or deteriorate. Since moisture is not present in the oven, a temperature of 165-170 C (329-338 F) is required for a 2 hour period.

3.3 Incineration: Contaminated materials that are combustible may be disposed of by burning. This method is also used for sterilizing inoculating needles and loops, and flaming the lips of test tubes and flasks.

3.4 Filtration. Filtration is used to sterilize liquids that are heat-sensitive. Filters include those composed of: asbestos-cellulose (Seitz filter), cellulose esters (0.22 μm MF or molecular filter), unglazed porcelain (Chamberland-Pasteur filter), or diatomaceous earth (Berkfeld filter). Although filters are normally used to remove bacteria, porosities small enough for the removal of viruses are available.

3.5 Ultraviolet Radiation: Ultraviolet light includes radiations between 150 and 4000 Angstrom units (A.U.), but radiations less than 1800 A.U. are absorbed by atmospheric oxygen. The greatest killing effect on microorganisms occurs at 2600 A.U. Commercial germicidal ultraviolet lamps emit primarily 2537 A.U. which has 85 percent of the germicidal ability of 2600 A.U. Ultraviolet radiation at the germicidal wavelength is used to sterilize laboratory equipment such as membrane filter units, inoculating rooms, bacteriological hoods and glove boxes. See Part IV-A, 4.2, for monitoring procedures.

3.6 Ethylene Oxide Chemical Sterilization: Low temperature ethylene oxide

TABLE II-B-1

Relationship of Steam Pressure to Temperature in the Autoclave

Pounds of pressure of steam per square inch	Corresponding Temperature In Degrees*	
	Celsius	Fahrenheit
0	100.0	212
5	108.3	227
10	115.5	240
15	121.6	250
20	126.6	260
25	130.5	267
30	134.4	274

*Correct at sea level and only if all air is evacuated from the sterilizing chamber since a mixture of steam and air at a given pressure gives a temperature that is less than that of pure steam only.

gas sterilization is used to sterilize plastics, rubber goods, delicate instruments and other materials that would be damaged by the high temperature of the steam pressure autoclave. Vent sterilized materials according to operating instructions.

4. Preparation and Use of Culture Media

The preparation of culture media and solutions is a critical aspect of water quality testing. In many laboratories media are prepared by nonprofessional personnel. If such personnel are properly trained and guided, they can perform the required tasks efficiently and reliably. However, the supervisor should monitor media preparation to maintain quality control. Use commercial dehydrated media which require only weighing and dissolving of the powder in laboratory pure water for preparation and are much more likely to have uniform high quality than media compounded in the laboratory. See IV-A, 7.3-7.5 for quality control on preparation of media.

4.1 Supplies of Dehydrated Media: The laboratory worker should keep a record of the lot numbers of commercial media in use. The data of receipt of media and the date of opening should be recorded in the quality control log. It is suggested that supplies of dehydrated media be purchased for anticipated use over the next year. Whenever practical, one-quarter pound bottles should be purchased to insure minimal exposure of contents to atmosphere. When a new lot number of medium is used, the contents should be tested for expected performance characteristics (see Part IV-A for details). Stocks of dehydrated media should be stored in a cool, dry place away from sunlight.

4.2 Rehydration of Media: In this Manual, dry ingredients are added to 1 liter (1000 ml) of laboratory pure water. However, liquids such as ethanol are added to graduate and brought up to volume with laboratory pure water. Care must be taken to completely dissolve and mix the ingredients before dispensing the medium into bottles, tubes or flasks. If heat is necessary, it must be applied with caution and for the shortest possible time. Direct heat, boiling water bath and flowing steam are used selectively according to the type and volume of medium as described below:

4.2.1 MF Broths and Agars: Small volumes of broth are rehydrated in a boiling water bath for 5 minutes. Agars and larger volumes of broth require direct heating to the first bubble of boil. Such heating must be applied with stirring and constant attention until agar is dissolved.

4.2.2 Other Broths: Some heat may be required to dissolve the ingredients prior to autoclaving. Heat can be applied by flame or waterbath.

4.2.3 Other Agars: Heat is required for complete solution. For large volumes, direct heat will effect solution more rapidly but must be applied with stirring and constant attention to prevent scorching.

4.3 Sterilization: Media must be dissolved before autoclaving, to insure timing for complete sterilization. The specific recommendations for sterilization are described in subsection 5, Composition of Media, in This Section.

The following general recommendations can be made:

Group	Sterilization
MF and *Salmonella* Media	Heat to boiling only
Media Containing Blood	Autoclave, cool, then add blood
Litmus Milk and Other Milk-containing Media	115 C for 20 minutes (10 lbs. pressure)
Most Carbohydrate-Containing Media for Fermentation Reactions such as Phenol Red Broth, Triple Sugar Iron Agar	118 C for 15 minutes (12 lbs. pressure) or 121 C for 12 minutes

MPN, some *Salmonella* Media and General Use Media such as Trypticase Soy Broth and Agar, Nutrient Broth and Agar	121 C for 12-15 minutes (15 lbs. pressure)

As soon as the autoclave pressure has fallen to zero, the sterilized media should be removed from the autoclave for cooling before use or storage. Refrigerated media should be allowed to come to room temperature before use. Incubate MPN tubed media overnight and discard tubes showing bubbles. Media that have been poured into petri dishes should be used on the day of preparation or refrigerated. Water loss from evaporation can be prevented by storing plates in plastic bags. Plates should be stored inverted. See Part IV-A, 7.9 for storage limits.

Autoclaves should be inspected routinely to insure proper functioning of pressure gauges and thermometers. The use of commercially available temperature indicator devices within the sterilizer (heat-resistant spore preparations, chemically-impregnated tapes, vials containing chemicals, etc.) is recommended to insure sterility (see Part IV-A for details).

5. Composition of Media

The formulas for media used in this Manual and for other more commonly-used media are given in detail in this Section. Normally, these media should not be prepared from basic ingredients when suitable commercial media are available.

The catalogue numbers cited are those for ¼ lb. size if available, from the two U.S. manufacturers whose media were used in most of the method development work. This listing is not restrictive; other sources can be utilized if the user confirms that the formulas are the same as those cited here and that the media produce comparable results.

Culture media used for the examination of water and wastewaters are described separately for:

5.1 General Use Media
5.2 MF Media for Coliforms
5.3 MPN Media for Coliforms
5.4 Media for Fecal Streptococci
5.5 Media for *Salmonella* and Other Enterics
5.6 Medium for Actinomycetes

5.1 General Use Media

5.1.1 Nutrient Agar (Difco 0001–02, BBL 11471)

Use: This medium is used to cultivate pure culture isolates for subsequent gram stain and other examinations and for general cultivation of microorganisms that are not fastidious.

Composition:

Peptone	5.0 g
Beef Extract	3.0 g
Agar	15.0 g

Final pH: 6.8 ± .1

Preparation: Add 23 grams of nutrient agar per liter of laboratory pure water and mix well. Heat in boiling water bath to dissolve the agar completely. Dispense in screw-cap tubes, bottles or flasks and sterilize for 15 minutes at 121 C (15 lbs. pressure). Remove tubes and slant.

5.1.2 Nutrient Broth (Difco 0003–02, BBL 11478)

Use: General laboratory use for the cultivation of non-fastidious microorganisms.

Composition:

Peptone	5.0 g
Beef Extract	3.0 g

Final pH: 6.8 ± .1

Preparation: Add 8 grams of nutrient broth per liter of laboratory pure water and warm to dissolve the medium completely. Dispense in containers and sterilize for 15 minutes at 121 C (15 lbs. pressure).

5.1.3 Trypticase Soy Agar (BBL 11042) Tryptic Soy Agar (Difco 036902)

Use: A general purpose medium for the cultivation of fastidious microorganisms. An excellent blood agar is prepared by adding sheep blood.

Composition:

Trypticase Peptone or Tryptone	15.0 g
Phytone Peptone or Soytone	5.0 g
Sodium Chloride	5.0 g
Agar	15.0 g

Final pH: 7.3 ± .2

Preparation: Add 40 grams of TS agar per liter of laboratory pure water and mix well. Heat in a boiling water bath to dissolve agar completely. Dispense into tubes, bottles or flasks and sterilize for 15 minutes at 121 C (15 lbs. pressure).

For blood agar, cool the sterile, melted agar to 45-46 C and add aseptically 5 ml of sterile defibrinated sheep blood for each 100 ml of agar. Mix flask of agar by swirling and dispense into petri dishes. Blood from other species may be used for particular purposes.

5.1.4 Tryptic Soy Broth (Difco 0370-02 Trypticase Soy Broth (BBL 11767)

Use: A general purpose medium for the cultivation of fastidious microorganisms.

Composition:

Tryptone or Trypticase Peptone	17.0 g
Soytone or Phytone Peptone	3.0 g
Sodium Chloride	5.0 g
Dextrose	2.5 g
Dipotassium Phosphate	2.5 g

Final pH: 7.3 ± .2

Preparation: Add 30 grams of TS broth per liter of laboratory pure water. Warm the broth and mix gently to dissolve the medium completely. Dispense and sterilize for 15 minutes to 121 C (15 lbs. pressure).

5.1.5 Standard Methods Agar (BBL 11637) Plate Count Agar (Difco 0479-02) (Tryptone Glucose Yeast Agar)

Use: Standard Plate Counts in water and in general pour plate procedures.

Composition

Tryptone or Trypticase Peptone	5.0 g
Yeast Extract	2.5 g
Dextrose	1.0 g
Agar	15.0 g

Final pH: 7.0 ± .2

Preparation: Add 23.5 grams of tryptone glucose yeast agar per liter of laboratory pure water. Mix well and heat in boiling water bath to dissolve agar completely. Dispense into screw-cap tubes, flasks or bottles and sterilize for 15 minutes at 121 C (15 lbs. pressure).

5.1.6 Phenol Red Broth Base (Difco 0092-02, BBL 11505)

Use: Phenol red broth base with the addition of carbohydrates is used in fermentation studies of microorganisms because its pH range of 6.9-8.5 indicates slight changes toward acidity. Although 0.5-1.0% carbohydrates have been used, the 1.0 percent level is recommended to prevent reversal of the reactions.

Composition:

Difco (0092–02)

Beef Extract	1.0	g
Proteose Peptone No. 3	10.0	g
Sodium Chloride	5.0	g
Phenol Red	0.018	g

BBL (11505)

Trypticase Peptone	10.0	g
Sodium Chloride	5.0	g
Phenol Red	0.018	g

Final pH: 7.4 $\pm$.2

Preparation: Add 15–16 grams (depending on manufacturer) of phenol red broth base per liter of laboratory pure water. Mix well to dissolve. To this solution, add 10 grams of test carbohydrate if heat-stable. Mix to complete solution. Distribute the medium in fermentation tubes and sterilize not more than 15 minutes at 118 C (12 lbs. pressure). Heat-sensitive carbohydrates or alcohols are filter-sterilized and added to the sterile medium tubes. Check pH and adjust if necessary with 0.1 N NaOH after addition and solution of the carbohydrate.

5.1.7 Purple Broth Base (Difco 0227-02, BBL 11558)

Use: For preparation of carbohydrate broths in fermentation studies, for the cultural identification of pure cultures of microorganisms, particularly the fecal streptococci. Although 0.5-1.0 percent carbohydrate has been used, 1.0 percent is recommended rather than 0.5 percent to prevent reversal of the reaction. BBL product does not contain beef extract.

Composition:

Proteose Peptone No. 3 or Peptone	10.0	g
Beef Extract	1.0	g
Sodium Chloride	5.0	g
Brom Cresol Purple	0.015	g

Final pH: 6.8 $\pm$.2

Preparation: Add 15-16 grams (depending on manufacturer) of purple broth base per liter of laboratory pure water and mix to dissolve. Add 10 grams of the test carbohydrate and dissolve. Dispense in screw-cap fermentation tubes and sterilize for not more than 15 minutes at 118 C (12 lbs. pressure).

Ten percent solutions of the following carbohydrates are prepared for differentiation of fecal streptococci.

- L-Arabinose (Difco 0159, BBL 11960)
- Raffinose (Difco 0174, BBL 12060)
- D-Sorbitol (Difco 0179)
- Glycerol (Difco 0282)
- Lactose (Difco 0156, BBL 11881)
- Inositol (Difco 0164)
- L-Sorbose (Merck 7760)

Sterilize heat-labile carbohydrates and alcohols by passage through a sterile 0.22 μm membrane filtration unit. The carbohydrates made up as 10% solutions are sterilized and added to the sterile base medium at a 1% concentration, wt/unit volume. Check pH and adjust if necessary with 0.1 N NaOH after addition of carbohydrate. Careful use of aseptic techniques is necessary to prevent contamination. As a QC check for contamination, incubate the prepared tubes for 24 hours at 35 C.

5.1.8 Purple Broth Base with Sorbose, pH 10

(Medium may not be commercially available).

Use: To determine the presence of Group Q Streptococci.

Composition:

Same as purple broth base but add 1% sorbose and adjust pH to 10.0.

Final pH: 10.0 $\pm$.2

Preparation: Prepare and sterilize purple broth base as in 5.1.7. Add sufficient volume of a 10% filter-sterilized solution of sorbose to produce a 1% final concentration of sorbose. Adjust pH to 10.0 with sterile 38% sodium phosphate solution ($Na_3PO_4 \cdot 12\ H_2O$).

5.1.9 IMViC Test Media

Use: Differentiation of the coliform group based on indole production from tryptophane broth, acid production in a glucose broth indicated by methyl red color change, formation of acetoin (actylmethylcarbinol) in salt peptone glucose broth and use of citrate as the sole carbon source.

(a) Tryptone 1% (Difco 0123–02) Tryptophane Broth (BBL 11920)

Use: For the detection of indole as a by-product of the metabolism of tryptophane and for the identification of bacteria.

Composition:

Tryptone or Trypticase Peptone	10.0 g

Final pH: 7.2 ± .2

Preparation: Add 10 grams of Tryptone or Trypticase to 900 ml of laboratory pure water and heat with mixing until dissolved. Bring solution to 1000 ml in a graduate or flask. Dispense in five ml volume tubes and sterilize for 15 minutes at 121 C (15 lbs. pressure).

(b) MR-VP Broth (Buffered Glucose) (BBL 11382, Difco 0016–02)

Use: For the performance of the Methyl Red and Voges-Proskauer Tests in differentiation of the coliform group.

Composition:

Buffered Peptone or Polypeptone	7.0 g
Dextrose	5.0 g
Dipotassium Phosphate	5.0 g

Final pH: 6.9 ± .2

Preparation: Add 17 grams of MR-VP medium to 1 liter of laboratory pure water. Mix to dissolve. Dispense 10 ml volumes into tubes and sterilize for 15 minutes at 121 C (15 lbs. pressure).

(c) Simmons Citrate Agar (BBL 11619, Difco 0091–02)

Use: Differentiation of gram-negative enteric bacteria on the basis of citrate utilization.

Composition:

Magnesium Sulfate	0.2 g
Monoammonium Phosphate	1.0 g
Dipotassium Phosphate	1.0 g
Sodium Citrate	2.0 g
Sodium Chloride	5.0 g
Brom Thymol Blue	0.08 g
Agar	15.0 g

Final pH: 6.8 ± .2

Preparation: Add 24.2 grams of Simmons Citrate agar per liter of laboratory pure water. Heat in boiling water bath with mixing for complete solution. Dispense into screw-cap tubes and sterilize for 15 minutes at 121 C (15 lbs. pressure). Cool tubes as slants.

5.1.10 Motility Test Medium (Edwards and Ewing)(BBL 11435)

Use: Detection of motility of gram-negative enteric bacteria.

Composition:

Beef Extract	3.0 g
Peptone	10.0 g
Sodium Chloride	5.0 g
Agar	4.0 g

Final pH: 7.3 ± .2

Preparation: Add 22 grams of dry medium to 1 liter of laboratory pure water. Add 0.05 grams of triphenyl tetrazolium chloride/liter. Heat with mixing to boil for 1 minute. Dispense 10 ml volumes into tubes and sterilize for 15 minutes at 121 C (15 lbs. pressure).

5.2 MF Media for Coliforms

Prepare heat-sensitive broths in sterile flasks.

5.2.1 M-FC Broth Base (Difco 0883-02) M-FC Broth (BBL 11364)

Use: Detection and enumeration of fecal coliform microorganisms by the membrane filter procedure.

Composition:

Tryptose or Biosate Peptone	10.0 g
Proteose Peptone No. 3 or Polypeptone	5.0 g
Yeast Extract	3.0 g
Sodium Chloride	5.0 g
Lactose	12.5 g
Bile Salts No. 3 or Bile Salts Mixture	1.5 g
Aniline Blue	0.1 g

Final pH: 7.4 ± .2

Preparation:

Rosolic Acid

Dissolve 1 gram of rosolic acid in 100 ml of 0.2 N sodium hydroxide to make a rosolic acid solution. Note: The quality of present supplies is quite variable. Performance of new batches should be compared against previous batch before it is exhausted.

Autoclaving will decompose rosolic acid reagent. Stock solutions should be stored in the dark at 4 C. Discard after 2 weeks or sooner if the color changes from dark red to muddy brown. The rosolic acid may be omitted from testing samples with stressed organisms and low background count.

M-FC Broth

Add 37 grams of M-FC medium to 1 liter of laboratory pure water containing 10 ml of the rosolic acid solution. Heat in a boiling water bath to dissolve before use. Store the prepared medium at 4 C in a refrigerator. Discard unused medium after 96 hours.

M-FC Agar

Prepare by adding 15 grams of agar per liter of M-FC broth. Heat in boiling water bath to solution then cool to about 45 C and add to 50 mm diameter glass or plastic petri dishes to a minimal agar depth of 2-3 mm. Allow to solidify. Protect the prepared medium from light. It can be stored at 4 C for up to 2 weeks.

Caution: Do not autoclave M-FC Broth or Agar.

5.2.2 M-Coliform Broth (BBL 11119) M-Endo Broth MF (Difco 0749–02)

Use: A selective and differential medium for enumeration of members of the coliform group by the membrane filter technique.

Composition:

Tryptose or Polypeptone	10.0 g
Thiopeptone or Thiotone	5.0 g
Casitone or Trypticase	5.0 g
Yeast Extract	1.5 g
Lactose	12.5 g
Sodium Chloride	5.0 g
Dipotassium Hydrogen Phosphate	4.375 g
Potassium Dihydrogen Phosphate	1.375 g
Sodium Lauryl Sulfate	0.050 g
Sodium Desoxycholate	0.1 g
Sodium Sulfite	2.1 g
Basic Fuchsin	1.05 g

Final pH: 7.2 ± .2

Preparation: Add 48 grams of M-Endo medium to 1 liter of laboratory pure water containing 20 ml of 95% ethanol. Denatured alcohol should not be used. Heat in boiling water bath for solution. Store prepared medium in the dark at 4 C. Discard the unused medium after 96 hours.

Prepare M-Endo agar by adding 15 grams of agar per liter of M-Endo medium. Heat to boiling, cool to about 45 C and dispense into glass or plastic petri dishes to provide minimal agar depth of 2-3 mm and allow to solidify. Protect prepared medium from light. It can be stored at 4 C for up to 2 weeks.

5.2.3 M-Endo Holding Medium

Use: Holding Medium in the delayed incubation total coliform procedure.

Composition:

Same as M-Endo Broth but add 0.384 grams of sodium benzoate per 100 ml.

Preparation:

Prepare M-Endo Broth as described in 5.2.2 and add 3.2 ml of a 12% solution of sodium benzoate per 100 ml of medium. Add cycloheximide if needed.

Sodium Benzoate Solution

Final pH: 7.2 ± .2

Dissolve 12 grams of sodium benzoate in about 85 ml of laboratory pure water, then bring to 100 ml final volume. Sterilize by autoclaving or filtration. Discard the solution after 6 months.

Cycloheximide Solution (Optional)

Cycloheximide is used for samples that have shown problems of overgrowth with fungi. Prepare an aqueous solution containing 1.25 grams of cycloheximide/100 ml of laboratory pure water. Store solution in refrigerator. Discard and remake after 6 months. Add 4 ml of the aqueous solution of cycloheximide per 100 ml of M-Endo Holding Medium.

Caution: Cycloheximide is a powerful skin irritant that should be handled with care. See Manufacturer's Directions (Actidione, Upjohn, Kalamazoo, MI).

5.2.4 M-Endo Agar LES (Difco 0736-02, BBL 11203)

Use: Determination of total coliforms using a two-step membrane filter technique with lauryl tryptose broth as the preliminary enrichment.

Composition:

Yeast Extract	1.2	g
Casitone or Trypticase Peptone	3.7	g
Thiopeptone or Thiotone	3.7	g
Tryptose or Biosate Peptone	7.5	g
Lactose	9.4	g
Dipotassium Hydrogen Phosphate	3.3	g
Potassium Dihydrogen Phosphate	1.0	g
Sodium Chloride	3.7	g
Sodium Desoxycholate	0.1	g
Sodium Lauryl Sulfate	0.05	g
Sodium Sulfite	1.6	g
Basic Fuchsin	0.8	g
Agar	14–15.0	g

Final pH: 7.2 ± .2

Preparation: Add 50 or 51 grams, depending on manufacturer, of agar per liter of laboratory pure water to which has been added 20 ml of 95% ethanol. Heat in boiling water bath to dissolve completely. Cool to about 45 C and dispense into 60 mm glass or plastic petri dishes to provide a minimal agar depth of 2-3 mm. Allow to solidify. If larger dishes are used, dispense sufficient agar to give equivalent depth. Protect prepared medium from light. It can be stored at 4 C for up to 2 weeks.

Caution: Do not autoclave.

5.2.5 M-Coliform Holding Broth (Difco 0842-02) (LES Holding Medium)

Use: Holding medium in the delayed-incubation total coliform procedure.

Composition:

Tryptone or Trypticase Peptone	3.0	g
M-Endo Broth MF	3.0	g
Dipotassium Hydrogen Phosphate	3.0	g
Sodium Benzoate	1.0	g
Sulfanilamide	1.0	g
Paraminobenzoic Acid	1.2	g
Cycloheximide	0.5	g

Final pH: 7.1 ± .2

Preparation: Add 12.7 grams per liter of laboratory pure water and mix to dissolve. Do not heat to dissolve medium.

5.2.6 M-VFC Holding Medium (3)

Use: Holding medium in the delayed incubation test for fecal coliform microorganisms.

Composition:

Casitone, Vitamin Free	0.2 g
Sodium Benzoate	4.0 g
Sulfanilamide	0.5 g

Final pH: 6.7 ± .2

Preparation: Add 4.7 grams of medium per liter of laboratory pure water containing 10 ml of 95% ethanol. Denatured alcohol should not be used. Heat slightly to dissolve ingredients, then sterilize by membrane filtration (0.22 µm). Store prepared medium at 4 C. Discard after 1 month.

5.3 MPN Media for Coliforms

5.3.1 Lauryl Sulfate Broth (BBL 11338) Lauryl Tryptose Broth (Difco 0241-02)

Use: Primary medium for the Presumptive Test for the total coliform group.

Composition:

Tryptose or Trypticase Peptone	20.0 g
Lactose	5.0 g
Dipotassium Hydrogen Phosphate	2.75 g
Potassium Dihydrogen Phosphate	2.75 g
Sodium Chloride	5.0 g
Sodium Lauryl Sulfate	0.1 g

Final pH: 6.8 ± .2

Preparation: Add 35.6 grams of the medium to 1 liter of laboratory pure water and mix to dissolve. Dispense 10 ml volumes in fermentation tubes (150 × 20 mm tubes containing 75 × 10 mm tubes) for testing 1 ml or less of samples. For testing 10 ml volumes of samples, add 71.2 grams of the medium per liter of laboratory pure water and mix to dissolve. Dispense in 10 ml amounts in fermentation tubes (150 × 25 mm tubes containing 75 × 10 mm tubes). Sterilize for 15 minutes at 121 C (15 lbs. pressure). The concentration of the medium should vary with the size of the sample according to the table below.

Compensation in Lauryl Tryptose Broth for Diluting Effects of Samples

LTB Medium /Tube in ml	Sample Size /Dilution in ml	Medium Concentration	Dehydrated LTB in grams/liter
10	0.1 to 1.0	1x	35.6
10	10	2x	71.2
20	10	1.5x	53.4

5.3.2 Brilliant Green Bile 2% (Difco 0007–02) Brilliant Green Bile Broth 2% (BBL 11079)

Use: Recommended for the confirmation of MPN Presumptive Tests for members of the coliform group.

Composition:

Peptone	10.0 g
Lactose	10.0 g
Oxgall or Bile	20.0 g
Brilliant Green	0.33 g

Final Ph: 7.2 ± .2

Preparation: Add 40 grams of brilliant green bile broth to 1 liter of laboratory pure water. Dispense 10 ml volumes of the broth in fermentation tubes (150 × 20 mm tubes containing 75 × 10 mm tubes). Sterilize for 15 minutes at 121 C (15 lbs. pressure).

5.3.3 Levine's Eosin Methylene Blue Agar (Difco 0005-02, BBL 11220)

Use: Isolation of coliform-like colonies as a preliminary to total coliform Completed Test Procedure.

Composition:

Peptone	10.0	g
Lactose	10.0	g
Dipotassium Phosphate	2.0	g
Agar	15.0	g
Eosin Y	0.4	g
Methylene Blue	0.065	g

Final pH: 7.1 ± .2

Preparation: Add 37.5 grams of Levine's E.M.B. agar to 1 liter of laboratory pure water and heat in a boiling water bath until dissolved completely. Dispense into tubes, flasks or bottles and sterilize for 15 minutes at 121 C (15 lbs. pressure). A flocculant precipitate may form after autoclaving. Resuspend the precipitate by gently shaking the flask prior to pouring the medium into sterile petri dishes.

5.3.4 EC Medium (Difco 0314-02) EC Broth (BBL 11187)

Use: Detection and enumeration of fecal coliform bacteria.

Composition:

Tryptose or Trypticase Peptone	20.0	g
Lactose	5.0	g
Bile Salts No. 3 or Bile Salts Mixture	1.5	g
Dipotassium Phosphate	4.0	g
Monopotassium Phosphate	1.5	g
Sodium Chloride	5.0	g

Final pH: 6.9 ± .2

Preparation: Add 37 grams of EC medium to 1 liter of laboratory pure water. Dispense into fermentation tubes (150 × 20 mm tubes containing 75 × 10 mm tubes). Sterilize for 15 minutes at 121 C (15 lbs. pressure).

5.4 Media for Fecal Streptococci

5.4.1 KF Streptococcus Agar (Difco 0496-02) KF Streptococcal Agar (BBL 11313)

Use: Selective cultivation and enumeration of fecal streptococci by direct plating or the membrane filter technique.

Composition:

Proteose Peptone No. 3 or Polypeptone	10.0	g
Yeast Extract	10.0	g
Sodium Chloride	5.0	g
Sodium Glycerophosphate	10.0	g
Maltose	20.0	g
Lactose	1.0	g
Sodium Azide	0.4	g
Brom Cresol Purple (in Difco medium only)	0.015	g
Agar	20.0	g

Final pH: 7.2 ± .2

Preparation: Add 76.4 grams of the medium per liter of laboratory pure water. Dissolve by heating in a boiling water bath with agitation. Heat in boiling water bath for 5 minutes after solution is complete. Caution: Do not autoclave. Cool to 60 C and add 1 ml of a filter-sterilized 1% aqueous solution of 2, 3, 5-triphenyl tetrazolium chloride per 100 ml of agar. If necessary, adjust pH to 7.2 with 10% Na_2CO_3. Do not hold the completed medium (with indicator) at 44–46 C for more than 4 hours before use. Store prepared medium (without indicator) in the dark for up to 30 days at 4 C. TTC solution is light-sensitive. It should be stored in the refrigerator and protected from light.

5.4.2 Azide Dextrose Broth (Difco 0837-02, BBL 11000)

Use: Primary inoculation medium for Fecal Streptococci Presumptive Test.

Composition:

Beef Extract	4.5	g
Tryptose or Polypeptone	15.0	g

Dextrose	7.5 g
Sodium Chloride	7.5 g
Sodium Azide	0.2 g

Final pH: 7.2 ± .2

Preparation; Add 34.7 grams of azide dextrose broth to 1 liter of laboratory pure water. Dissolve and dispense into tubes without inner vials. Note: Azide dextrose broth should be sterilized at 118 C for 15 minutes (12 lbs. pressure). Prepare the medium in multiple strength for larger inocula to preserve the correct concentration of ingredients. For example, if 10 ml of inoculum is to be added to 10 ml of medium, the medium should be prepared double strength.

5.4.3 Ethyl Violet Azide Broth (BBL 11226) EVA Broth (Difco 0606-02)

Use: Confirmed Test for fecal streptococci.

Composition:

Tryptose or Biosate Peptone	20.0 g
Dextrose	5.0 g
Dipotassium Phosphate	2.7 g
Monopotassium Phosphate	2.7 g
Sodium Chloride	5.0 g
Sodium Azide	0.4 g
Ethyl Violet	0.83 mg

Final pH: 7.0 ± .2

Preparation: Add 35.8 grams of the medium to 1 liter of laboratory pure water. Dissolve and dispense in 10 ml amounts into tubes. Sterilize for 15 minutes at 121 C (15 lbs. pressure).

5.4.4 PSE Agar (Pfizer Selective Enterococcus Agar) 224B, formerly from Pfizer Diagnostics Division. Now available from Grand Island Biological Company (GIBCO), 3175 Staley Road, Grand Island, NY 14072.

Use: Isolation of fecal streptococci.

Composition:

Pfizer Peptone C	17.0 g
Pfizer Peptone B	3.0 g
Pfizer Yeast Extract	5.0 g
Pfizer Bile	10.0 g
Sodium Chloride	5.2 g
Esculin	1.0 g
Sodium Citrate	1.0 g
Ferric Ammonium Citrate	0.5 g
Sodium Azide	0.25 g
Agar	15.0 g

Final pH: 7.1 ± .2

Preparation: Add 58 grams of PSE agar to 1 liter of laboratory pure water. Heat in a boiling water bath to complete solution. Dispense into tubes or flasks and sterilize for 15 minutes at 121 C (15 lbs. pressure).

5.4.5 Brain Heart Infusion Broth (BHI) (Difco 0037-02, BBL 11058)

Use: For separation of enterococcus group organisms from *S. bovis* and *S. equinus* by testing for growth at 10 C and 45 C. For general cultivation of fastidious microorganisms.

Composition:

Calf Brain Infusion	200.0 g
Beef Heart Infusion	250.0 g
Peptone	10.0 g
Sodium Chloride	5.0 g
Disodium Phosphate	2.5 g
Dextrose	2.0 g

Final pH: 7.4 ± .2

Preparation: Dissolve 37 grams of brain heart infusion broth in 1 liter of laboratory pure water. Dispense in 8–10 ml volumes in screw-cap tubes and sterilize for 15 minutes at 121 C (15 lbs. pressure). If the medium is not used the same day as prepared and sterilized, heat at 100 C for several minutes to remove absorbed oxygen, and cool quickly without agitation, just prior to inoculation.

5.4.6 Brain Heart Infusion Agar (Difco 0418-02, BBL 11064)

Use: Cultivation of streptococci isolates or other fastidious bacteria.

Composition: Brain heart infusion agar contains the same components as BHI broth (see 5.4.5 above) with the addition of 15.0 grams agar.

Preparation: Heat in boiling water bath until dissolved. Dispense 10–12 ml of medium in screw-cap test tubes and slant after sterilization. Sterilize for 15 minutes at 121 C (15 lbs. pressure).

Final pH: 7.4 ± .2

5.4.7 Brain Heart Infusion (BHI) Broth with 6.5% NaCl

Use: Identification of fecal streptococci.

Composition: Brain heart infusion broth with 6.5% NaCl is the same as BHI broth in 5.4.5 above with addition of 60.0 grams NaCl per liter of medium.

Since most commercially available dehydrated media already contain sodium chloride, this amount is taken into consideration for determining the final NaCl percentage above.

5.4.8 Brain Heart Infusion Broth (BHI), pH 9.6

Use: Identification of fecal streptococci.

Composition: Same as for BHI broth above in 5.4.5 with addition of sterile 38% sodium phosphate solution ($Na_3PO_4 \cdot 12\ H_2O$) to produce a final pH of 9.6.

5.4.9 Brain Heart Infusion Broth with 40% Bile

Use: Verification of fecal streptococci.

Composition: Same as for BHI broth above in 5.4.5 with addition of 40 ml of sterile 10% oxgall to 60 ml of basic medium or 668 ml to each liter of medium.

Final pH: 7.4 ± .2

Preparation of Medium: Add 37 grams of BHI broth to 1 liter of laboratory pure water and heat gently with agitation to dissolve. Dispense 60 ml amounts of the medium into screw-cap flasks. Sterilize for 15 minutes at 121 C (15 lbs. pressure).

Preparation of 10% oxgall: Add 10 grams of oxgall per 100 ml of laboratory pure water. After dissolving and mixing, filter-sterilize the solution.

Preparation of Final Medium: Cool the BHI broth and add 40 ml of the sterile 10% oxgall solution to each 60 ml of sterile, cool BHI broth, resulting in a 40% bile concentration. Dispense as needed aseptically in 10 ml volumes into sterile culture tubes.

5.4.10 Starch Agar

(Medium may not be commercially available).

Use: Starch hydrolysis tests for separation and confirmation of fecal streptoccal species.

Composition:

Peptone	10.0 g
Yeast Extract	5.0 g
Sodium Chloride	5.0 g
Starch (Soluble)	2.0 g
Agar	15.0 g

Final pH: 6.8 ± .2

Preparation: Add 37 grams of starch agar in 1 liter of cold laboratory pure water. Heat to boiling to dissolve and dispense into tubes, flasks or bottles. Sterilize for 15 minutes at 121 C (15 lbs. pressure). Cool medium after sterilization and pour into petri dishes. Allow to solidify.

5.4.11 Starch Liquid Medium

(Medium may not be commercially available).

Use: Starch hydrolysis for speciation of fecal streptococci.

Composition:

Tryptone or Trypticase Peptone	10.0 g
Yeast Extract	3.0 g
Dipotassium Phosphate	2.0 g
Glucose	0.5 g
Starch (Soluble)	5.0 g

Final pH: 6.8 ± .2

Preparation: Add 30.5 grams of dry ingredients to 1 liter of laboratory pure water and heat to boiling. Dispense into tubes and sterilize for 15 minutes at 121 C (15 lbs. pressure).

5.4.12 Nutrient Gelatin (BBL 11481, Difco 0011-02)

Use: Detection of gelatin liquefaction for identification of the fecal streptococci and other microorganisms.

Composition:

Peptone	5.0 g
Beef Extract	3.0 g
Gelatin	120.0 g

Final pH: 6.8 ± .2

Preparation: Add 128 grams of nutrient gelatin to 1 liter of cold laboratory pure water and warm to about 50 C to dissolve the medium. Dispense 5 ml in screw-cap test tubes and sterilize for 15 minutes at 121 C (15 lbs. pressure). Store tubes in refrigerator until use.

5.4.13 Litmus Milk (Difco 0107-02, BBL 11343)

Use: To separate and identify fecal streptococci and generally to determine the action of bacteria on milk.

Composition:

Skim Milk	100.0 g
Litmus	0.75 g

Final pH: 6.8 ± .2

Preparation: Add 100 grams of litmus milk to 1 liter of laboratory pure water and warm to about 50 C to dissolve the medium. Dispense 10 ml volumes into screw-cap tubes and sterilize for 20 minutes at 115 C (10 lbs. pressure). Do not overheat. Control pressure and time carefully since overheating or prolonged heating during sterilization can caramelize the milk sugar.

5.4.14 Skim Milk with 0.1% Methylene Blue

Use: Identification of fecal streptococci by reduction of methylene blue.

Composition:

Skim Milk Powder	100.0 g
Methylene Blue Powder	1.0 g

Final pH: 6.4 ± .2

Preparation: Add 100 grams of skim milk powder and 1 gram of methylene blue to 1 liter of distilled water, and warm to 50 C to dissolve the medium. Dispense 10 ml volumes into screw-cap tubes and sterilize for 20 minutes at 115 C (10 lbs. pressure). Do not overheat. Prolonged heating or overheating during sterilization results in caramelization of the milk sugar.

5.4.15 Brain Heart Infusion Agar with 0.04% Potassium Tellurite

Use: Identification of fecal streptococci by tellurite reduction.

Composition:

Calf Brain Infusion	200.0 g
Beef Heart Infusion	250.0 g
Proteose Peptone	10.0 g
Dextrose	2.0 g
Sodium Chloride	5.0 g
Disodium Phosphate	2.5 g
Agar	15.0 g

Final pH: 7.4 ± .2

Preparation: Add 52 grams of brain heart infusion agar to 1 liter of cold laboratory pure water. Heat in a boiling water bath to dissolve the agar and dispense 100 ml volumes in screw-cap flasks and sterilize for 15 minutes at 121 C (15 lbs. pressure). Cool to 50 C and add 1 ml of sterile warm (50 C) 4% potassium tellurite to each 100 ml flask of brain heart infusion agar. This should produce a final potassium tellurite concentration of 0.04%. Dispense melted sterile medium into sterile petri dishes.

5.4.16 Blood Agar with 0.04% Potassium Tellurite

Use: Identification of fecal streptococci by tellurite reduction.

Composition:

Evans Peptone	10.0 g
Sodium Chloride	5.0 g
Meat Extract (Lab Lemco)	10.0 g
Yeast Extract	3.0 g
Agar	15.0 g

Final pH: 7.2 ± .2

Preparation: Add 43 grams of blood agar base to 1 liter of laboratory pure water. Heat in a boiling water bath to dissolve the agar and dispense 100 ml volumes in screw-cap flasks, sterilize for 15 minutes at 121 C (15 lbs. pressure) and cool to 44-46 C. Aseptically add 10% sterile defibrinated horse blood to the medium. The mixture is heated at 70 C for 10 minutes, then cooled to 45 C. A filter-sterilized 4.0% solution of potassium tellurite is added aseptically to give a final concentration of 0.04%. Dispense the completed medium into petri dishes.

5.4.17 Tetrazolium Glucose (TG) Agar or 2, 3, 5-Triphenyl Tetrazolium Chloride (TTC Agar) (Medium may not be commercially available).

Use: Identification of fecal streptococci by tetrazolium reduction.

Composition:

Peptone	10.0 g
Beef Extract	10.0 g
Sodium Chloride	5.0 g
Agar	14.0 g

Final pH: 7.0–7.3

Preparation of 50% Glucose Solution: Weigh out 50 grams of reagent grade glucose. Add to 50 ml laboratory pure water in a 100 ml volumetric flask. Dissolve glucose and bring up to volume. Filter-sterilize solution and store in a screw-cap flask.

Preparation of 1% TTC Solution: Weigh out 1 gram of 2, 3, 5 triphenyl tetrazolium chloride. Add to 50 ml laboratory pure water in a 100 ml volumetric flask. Dissolve and bring up to volume. Filter-sterilize solution and store in a screw-cap flask.

Preparation of Final Medium: Add 39 grams of TG agar to 1 liter of cold laboratory pure water and heat in a boiling water bath to dissolve the agar. Sterilize for 15 minutes at 121 C (15 lbs. pressure) and cool to about 50 C. To 970 ml of liquid TG agar, aseptically add 20 ml of 50% glucose solution and 10 ml of 1% TTC solution. Mix well and pour into sterile petri dishes.

5.4.18 Blood Agar Base (Optional – 10% Blood) (Difco 0045–02, BBL 11036)

Use: Identification of hemolytic properties of fecal streptococci.

Composition:

BBL 11036

Beef Heart Infusion	375.0 g
Tryptose or Thiotone (Peptone)	10.0 g
Sodium Chloride	5.0 g
Agar	15.0 g

Difco 0045–02

Beef Heart Infusion	500.0 g
Tryptose or Thiotone (Peptone)	10.0 g
Sodium Chloride	5.0 g
Agar	15.0 g

Final pH: 7.4 ± .2

Preparation: Add 40 grams of blood agar base to 1 liter of cold laboratory pure water. Heat in a boiling water bath to dissolve the agar. Dispense in 90 ml volumes into screw-cap flask. Sterilize for 15 minutes at 121 C (15 lbs. pressure). Store at 4 C for later use. When ready to use, heat/cool the blood agar base to 45–50 C and add 10% by volume of fresh sterile defibrinated horse blood. Mix and pour into sterile petri dishes.

5.5 Media for *Salmonella*

5.5.1 Selenite F Broth (BBL 11607, Difco 0275–02)

Use: Primary enrichment of salmonellae.

Composition:

Tryptone or Polypeptone	5.0 g
Lactose	4.0 g
Disodium Hydrogen Phosphate	10.0 g
Sodium Selenite	4.0 g

Final pH: 7.0 $\pm$ .2

Preparation: Add 23 grams of selenite broth to 1 liter of laboratory pure water. Mix and warm gently until dissolved. Dispense in tubes to a depth of 6 cm and expose to flowing steam for 15 minutes. Avoid excessive heating. Do not autoclave. Sterilization is unnecessary if broth is used immediately.

5.5.2 Tetrathionate Broth Base (Difco 0104-02, BBL 11706)

Use: Primary enrichment of salmonellae.

Composition:

Proteose Peptone or Polypeptone	5.0 g
Bile Salts	1.0 g
Calcium Carbonate	10.0 g
Sodium Thiosulfate	30.0 g

Final pH: 7.8 $\pm$ .2

Preparation: Add 46 grams of tetrathionate broth base to 1 liter of laboratory pure water and heat to boiling. Cool to less than 45 C and add 20 ml of iodine solution.* Mix and dispense in 10 ml volumes into screw-cap tubes. Do not heat after the addition of the iodine. Do not autoclave. The tetrathionate broth base without iodine may be stored for later use. The complete medium (with iodine) should be used on the day it is prepared.

*The iodine-iodide solution is prepared by dissolving 6 grams iodine crystals and 5 grams potassium iodide in 20 ml of distilled water.

5.5.3 Dulcitol Selenite Broth

(Medium may not be commercially available).

Use: Primary enrichment of salmonellae.

Composition:

Proteose Peptone	4.0 g
Yeast Extract	1.5 g
Dulcitol	4.0 g
Sodium Selenite	5.0 g
Disodium Hydrogen Phosphate	1.25 g
Potassium Dihydrogen Phosphate	1.25 g

Final pH: 6.9 $\pm$ .2

Preparation: Add 16.5 grams of dulcitol selenite broth to 1 liter of laboratory pure water and heat carefully to dissolve ingredients. Do not boil. The prepared medium should be buff-colored. Dispense into screw-cap tubes to a depth of 6 cm. Do not autoclave.

5.5.4 Tetrathionate Brilliant Green Broth

Use: Primary enrichment for salmonellae.

Composition: Same as tetrathionate broth base (5.5.2) with addition of 0.01 gram of brilliant green per liter.

5.5.5 Brilliant Green Agar (BBL 11072, Difco 0285–02)

Use: As a primary plating medium for the isolation of salmonellae.

Composition:

Yeast Extract	3.0	g
Polypeptone or Proteose Peptone	10.0	g
Sodium Chloride	5.0	g
Lactose	10.0	g
Saccharose (Sucrose)	10.0	g
Phenol Red	0.08	g
Brilliant Green	0.0125	g
Agar	20.0	g

Final pH: 6.9 ± .2

Preparation: Add 58 grams of brilliant green agar to 1 liter of cold laboratory pure water and heat to boiling. Dispense in screw-cap flasks and sterilize for 15 minutes at 121 C (15 lbs. pressure). Pour into sterile petri dishes.

Warning: A longer period of sterilization will reduce the selectivity of the medium.

5.5.6 Xylose Lysine Brilliant Green (XLBG) Agar

Use: *Salmonella* Differentiation.

Composition of XL Agar Base:

BBL 11836

Xylose	3.5	g
L-Lysine	5.0	g
Lactose	7.5	g
Saccharose (Sucrose)	7.5	g
Sodium Chloride	5.0	g
Yeast Extract	3.0	g
Phenol Red	0.08	g
Agar	13.5	g

Difco 0555–02

Xylose	3.75	g
L-Lysine	5.0	g
Lactose	7.5	g
Saccharose (Sucrose)	7.5	g
Sodium Chloride	5.0	g
Yeast Extract	3.0	g
Phenol Red	0.08	g
Agar	15.0	g

Final pH: 7.4 ± .2

Preparation: Add 45 or 47 grams of XL agar base to 1 liter of cold laboratory pure water. Heat in a boiling water bath to dissolve the agar. Prior to sterilization, add 1.25 ml of 1% aqueous brilliant green. Sterilize for 15 minutes at 121 C (15 lbs. pressure). Cool the sterilized medium to about 45–50 C and add 20 ml of a solution containing 34% sodium thiosulfate and 4% ferric ammonium citrate. Pour into sterile petri dishes.

5.5.7 Xylose Lysine Desoxycholate (XLD) Agar

Use: Differentiation of *Salmonella*.

Composition:

BBL 11837

Xylose	3.5	g
L-Lysine	5.0	g
Lactose	7.5	g
Saccharose (Sucrose)	7.5	g
Sodium Chloride	5.0	g
Yeast Extract	3.0	g
Phenol Red	0.08	g
Agar	13.5	g
Sodium Desoxycholate	2.5	g
Sodium Thiosulfate	6.8	g
Ferric Ammonium Citrate	0.8	g

Difco 0788–02

Xylose	3.75	g
L-Lysine	5.0	g
Lactose	7.5	g
Saccharose (Sucrose)	7.5	g
Sodium Chloride	5.0	g
Yeast Extract	3.0	g
Phenol Red	0.08	g
Agar	15.0	g
Sodium Desoxycholate	2.5	g
Sodium Thiosulfate	6.8	g
Ferric Ammonium Citrate	0.8	g

Final pH: 7.4 ± .2

Preparation: Add 55 or 57 grams of XLD agar in 1 liter of cold laboratory pure water, heat to boiling with mixing. Do not overheat and do not autoclave. Pour into sterile petri dishes.

Note: Taylor and Schelhart report better recoveries by using XL Agar Base (BBL 11835 or Difco 9555), see 5.5.6, and adding separately, sterile solutions of the last three ingredients (4).

5.5.8 Bismuth Sulfite Agar (Difco 0073–02, BBL 11030)

Use: Differentiation of salmonellae, especially *S. typhosa*.

Composition:

Polypeptone or Proteose Peptone	10.0	g
Beef Extract	5.0	g
Dextrose	5.0	g
Disodium Hydrogen Phosphate	4.0	g
Ferrous Sulfate	0.3	g
Bismuth Sulfite Indicator	8.0	g
Brilliant Green	0.025	g
Agar	20.0	g

Final pH: 7.6 ± .2

Preparation: Add 52 grams of bismuth sulfite agar to 1 liter of cold laboratory pure water and heat to boiling. Do not autoclave or overheat. Twirl the flask prior to pouring plates to evenly dispense the characteristic precipitate. Use the plated medium on the day prepared.

5.5.9 Triple Sugar Iron (TSI) Agar

Use: Differentiation of gram negative enterics by their differing ability to ferment dextrose, lactose and sucrose and ability to produce hydrogen sulfide.

Composition:

Difco 0265–02

Beef Extract	3.0	g
Yeast Extract	3.0	g
Peptone	15.0	g
Proteose Peptone	5.0	g
Lactose	10.0	g
Saccharose (Sucrose)	10.0	g
Dextrose	1.0	g
Ferrous Sulfate	0.2	g
Sodium Chloride	5.0	g
Sodium Thiosulfate	0.3	g
Agar	12.0	g
Phenol Red	0.024	g

BBL 11748

Peptone	20.0	g
Lactose	10.0	g
Saccharose (Sucrose)	10.0	g
Dextrose	1.0	g
Ferrous Sulfate	0.2	g
Sodium Chloride	5.0	g
Sodium Thiosulfate	0.2	g
Agar	13.0	g
Phenol Red	0.025	g

Final pH: 7.3 ± .2

Preparation: Add 65 grams or 59.4 grams, depending on manufacturer, of triple sugar iron agar to 1 liter of cold laboratory pure water and heat in a boiling water bath to dissolve the agar. Dispense into screw-cap tubes and sterilize for 15 minutes at 118 C (12 lbs. pressure). Slant tubes for a generous butt. Inoculated TSI slants must be incubated with loosened caps to prevent complete blackening of the medium from H_2S.

5.5.10 Lysine Iron Agar

Use: Differentiation of *Proteus, Citrobacter* and *Shigella* from *Salmonella* based on deamination of lysine and hydrogen sulfide production. *Salmonella* cultures produce large amounts of hydrogen sulfide and lysine decarboxylase.

Composition:

Difco 0849–02

Peptone	5.0	g

Yeast Extract	3.0 g
Dextrose	1.0 g
L-Lysine	10.0 g
Ferric Ammonium Citrate	0.5 g
Sodium Thiosulfate	0.04 g
Brom Cresol Purple	0.02 g
Agar	15.0 g

BBL 11362

Peptone	5.0 g
Yeast Extract	3.0 g
Dextrose	1.0 g
L-Lysine	10.0 g
Ferric Ammonium Citrate	0.5 g
Sodium Thiosulfate	0.04 g
Brom Cresol Purple	0.02 g
Agar	13.5 g

Final pH: 6.7 $\pm$.2

Preparation: Add 34.5 or 33 grams, depending on manufacturer, of lysine iron agar to 1 liter of laboratory pure water. Heat in a boiling water bath to dissolve the agar. Dispense in 4 ml amounts in screw-cap tubes and sterilize for 12 minutes at 121 C (15 lbs. pressure). Cool to give a deep butt and short slant. Inoculated LIA slants must be incubated with loosened caps.

5.5.11 Urea Agar Base (BBL 11794, Difco 0283-02)

Use: To differentiate enteric microorganisms, especially *Proteus sp.* on basis of urease activity.

Composition:

Peptone	1.0 g
Dextrose	1.0 g
Sodium Chloride	5.0 g
Monopotassium Phosphate	2.0 g
Urea	20.0 g
Phenol Red	0.012 g

Final pH: 6.8 $\pm$.2

Preparation: Add 29 grams of urea agar base to 100 ml of laboratory pure water. Dissolve and filter-sterilize. Add 15 grams of agar to 900 ml laboratory pure water and boil to dissolve. Sterilize for 15 minutes at 121 C (15 lbs. pressure). Cool to 50–55 C and add aseptically 100 ml of filter-sterilized urea agar base. Mix and dispense in sterile tubes. Slant tubes to form a 2 cm butt and 3 cm slant and cool.

Urea Agar Base 10X Concentrate (Difco 0284–60)

Use: Same as urea agar base, for preparation of small volumes of urea agar.

Composition: A filter-sterilized 10X solution of urea agar base, 10 ml volumes in tubes. Refrigerate to store.

Preparation: Add 1.5 grams of agar to 90 ml of laboratory pure water and dissolve by boiling. Sterilize for 15 minutes at 121 C (15 lbs. pressure). Cool the agar to 50–55 C and aseptically add a 10 ml tube urea agar base concentrate. Mix agar and concentrate. Dispense aseptically into sterile tubes and slant.

5.5.12 Phenylalanine Agar (BBL 11536, Difco 0745–02)

Use: Differential tube medium for the separation of members of the *Proteus* and *Providencia* genera from other members of the Enterobacteriaceae based on deaminase activity.

Composition:

Yeast Extract	3.0 g
DL-Phenylalanine	2.0 g
Disodium Phosphate	1.0 g
Sodium Chloride	5.0 g
Agar	12.0 g

Final pH: 7.3 $\pm$.2

Preparation: Add 23 grams of phenylalanine agar to 1 liter of cold laboratory pure water. Heat in a boiling water bath to dissolve the agar. Dispense in screw-cap tubes and sterilize in the autoclave for 15 minutes at 121 C (15 lbs. pressure). Slant and cool tubes.

5.5.13 Malonate Broth (Modified) (BBL 11398, Difco 0569–02)

Use: Differentiation of enteric organisms based on utilization of malonate. Described by Leifson and modified by Ewing, the medium is used in differentiation of *Salmonella*.

Composition:

Yeast Extract	1.0	g
Ammonium Sulfate	2.0	g
Dipotassium Phosphate	0.6	g
Monopotassium Phosphate	0.4	g
Sodium Chloride	2.0	g
Sodium Malonate	3.0	g
Dextrose	0.25	g
Brom Thymol Blue	0.025	g

Final pH: 6.7 $\pm$ .2

Preparation: Dissolve 9.3 grams in 1 liter of laboratory pure water. Dispense into tubes and sterilize for 15 minutes at 121 C (15 lbs. pressure).

5.5.14 Decarboxylase Medium Base (Difco 0872-02, BBL 11429)

Use: Differentiation of microorganisms based on decarboxylase activity in presence of L-lysine HCl, L-arginine HCl, L-ornithine HCl, glutamic acid or other amino acids.

Composition:

Peptone	5.0	g
Yeast Extract	3.0	g
Dextrose	1.0	g
Brom Cresol Purple	0.02	g

Final pH: 6.5 $\pm$ .2

Preparation: Add 9 grams of base to 1 liter of cold laboratory pure water and warm to dissolve completely. Add 5 grams L-lysine, L-ornithine, L-arginine or other L-amino acids as desired per liter of medium and warm to dissolve completely. When D/L amino acids are used add 10 g rather than 5 g. If ornithine HCl is used, adjust pH with 10 N NaOH. (About 2.1 ml required for 1 liter of medium containing 5 grams of ornithine HCl). Lysine or arginine do not require pH adjustment. Dispense in 5 ml volumes into screw-cap tubes and sterilize for 15 minutes at 121 C (15 lbs. pressure). The proper pH for the complete medium (6.5) is indicated by purple color of broth.

5.5.15 Motility Test Medium (BBL 11435)

Use. Recommended for detection of motility of gram-negative enteric bacilli.

Composition:

Beef Extract	3.0	g
Peptone	10.0	g
Sodium Chloride	5.0	g
Agar	4.0	g

Final pH: 7.3 $\pm$ .2

Preparation: Add 22 grams of motility test medium to 1 liter of cold laboratory pure water and heat in a boiling water bath to dissolve the agar. Dispense in tubes and sterilize for 15 minutes at 121 C (15 lbs. pressure).

To aid in recognizing motility, add 0.05 grams of triphenyl tetrazolium chloride/liter after sterilization.

5.5.16 Motility Sulfide Medium (Difco 0450–17)

Use: Determination of motility and the production of hydrogen sulfide from l-cystine.

Composition:

Beef Extract	3.0	g
Peptone No. 3	10.0	g
L-Cystine	0.2	g
Ferrous Ammonium Citrate	0.2	g
Sodium Citrate	2.0	g
Sodium Chloride	5.0	g
Gelatin	80.0	g
Agar	4.0	g

Final pH: 7.3 $\pm$ .2

Preparation: Add 104 grams of Motility Sulfide Medium to 1 liter of cold laboratory pure water. After wetting powder, heat carefully to boiling on a hot plate to dissolve completely. Dispense in 4 ml amounts in tubes, cap loosely and sterilize for 15 minutes at 117 C (10 lbs. pressure).

5.5.17 H Broth (Difco 0451-02)

Use: Preparation of H agglutinating antigens of members of genus, *Salmonella*.

Composition:

BBL 11288

Thiotone (peptone)	10.0 g
Beef Extract	3.0 g
Dextrose	1.0 g
Sodium Chloride	5.0 g
Dipotassium Phosphate	2.5 g

Difco 0451-02

Thiotone (Peptone)	5.0 g
Tryptone	5.0 g
Beef Extract	3.0 g
Dextrose	1.0 g
Sodium Chloride	5.0 g
Dipotassium Phosphate	2.5 g

Final pH: 7.2 ± .2

Preparation: Add 21.5 grams of H broth to 1 liter of laboratory pure water, mix well and dissolve by warming. Dispense 5 ml amounts in screw-cap test tubes. Sterilize for 15 minutes at 121 C (15 lbs. pressure).

5.6 Medium for Actinomycetes

Starch-Casein Agar

(Medium may not be commercially available).

Use: Isolation of actinomycetes from water or soil.

Composition:

Soluble Starch (or Glycerol)	10.0 g
Casein (Vitamin-free)	0.3 g
Sodium Nitrate	2.0 g
Sodium Chloride	2.0 g
Dipotassium Phosphate	2.0 g
Magnesium Sulfate•7 H_2O	0.05 g
Calcium Carbonate	0.02 g
Iron Sulfate•7 H_2O	0.01 g
Agar	15.0 g

Final pH: 7.0–7.2

Preparation: Weigh out ingredients and add in turn to 1 liter of laboratory pure water in a two liter flask. Dissolve ingredients using gentle heat. Add the agar last and place in a boiling water bath. Heat and stir occasionally until dissolved. Dispense 250 ml volumes in 500 ml screw-cap flasks and 17 ml volumes in screw-cap tubes. Sterilize for 15 minutes at 121 C (15 lbs. pressure).

6. Laboratory Pure Water

Distilled or deionized water free of nutritive and toxic materials is required for preparation of media and dilution/rinse water.

6.1 Distilled Water Systems: Water distillation units can produce good grades of pure water. Stills are dependable and long-lived if maintained and cleaned properly. Use of softened water as the source water increases the interval between cleanings of the still. Stills characteristically produce a good grade of water which gradually deteriorates as corrosion, leaching and fouling set in. There is no sudden loss of water quality unless a structural failure occurs. Stills are efficient in removing dissolved chemicals but not dissolved gases. Fresh laboratory pure water may contain chlorine and ammonia. On storage, ammonia will increase and CO_2 will appear from air contamination. Distilled water systems should be monitored continuously for conductance and analyzed monthly for chlorine, ammonia and standard plate count and at least annually for trace metals. See Table IV-A-3.

6.2 Deionized Water Systems: Deionization systems produce a good grade of pure water and can produce an ultrapure water when combined with filtration and activated carbon in a recirculating system.

In contrast to distilled water systems, deionization systems do not gradually deteriorate. Rather, they continue to produce the same quality water for a long period of time until the resins and/or carbon are exhausted, whereupon the quality deteriorates quickly. Deionized water systems should be monitored continually with a conductance meter and analyzed monthly for ammonia/amines, total organic carbon, specific organic pollutants and Standard Plate Count and at least annually for metals. Amines may elute from the resin. Organic carbon results from organic chemicals in the water or from bacterial growth in the columns. Use of a 0.22 μm final filter is recommended to remove bacterial contamination. See Table IV-A-3.

Avoid the sudden loss of good quality water by continuously monitoring performance of the system, anticipating the remaining life of cartridges and replacing them before failure occurs.

6.3 Quality Control

Pure water systems should be monitored carefully as a part of the intralaboratory QC program. The water quality should meet the standards set in this Manual, Part IV-A, 5.2-5.4.

7. Dilution Water

Dilution water is used to reduce the number of microbial cells/unit volume of sample so that the density of cells is low enough to permit enumeration or manipulation by the technique of choice: pour-plate, spread plate, MF or MPN.

The ideal dilution water is neutral in effect. It maintains bacterial populations without stimulating cell growth and reproduction, damaging cells or reducing their ability to survive, grow or reproduce. Its basic purpose is to simulate the chemical conditions of the natural environment which are favorable to cell stability.

Microbiologists have tried different approaches to obtain an ideal diluent. Some workers have copied the natural environment by use of sterile fresh or marine waters as diluents, but these are non-standard. Other workers have used tap waters with the same lack of uniformity and an added potential for toxicity. Certainly for comparability of microbiological data the dilution water must be uniform between laboratories. The chemical elements and compounds required in natural conditions to insure a balance of cell solutes and maintain cell turgidity must be reproduced in the laboratory.

Inorganic constituents such as sodium, potassium, magnesium, phosphate, chloride and sulfate, and soluble organics such as peptone are used in synthetic dilution waters. pH is usually held to a near-neutral reaction. Two standard dilution waters are:

7.1 Phosphate Buffered Dilution Water

7.1.1 Stock Phosphate Buffer Solution

Potassium Dihydrogen Phosphate (KH_2PO_4)	34.0	g
Laboratory Pure Water	500	ml

Adjust the pH of the solution to 7.2 with 1 N NaOH and bring volume to 1000 ml with laboratory pure water. Sterilize by filtration or autoclave for 15 minutes at 121 C (15 lbs. pressure).

7.1.2 Storage of Stock Solution: After sterilization of the stock solution store in the refrigerator until used. Handle aseptically. If evidence of mold or other contamination appears, the stock solution should be discarded and a fresh solution prepared.

7.1.3 Working Solution

Stock Phosphate Buffer	1.25	ml
$MgCl_2$ Solution (38 g/liter)	5	ml
Laboratory Pure Water	1000	ml

Final pH: 7.2 ± 0.1

7.2 Peptone Dilution Water

Composition:

Peptone	1.0	g

Laboratory Pure Water 1000 ml

Final pH: 7.0 ± 0.1

7.3 Preparation of Dilution and Rinse Water

Dispense 102 ml volumes of phosphate buffer or peptone dilution water into borosilicate glass, screw-cap dilution bottles scribed at 99 ml. Loosen screw-caps on bottles and sterilize at 121 C for 15 minutes (15 lbs. pressure). Final volumes after sterilization should be 99 ± 2 ml. Cool and separate bottles outside of the 99 ml ± 2 ml limit. Tighten screw-caps after sterilization and store in a cool place.

Prepare dilution water for rinsing in 500 ml or larger volumes and autoclave for 30 minutes or more. Bottles or flasks must be separated sufficiently in the autoclave to permit easy access for steam.

REFERENCES

1. Songer, J. R., J. F. Sullivan and J. W. Monroe, 1971. Safe, convenient pipetting device. Appl. Microbiol. 21:1097.

2. Songer, J. R., D. T. Braymen and R. G. Mathis, 1975. Safe, convenient pipetting station. Appl. Microbiol. 30:887.

3. Taylor, R., R. Bordner and P. Scarpino, 1973. Delayed incubation membrane-filter test for fecal coliforms. Appl. Microbiol. 25:363.

4. Taylor, W. I., 1965. Isolation of Shigellae. I. Xylose lysine agars; new media for isolation of enteric pathogens. Am. J. Clin. Path. 14:471.

GENERAL REFERENCES

Difco Laboratories. General Conditions Pertaining to the Cultivation of Microorganisms; and, Preparation of Media from Dehydrated Culture Media. Difco Manual. 9th Edition. Difco Laboratories, Detroit, MI. p. 16 (1953).

Baltimore Biological Laboratories. General Suggestions for Use of Media: Dehydrated-Prepared. BBL Manual of Products and Laboratory Procedures. 5th Edition, BBL, Division of Becton-Dickinson and Co., Cockeysville, MD. p. 88 (1968).

Geldreich, E. E., 1975. Handbook for Evaluating Water Laboratories (2nd Edition). EPA-670/9-75-006, U.S. Environmental Protection Agency, Cincinnati, OH p. 83.

American Public Health Association, 1976. Standard Methods for the Examination of Water and Wastewater (14th Edition). American Public Health Association, Inc. p. 892.

PART II. GENERAL OPERATIONS

Section C Techniques for Isolation and Enumeration of Bacteria of Sanitary Significance

This Section describes the fundamental laboratory procedures needed for microbiological analyses of water and wastewater. *Although experienced microbiologists and technicians may not require the depth of information and the degree of detail given in this Section, it is provided to serve the technical personnel who are new to environmental microbiology.* The procedures included are:

1. **Preparation for Analyses**
2. **Streak Plate, Pour Plate and Spread Plate Methods**
3. **Membrane Filtration Method**
4. **Most Probable Number Method**
5. **Staining Procedures**
6. **Shipment of Cultures**

The specific details unique to tests are described in separate Sections of Part III: Standard Plate Count, Total Coliform, Fecal Coliform, Fecal Streptococci, *Salmonella* and Actinomycetes. Refer to Part IV, A & C for details of quality control on these analyses.

1. Preparation for Analyses

1.1 Preparation of Data Sheets

Select a standard format for bench sheet or card and use to record pertinent data: sample identification, sampling conditions, chlorine residual, temperature, pH, sampling site, station number, date and time of collection, sample collector, required chain of custody information, data and time received and analyzed, time elapsed from sample collection, analyses performed, sample volume(s), analyst, and laboratory identification numbers. (See Figures II-C-1, 2 and 3).

1.2 Disinfection of Work Area

Wipe the work area before and after use with laboratory-strength disinfectant and allow surface to dry before use. See Table V-C-3.

Keep a covered container of iodophor or quaternary ammonium disinfectant available for emergency use. Phenolics are acceptable if analyses for these compounds are not performed as part of laboratory work.

1.3 Pretreatment of Samples

Prior to dilution of samples for analyses, the analyst should examine the sample for free chlorine residual and possible uneven distribution of microorganisms.

MICROBIOLOGY LABORATORY RECORD
MEMBRANE FILTER ANALYSIS

PROJECT: ______ NO: ______
SAMPLE SOURCE: ______
SAMPLING DATE: ______ TIME: ______ FILTRATION DATE: ______ TIME: ______
COLLECTOR: ______ TEMP: ______ pH: ______ ANALYST: ______

VOL.	COUNT	COUNT/100 ml		VOL.	COUNT	COUNT/100 ml		VOL.	COUNT	COUNT/100 ml	

Remarks: ______ Remarks: ______ Remarks: ______

EPA-209 (Cin)
(Rev- 4-76)

FIGURE II-C-1. Microbiological Bench Card for MF Analyses.

MICROBIOLOGY LABORATORY RECORD
COLIFORM MPN ANALYSIS

Sample # ______ Date ______ Time ______ EST/DST River ______
Location ______
Tide ______ Rain past 24 Hrs. ______ Depth ______ Water Temp. °C ______ Secchi Disc. ______

Dilution					MPN/100 ml
Coliform Pres. LST 24 Hrs.					
48 Hrs.					
Coliform Conf. BGLB 48 Hrs.					
Fecal Coliform EC 24 Hrs.					

AFO, Region III, USEPA

FIGURE II-C-2. Bench Card for MPN Analyses.

BACTERIAL INDICATOR ORGANISMS OF POLLUTION

STATION ______________________ BENCH NO ________

LOCATION ______________________

	DATE	HOURS	SAD NO. ________
COLLECTED	. . . .	. . . .	ANALYST
EXAMINED	. . . .	. . . .	________

M P N

DiL. ml.	PRESUMPTIVE LTB 35°C		CONFIRMED B.G.L.B. 35°C		EC 44.5°C	MEMBRANE FILTER	
	24 hrs.	48 hrs.	24 hrs.	48 hrs.	24 hrs.	VOLUME FILTERED	COLONY COUNT
10	—						TOTAL COLIFORMS
	—						
	—						
	—						
	—						
10⁻	—						FECAL COLIFORMS
	—						
	—						
	—						
	—						
10⁻	—						
	—						
	—						
	—						
	—						
10⁻	—					ORGANISMS PER 100 ML.	
	—					TOTALS ________	
	—					FECALS ________	
	—					________	
	—						
10⁻	—					SALMONELLA ________	
	—					________	
	—					ISOLATION ________	
	—					SEROTYPE(S) ________	
	—						

CONFIRMED MPN . . . . PER 100 ML.

FECAL COLIFORM MPN . . . . PER 100 ML.

REMARKS: ______________________

____

____

____

SERL, Region IV, US EPA

FIGURE II-C-3. Combined Microbiological Bench Card

1.3.1 Water Samples with High Solids

(a) Blending of sediments, primary effluents, sludge and highly turbid waters is essential for representative subsampling.

(b) Blend the entire water sample containing particulates in a Waring-type blender. Use only autoclavable pyrex glass, stainless steel or plastic blender containers with safety screw covers to prevent release of aerosols.

(c) Limit blending to no more than 30 seconds at about 5000 RPM to avoid overheating or shearing damage.

(d) Dilute sediments or soils containing limited amounts of water at a 1:1, 1:2 ratio or more with dilution water to ensure good blending action and to reduce heat generation. Use of a large blender container rather than smaller units also reduces heat.

1.3.2 Dry Solid Samples

(a) Mix sample thoroughly and weigh 50 grams aliquot in a tared weighing pan. Dry at 105-110 C to constant weight. The final weight is used in calculating numbers of organisms/gram dry solids.

(b) Prepare the initial dilution by weighing out a second aliquot of 11 grams of original sample. Add to a 99 ml volume of buffered dilution water for a 1:10 dilution and blend sample aseptically in a Waring-type blender at 5000 rpm for 30 seconds. Use only a pyrex glass, stainless steel or plastic blender container with safety screw lid to prevent release of aerosols.

(c) Transfer an 11 ml sample of the 1:10 dilution to a second dilution bottle containing 99 ml buffered dilution water and shake vigorously about 25 times. Repeat this process until the desired dilution is reached.

1.3.3 Analytical Method

Although high solids samples with low microbial densities may require MPN or pour plate procedures, high density samples such as polluted soils, sludges and feces are diluted so that the MF method is applicable.

1.4 Dilution of Samples

1.4.1 Necessity for Dilutions: Dilutions of the original sample of water, wastewater or other material are often necessary to reduce the number of bacterial cells to measureable levels or to isolate single cells for purification and identification (see Part II-B, 7 for details on dilution water).

1.4.2 Serial Dilutions: A known quantity of the sample (usually 1.0 ml, but other volumes can be used) is transferred through a series of known volumes (e.g., 9 or 99 ml) of dilution water. This procedure is repeated until the desired bacterial density is reached. After dilution of the sample, the bacteria are enumerated using the membrane filtration, pour plate, streak plate, or the most probable number technique.

For ease of calculation and preparation, serial dilutions are usually prepared in succeeding ten-fold volumes called decimal dilutions. The decimal dilution procedure is shown in Figure II-C-4.

1.4.3 Special Dilutions for Membrane Filtration Procedures: The normally accepted limits for colonies per plate in membrane filtration methods (20–60, 20–80 or 20–100) require that decimal dilution series be modified to assume an MF plate count within the accepted limit.

The recommended method for obtaining counts within these limits is to filter dilution volumes of the decimal series which have a factor of 3, 4 or 5 among them (see Table II-C-1 for details).

1.4.4 Filtration Volumes for Membrane Filter Analyses: For sample volumes of 1-10 ml, add 20 ml of dilution water to the funnel before adding sample, to evenly disperse cells.

1.4.5 Preparation of Dilutions: Shake sample bottle vigorously (about 25 times in 7 seconds) to evenly distribute the bacteria. Take care to secure the screw-cap and prevent leakage during shaking.

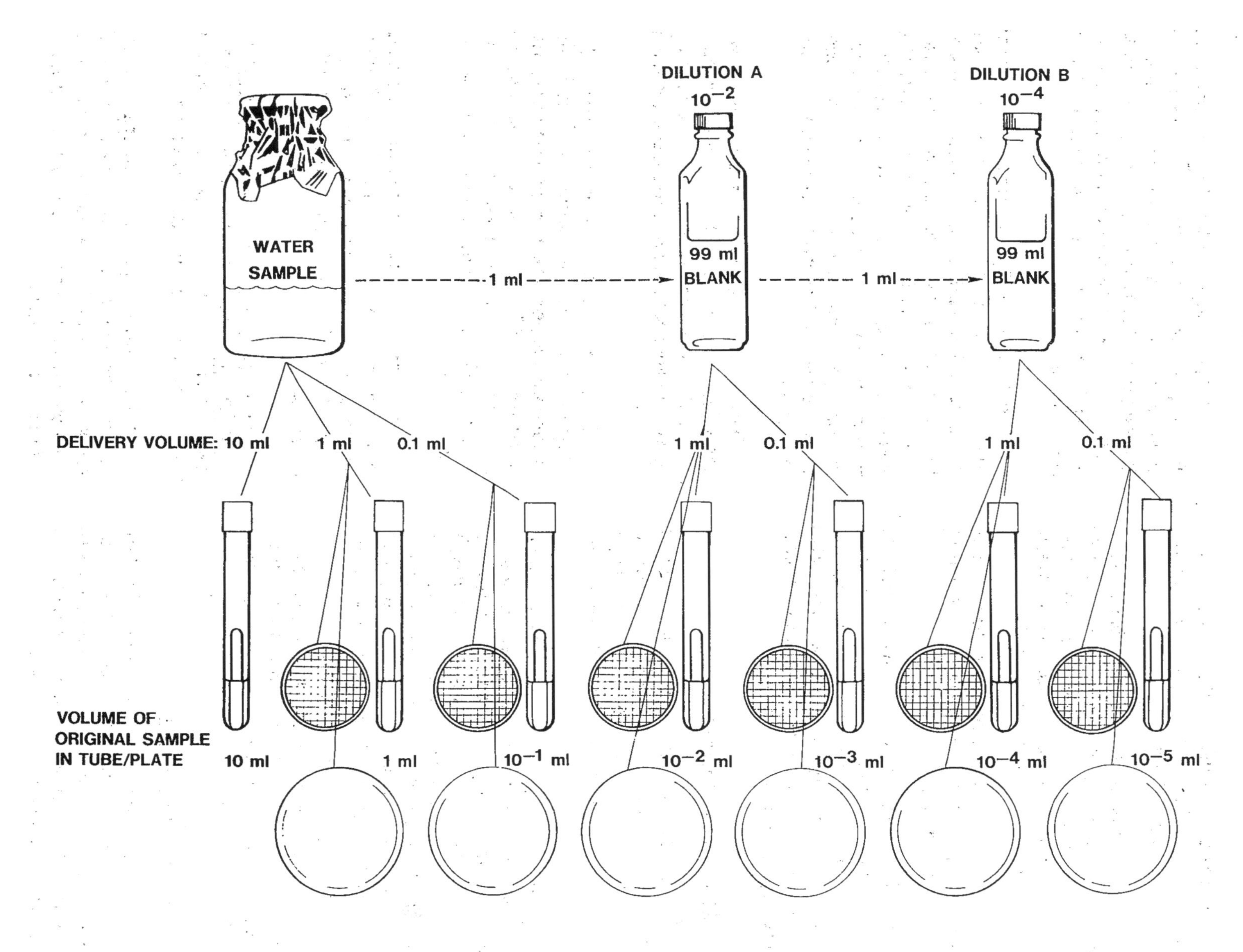

FIGURE II-C-4. Preparation of Decimal Dilution.

TABLE II-C-1

Recommended Filtration Volumes of Samples in MF Analyses

MF Count Range 20–60		MF Count Range 20–80		MF Count Range 20–100	
Sample vol., ml	added as:	Sample vol., ml	added as:	Sample vol., ml	added as:
.01	1 ml of 10^{-2}	.04	4 ml of 10^{-2}	.01	1 ml of 10^{-2}
.03	3 ml of 10^{-2}	.15	1.5 ml of 10^{-1}	.05	5 ml of 10^{-2}
.1	1 ml of 10^{-1}				
.3	3 ml of 10^{-1}	.5	5 ml of 10^{-1}	.25	2.5 of 10^{-1}
1.0	1 ml sample				
3	3 ml sample	2.0	2 ml sample	1.25	1.25 ml sample
10	10 ml sample	8.0	8 ml sample	6	6 ml sample
30	30 ml sample	30	30 ml sample	30	30 ml sample

(a) Withdraw 1.0 ml or 0.1 ml of original sample to test samples directly for a 1×10^0 ml and 1×10^{-1} ml volumes, respectively.

(b) Transfer a second 1.0 ml volume to a 99 ml dilution water bottle (Dilution A). Shake sample vigorously about 25 times and withdraw 1.0 ml or 0.1 ml of diluted sample for testing of 1×10^{-2} and 1×10^{-3} ml sample volumes. The resultant dilution is calculated as follows:

$$\frac{\text{Volume of Sample}}{\text{Volume of Dilution Blank} + \text{Volume of Sample}} = \text{Dilution Ratio}$$

or,

$$\frac{1.0 \text{ ml}}{99.0 + 1.0 \text{ ml}} = \frac{1}{100}$$

or 10^{-2} (Dilution A)

(c) When 1.0 ml is transferred from dilution bottle A to a second dilution bottle (B), the dilution ratio for bottle B dilution shown in Figure II-C-4 is the product of the individual dilutions as follows:

$$A \times B = \text{Final or Total Dilution Ratio}$$

or

$$\frac{1}{100} \times \frac{1}{100} = \frac{1}{10{,}000}$$

Volumes of 0.1 ml can be tested directly from each serial dilution to provide intermediate dilutions.

(d) Alternatively, if an initial sample volume of 11 ml is transferred into the first 99 ml volume dilution blank, an intermediate dilution can be obtained with the added precision resulting from measurements of 1 ml volumes serially as opposed to the measurement of 0.1 ml volumes in (c).

(e) Different dilutions can be obtained by varying sample and dilution preparations. For example, if 2.0 ml of the sample is transferred to a 98.0 ml dilution blank, the final dilution ratio is calculated as follows:

$$\frac{2.0 \text{ ml}}{98.0 + 2.0 \text{ ml}} = \frac{2}{100} = \frac{1}{50}$$

(f) If 0.5 ml is added to a dilution blank containing 99.5 ml of buffer, the dilution ratio is calculated as follows:

$$\frac{0.5 \text{ ml}}{99.5 + 0.5 \text{ ml}} = \frac{0.5}{100} = \frac{1}{200}$$

(g) Varying the final volume tested will also permit modification of dilutions without increasing the number of dilution bottles as follows:

A 1:200 dilution can be obtained by testing 0.5 ml of a 1:100 dilution.

A 1:500 dilution can be obtained by testing 2 ml of a 1:1000 dilution.

1.4.6 Prompt Use of Dilutions: The potential toxicity of phosphate dilution water and the stimulatory effect of peptone dilution water increase rapidly with time. Therefore, dilutions of samples should be tested as soon as possible after make-up and should be held no longer than 30 minutes after preparation.

2. Streak, Pour and Spread Plate Methods

2.1 Summary: There are three plate dilution techniques commonly used for isolating and/or enumerating single colonies of bacteria: the streak plate, pour plate and spread plate procedures. These techniques described herein use solid or melted agar plating media to dilute out the microorganisms so that individual species or cells can be selected or counted from mixed cultures. Because colonies can originate from more than one cell, results may be reported as colony-forming-units (CFUs).

2.1.1 Streak Plate Method: To obtain a pure culture the analyst dilutes and isolates bacteria from a mixed culture by drawing a small amount of the bacterial growth lightly across the surface of an agar plate in a pattern with an inoculating needle or loop.

In one suggested pattern, the plate is streaked in parallel lines over half of the surface, rotated a quarter turn, streaked again, rotated another quarter turn and streaked once more. The inoculum is progressively diluted with each successive streak, and eventually single cells are deposited on the agar surface. After suitable incubation, single isolated colonies develop in the path of the streak. (see Figure II-C-5).

2.1.2 Pour Plate Method: The analyst dilutes and isolates cells in a bacterial suspension by consecutively transferring a portion of the original sample through a series of dilution water blanks. After an appropriate series of dilutions, the original bacterial population is diluted out to a countable level, as described in 1.4. Aliquots of the diluted sample are added to sterile petri dishes and mixed with melted agar. After the agar solidifies, the plates are inverted and incubated for a predetermined time. Surface or subsurface colonies will develop in some of the agar plates. These colonies can be counted to provide a quantitative value for the bacterial density of the original sample, or they can be picked for further qualitative study.

2.1.3 Spread Plate Method: The analyst isolates bacterial cells by delivering a small volume of a diluted sample onto a solid agar plate and spreading the inoculum by use of a sterile glass rod bent at an angle of about 120°. The inoculum is spread uniformly by holding the stick at a set angle on the agar and rotating the agar plate or rotating the stick until the inoculum is distributed evenly.

2.2 Scope and Application

2.2.1 Streak and pour plate methods provide the means to separate individual bacteria so that each cell will develop into an isolated colony in or on a solid medium. The methods can isolate specific bacteria by the use of selective or differential media.

The streak plate is only qualitative but the pour plate procedure can be used to quantitate bacteria present in a sample as in the Standard Plate Count Method (see Part III-A). The spread plate method provides a quantitative method for aerobic surface growth of cultures against which other surface growth methods such as the MF technique can be compared.

2.2.2 Because the volumes tested with the spread plate technique are limited to 0.1–0.5 ml, the sample must be diluted to contain at least 40–2000 cells/ml in order to have a counting range of 20–200.

2.2.3 In the pour plate technique, test volumes are limited to 0.1–2 ml per 100 ml petri dish so the sample must be diluted to contain 60–3000 cells/ml to have a 30–300 counting range.

2.3 Apparatus and Materials

2.3.1 Incubator set at 35 ± 0.5 C.

2.3.2 Water bath set at 44–46 C for tempering agar.

2.3.3 Colony Counter, Quebec darkfield model or equivalent.

2.3.4 Hand tally or electronic counting device.

2.3.5 Thermometer which has been checked against a National Bureau of Standards-Certified Thermometer or one of equivalent accuracy.

2.3.6 Inoculating needle and loop.

2.3.7 Pipet containers of stainless steel, aluminum or pyrex for glass pipets.

2.3.8 Petri dish containers of stainless steel, aluminum or pyrex glass for glass petri dishes.

2.3.9 Glass spreader rods.

2.3.10 Sterile T.D. bacteriological or Mohr pipets, glass or plastic of appropriate volume.

2.3.11 Sterile 100 mm × 15 mm petri dishes, glass or plastic.

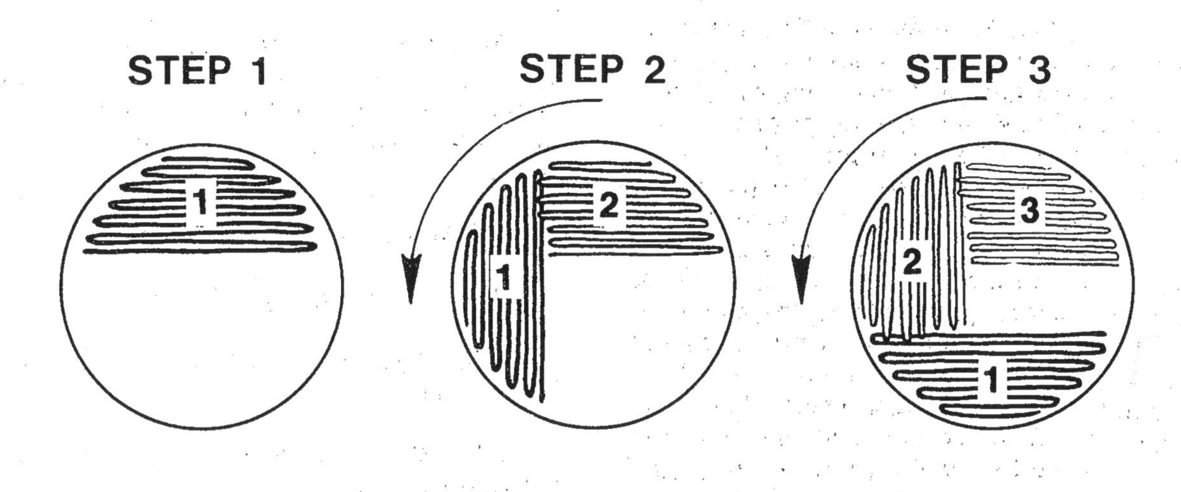

FIGURE II-C-5. Suggested Pattern for Preparing a Streak Plate.

2.3.12 Dilution (milk dilution) bottles, pyrex, marked at 99 ml volume, screw-cap with neoprene rubber liner.

2.3.13 Bunsen/Fisher type burner or electric incinerator.

2.4 Media

2.4.1 Sterile agar medium dispensed in bulk quantities in screwcapped bottles or flasks.

2.4.2 Sterile dilution water in bottles containing 99 $\pm$ 2 ml volumes.

2.5 Streak Plate Procedures

2.5.1 Melt the nonselective agar, such as nutrient agar or Trypticase soy agar, temper at 44-46 C, and add about 15 ml agar to each sterile petri dish. Allow to harden and dry for best results in streaking.

2.5.2 Bend an inoculation needle or loop at an angle about 1 cm from the needle tip to prevent cutting of the agar during streaking. Sterilize the needle by heating it to redness in a flame, and air-cool.

2.5.3 Remove screw cap and pick a small amount of growth from an isolated colony or from a mass of growth.

2.5.4 Draw the needle containing the bacteria back and forth across the surface of a previously-poured and hardened agar plate in a specific pattern, such as shown in Figure II-C-5. Streak 1/3–1/2 of the agar surface. Flame and cool needle after each step and inoculate plate further by drawing the needle across the area previously streaked.

2.5.5 Rotate the plate one quarter turn clockwise as each step is completed to make the streaking easier. Streaking patterns other than the model shown in Figure II-C-5 can be used; the objective is simply to deposit fewer and fewer cells along the streak until single cells are deposited on the agar surface. After incubation, these cells will develop into well-separated pure colonies of bacteria, each theoretically arising from a single bacterium.

2.5.6 After streaking, incubate the petri dishes at 35 C for 24 hours (or other appropriate conditions) in an inverted position to prevent condensation of water on the agar surface. Moisture interferes with development of isolated colonies by spreading bacterial growth over the agar surface.

2.5.7 For further purification, examine plates after incubation for single, well-isolated colonies. Pick typical colonies using a sterile inoculating needle, suspend cells in dilution water, and restreak on an agar plate, repeating steps 2.5.2–2.5.6. Isolated, single colonies from a plate containing like colonies may be considered to be pure.

2.5.8 Streaking may also be done on selective media, such as Endo or EMB agars or on selective/differential media e.g., in *Salmonella* testing.

2.6 Pour Plate Procedure

2.6.1 Shake the sample bottle vigorously about 25 times to disperse the bacteria. During shaking, close cap tightly to prevent leakage of sample and the danger of contamination.

2.6.2 Using a 1.1 ml bacteriological pipet, prepare the initial dilution by pipeting 1 ml of the sample into a 99 ml dilution water blank. This initial dilution represents a 10^{-2} dilution (see this Section, 1.4, Dilution of Samples).

2.6.3 Using the same pipet transfer 0.1 and 1.0 ml from the undiluted sample to two separate petri dishes.

2.6.4 Shake the 1:100 dilution bottle vigorously again and pipet 0.1 and 1.0 ml of the 1:100 dilution into two petri dishes using another sterile 1.1 ml pipet.

2.6.5 Pour aseptically 12–15 ml of the melted agar medium cooled to 44–46 C into each petri dish. Mix agar and inoculum by rotation, being careful to prevent spillover of agar. One recommended technique uses a sequence of five rotations to the left, five to the right and five forward and backward. Allow the agar to solidify on a level surface.

2.6.6 Invert the dishes and incubate at the specific temperature and time. After incubation, well-isolated surface and subsurface colonies should develop in some of the plates.

2.6.7 When the pour plate technique is used quantitatively, count plates containing between 30 and 300 colonies. This is the technique used in the Standard Plate Count Method described in Part III-A. Pour plate counts are reported as the count per ml.

$$\text{Count/ml} = \frac{\text{Sum of Plate Counts}}{\text{Total Volume of Sample in ml}}$$

2.7 Spread Plate Procedure

2.7.1 Prepare the appropriate melted agar. Pour about 15 ml of the melted agar into each 100 mm petri dish. Keep covers opened slightly until agars have hardened and moisture or condensation have evaporated. Close dishes and store in refrigerator. Warm at room temperature before use.

2.7.2 Prepare a series of dilutions based upon the estimated concentration of bacteria so that 0.1-0.5 ml of inoculum will give a 20-200 count (equivalent to a 40-2000 count/ml in the diluted sample). The dilutions should bracket the estimated density of bacteria. The analyst must remember that if only 0.1 ml volume is tested, it must be plated on the agar plate marked with the next higher dilution; for example, 0.1 ml of the 10^{-1} dilution onto the surface of the plate marked 10^{-2}. Inoculate agar plates.

2.7.3 Remove the glass spreader from alcohol and flame. Cool for 15 seconds. Test glass rod on edge of agar to verify safe temperature before use. This step can be simplified by making and sterilizing a number of glass spreaders.

2.7.4 Place cool glass spreader on agar surface next to inoculum. Position spreader so that the tip forms a radius from the center to the plate edge. Holding spreader motionless, rotate plate several revolutions, or hold plate and move the spreader in a series of sweeping arcs. The purpose is to spread the inoculum uniformly over the entire surface of the agar.

2.7.5 Lift the glass spreader from the agar and place in alcohol solution. Cover plate partially, leaving open slightly to evaporate excess moisture for 15-30 minutes.

2.7.6 When agar surfaces are dry, close dishes, invert them and incubate as required for the specific test.

2.7.7 After incubation at the proper time and temperature, isolated surface colonies should develop in one or more dilutions within the acceptable counting range of 20-200 colonies. The maximum recommended number of colonies/spread plate is fewer than for other plate techniques because surface colonies are larger than subsurface colonies and crowding can result at lower count levels.

2.7.8 Count the colonies by normal techniques and report on a count/ml or count/100 ml basis dependent on the use of the data.

2.8 Reporting Results

2.8.1 Significant Figures: To prevent false precision in the reporting of counts, the plate counts must be limited to the digit(s) known definitely plus one digit which is in doubt. These combined digits are termed the Significant Figures (S.F.).

(a) For example, if an analyst reports a plate count of 124 to three significant figures he is indicating that he is certain of the first two digits, 1 and 2, but is uncertain whether the last digit is 3, 4 or 5. If the analyst were reporting that same number to two significant figures, he would report the first figures, 1, as certain, the second figure, 2, as uncertain, and the third figure, 4, as unknown. Hence he would report it as 120, inserting the zero only as a spacer. Large counts of 1200, 12,000 and

12,000,000 only contain two significant figures. Of course, zeros can be significant in counts of 10, 60, 105, etc.

(b) In plate count and MF methods, the number of significant digits which can be reported are dictated by the method itself as follows: within the acceptable counting range of the method itself, i.e., 20–60, 20–80, 20–100 or 30–300 the actual number of colonies observed is the best estimate of the true density. The number of significant figures are equal to the number of colonies.

TABLE II-C-2

Number of Significant Figures (S.F.) Reported

Actual Colony Count	Pour Plate/ Spread Plate Method	Membrane Filtration Method
1 - 9	1 S.F.	1 S.F.
10 - 99	2 S.F.	2 S.F.
100 - 300	3 S.F.	—

2.8.2 Rounding Off Counts: Since plate counts must be limited to the number of significant figures obtainable by the method, the non-zero number which is not significant should be treated by the standard scientific convention:

(a) If the insignificant digit is less than five, replace it with a zero, e.g., 3530 becomes 3500.

(b) If the insignificant digit is five, round the preceding significant digit to the nearest even number, e.g., with two S.F., 3450 becomes 3400, and 3550 becomes 3600.

(c) If the insignificant digit is greater than five, drop the digit and increase the preceding significant number by one, e.g., 3480 becomes 3500.

3. Membrane Filtration Method

3.1 Summary: The membrane filter method provides a direct count of bacterial colonies on the surface of the filter. The sample is filtered as soon as possible after collection. After the sample is filtered, the membrane filter is placed on a nutrient medium formulated to encourage growth of the bacteria for which the test is designed and to suppress the growth of other microorganisms. After incubation under the specified conditions, the bacteria retained on the surface of the membrane develop into visible colonies. The medium and the temperature of incubation influence the kinds and appearance of bacteria that develop. Two-step enrichment and delayed incubation MF procedures can also be used. The two-step procedure involves an acclimation period on another medium before the selective growth step.

3.2 Scope and Application

Membrane filter methods are preferred over MPN or other techniques, where applicable because of the following advantages.

3.2.1 Advantages

(a) One of the primary advantages of this method is its speed. Definitive results for total and fecal coliforms can be obtained in 22–26 hours, whereas 48–96 hours are required for the multiple-tube fermentation method.

(b) Considerably larger, more representative water samples can be examined than with the MPN. With waters of low bacterial densities such as finished waters, larger sample aliquots can be used to enhance the reliability of the results.

(c) The precision is greater with the MF than the multiple-tube technique because the former makes a direct count of colonies/unit volume.

(d) The method represents savings in time, labor, space, supplies and equipment.

(e) Because of its portability, this procedure is also very practical for field studies.

3.2.2 Limitations: Although the majority of water samples can be tested by membrane filtration, there are limitations with certain samples and some problems with membrane filters themselves.

(a) Some samples contain large quantities of colloidal materials or suspended solids such as iron, manganese or alum flocs or clay (1). Other samples may contain algae. These substances can clog the filter pores and prevent filtration or can cause the development of spreading bacterial colonies. When the bacterial counts of such samples are high, a smaller volume or a higher sample dilution can be used to minimize the effect of sample turbidity. The membrane filter method may be used with samples containing turbidity by filtration of several smaller replicate sample volumes and compositing the results. However, with waters of high turbidity and low bacterial count, the membrane filter method may not be applicable. In the latter situation the multiple-tube procedure should be used.

(b) Large non-specific populations may mask the appearance of indicators on selective media such as M-Endo MF medium (4).

(c) Industrial wastewaters may contain zinc, copper, or other heavy metallic compounds (2) which adsorb onto the membrane surface and interfere with subsequent bacterial development (1, 2, 3).

(d) MF analyses require preparation of MPN tubed media for verification.

(e) Inhibition may result in seawater or from toxic materials such as chlorine or phenols.

(f) Indicator organisms stressed in the environment may be poorly recovered (5).

3.3 Apparatus and Materials

3.3.1 Incubator set at 35 ± 0.5 C for total coliform test (Part III-B) and fecal streptococci test (Part III-D).

3.3.2 Water bath or other type of incubators such as the aluminum heat sink incubator, or equivalent, set at 44.5 ± 0.2 C for fecal coliform test (Part III-C).

3.3.3 Stereoscopic microscope, with magnification of 10–15×, wide-field type.

3.3.4 A microscope lamp producing diffuse light from cool, white fluorescent lamps adjusted to give maximum color or sheen appearance.

3.3.5 Hand tally.

3.3.6 Pipet container of stainless steel, aluminum or pyrex glass for glass pipets.

3.3.7 Graduated cylinders covered with aluminum foil or kraft paper and sterilized.

3.3.8 Membrane filtration units (filter base and funnel), glass, plastic or stainless steel, see Figure II-C-6. These are wrapped with aluminum foil or kraft paper and sterilized. See Figure II-C-7 for an exploded view of a stainless steel MF assembly and filter.

3.3.9 Ultraviolet sterilizer for the filter funnel is optional (6).

3.3.10 Line vacuum, electric vacuum pump or aspirator is used as a vacuum source. In an emergency or in the field, a hand pump or a syringe can be used. Such vacuum-producing devices should be equipped with a check valve to prevent the return flow of air.

3.3.11 Vacuum filter flask, usually 1 liter, with appropriate tubing. Filter manifolds to hold a number of filter bases are optional.

3.3.12 Safety trap flask placed between the filter flask and the vacuum source.

3.3.13 Forceps, straight or curved, with smooth tips to permit easy handling of filters without damage.

3.3.14 Alcohol, ethanol or methanol, in small wide mouthed vials, for sterilizing forceps.

FIGURE II-C-6. Membrane Filtration Units Made by Various Manufacturers for Detection of Bacteria in Aqueous Suspensions.

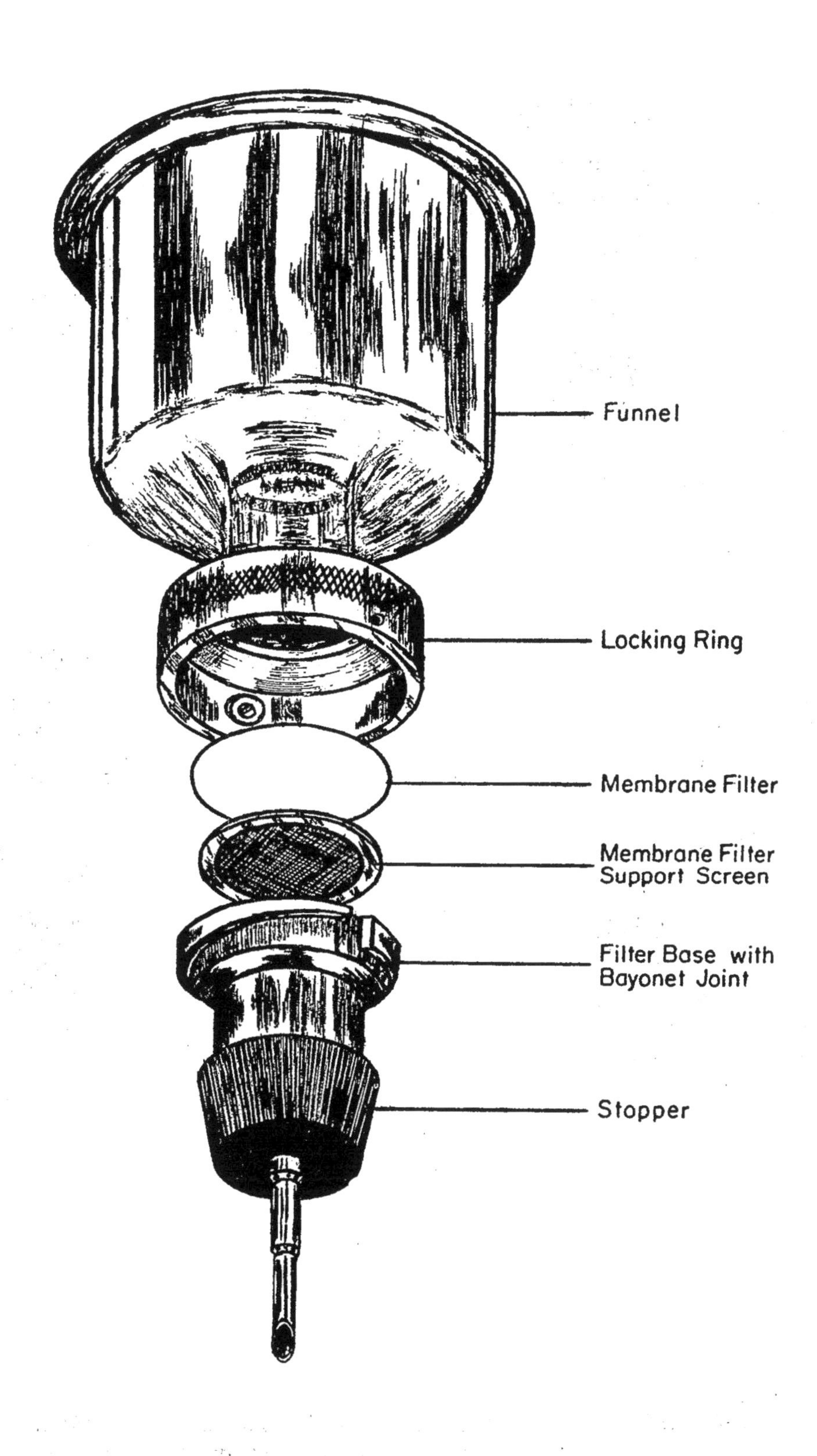

FIGURE II-C-7. Exploded View of a Stainless Steel Membrane Filtration Unit.

3.3.15 Bunsen or Fisher type burner or electric incinerator unit.

3.3.16 Sterile T.D. bacteriological or Mohr pipets, glass or plastic, of appropriate volume.

3.3.17 Sterile petri dishes, 50 × 12 mm, with tight-fitting lids and 60 × 15 mm, with loose-fitting lids glass or plastic.

3.3.18 Dilution bottles (milk dilution) pyrex, marked at 99 ml, screw-cap with neoprene liners.

3.3.19 Membrane filters, white, grid marked, 47 mm diameter, with 0.45 µm ± .02 µm pore size or other pore size recommended by the manufacturer for water analyses.

3.3.20 Absorbent pads of cellulosic paper, 47 mm diameter. The paper should be of high quality and free of sulfites or other substances that could inhibit bacterial growth.

3.3.21 Waterproof plastic bags.

3.3.22 Inoculation loops, at least 3 mm diameter, or needles, nichrome or platinum wire, 26 B&S gauge, in suitable holder. Disposable applicator sticks or plastic loops as alternatives to inoculation loops.

3.3.23 Media: Media required for a specific test should be prepared in pre-sterilized erlenmeyer flasks with metal caps, aluminum foil covers, or screw caps.

3.3.24 Dilution Water

(a) Sterile buffered or peptone dilution water dispensed in 99 ± 2 ml amounts in screw-capped dilution bottles.

(b) Sterile dilution water prepared in larger volumes for wetting membranes before addition of the sample and for rinsing the funnel after sample filtration.

3.4 Procedure

3.4.1 Prepare the required media as outlined in Part II-B. If the medium is an agar, cool to room temperature. Use sterile forceps for manipulation of absorbent pads and membrane filters, contacting the outer edges only, to avoid touching the filtering area or damaging the membrane filter surface. Sterilize forceps by immersing the tips in ethanol and flaming. Place absorbent pad in bottom of 50 or 60 mm petri dish. Add 1.8–2.0 ml broth to the sterile absorbent pad. Saturate but do not flood the pad. Tip the petri dish to drain off excess. If agar medium is used, add about 5–6 ml (to a depth of 2–3 mm) in the petri dish.

3.4.2 Arrange petri dishes in rows according to the dilution series. Mark each dish to identify the sample, volume or dilution to be filtered.

3.4.3 Using sterile forceps, place a membrane filter, grid-side up, on the porous plate of the filter base.

3.4.4 Attach the funnel to the base of the filter unit, taking care not to damage or dislodge the filter, The membrane filter is now fitted between the funnel and the base.

3.4.5 Shake the sample container vigorously about 25 times.

3.4.6 Prepare at least three sample increments according to 1.4.3, in this Section. Measure the desired volume of sample into the funnel with the vacuum turned off. To measure the sample accurately and to obtain good distribution of colonies on the filter surface, the following methods are recommended:

(a) Sample volumes of 20 ml or more: Measure the sample in a sterile graduated cylinder and pour it into the funnel. Rinse the graduate twice with sterile dilution water, and add the rinse water to the funnel. For potable waters, 100 ml volumes may be measured directly in a precalibrated funnel.

(b) Sample volumes of 10-20 ml: Measure the sample with a sterile 10 ml or 20 ml pipet into the funnel.

(c) Sample volumes of <10 ml: Pour about 10 ml of sterile dilution water into the funnel and add the sample to the sterile water using appropriate sterile pipet.

(d) Sample volumes of less than 0.1 ml: Prepare appropriate dilutions in sterile dilution water and proceed as applicable in steps (b) or (c) above.

(e) The time elapsing between preparation of sample dilutions and filtration should be minimal and never more than 30 minutes.

3.4.7 Turn on the vacuum to filter the sample. Leave the vacuum on and rinse down the funnel walls at least twice with 20–30 ml of sterile dilution water. Turn off vacuum.

3.4.8 Remove the funnel from the base of the filter unit. An ultraviolet sterilizer unit can be used to hold and sterilize the funnel between filtrations. At least 2 minutes exposure time is required for funnel decontamination (6). Protect eyes from UV irradiation with glasses, goggles, or an enclosed UV chamber (7).

3.4.9 Holding the membrane filter at its edge with a sterilized forceps gently lift and place the filter grid-side up in the culture dish. Slide the filter onto the absorbent pad or agar, using a rolling action to avoid trapping air bubbles between the membrane filter and the underlying pad or agar. Reseat the membrane if non-wetted areas occur due to air bubbles.

3.4.10 Invert the petri dishes and incubate at the appropriate temperature in an atmosphere with close to 100% relative humidity for the required time.

3.5 Counting Colonies: The grid lines are used in counting the colonies.

3.5.1 Count the colonies for the parameter of interest following a preset plan such as shown in Figure II-C-8. Some colonies will be in contact with grid lines. A suggested procedure to reduce error in counting these colonies is shown in Figure II-C-9. Count the colonies in the squares indicated by the arrows.

3.5.2 The fluorescent lamp should be nearly perpendicular to the membrane filter, Count colonies individually, even if they are in contact with each other. The technician must learn to recognize the difference between two or more colonies which have grown into contact with each other and single, irregularly shaped colonies which sometimes develop on membrane filters. The latter colonies are usually associated with a fiber or particulate material and the colonies conform to the shape and size of the fiber or particulates. Colonies which have grown together almost invariably show a very fine line of contact.

3.6 Calculation of Results: Select the membrane filter with the number of colonies in the acceptable range and calculate count per 100 ml according to the general formula:

$$\text{count per 100 ml} = \frac{\text{No. colonies counted}}{\text{Volume of sample filtered, in ml}} \times 100$$

3.6.1 <u>Counts Within the Acceptable Limits</u>

(a) The acceptable range of colonies which is countable on a membrane is a function of the parameter as shown in Table II-C-3.

(b) Assume that filtration of volumes of 50, 15, 5, 1.5, and 0.5 ml produced colony counts of 200, 110, 40, 10, 5, respectively.

(c) An analyst would <u>not</u> actually count the colonies on all filters. By inspection he would select the membrane filter(s) with 20–80 coliform colonies and then limit his actual counting to such membranes.

(d) After selecting the best membrane filter for counting, in this case the MF with a 40 colony count, the analyst counts colonies according to the counting procedures in 3.6 and applies the general formula as follows:

$$\text{Colonies per 100 ml} = \frac{40}{5} \times 100 = 800$$

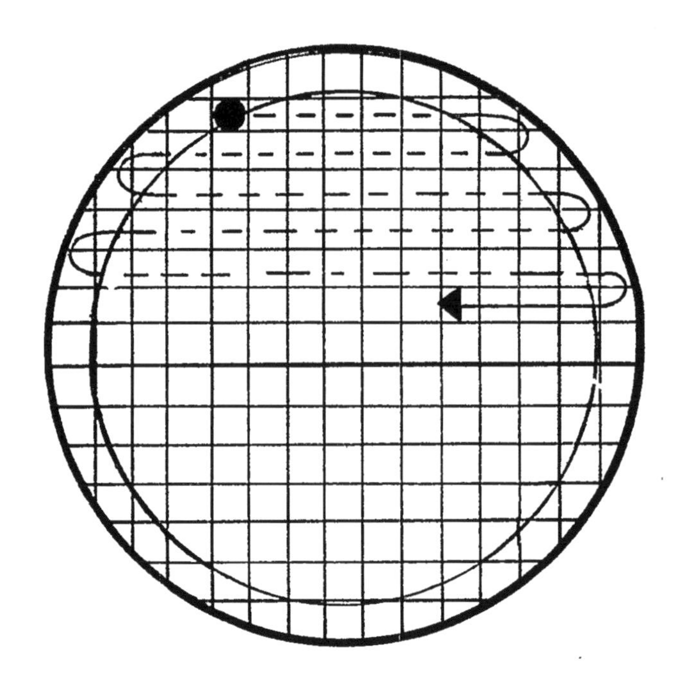

FIGURE II-C-8. Colony-Counting Pathway. (The inner circle indicates the effective filtering area, dashed line indicates the pathway.)

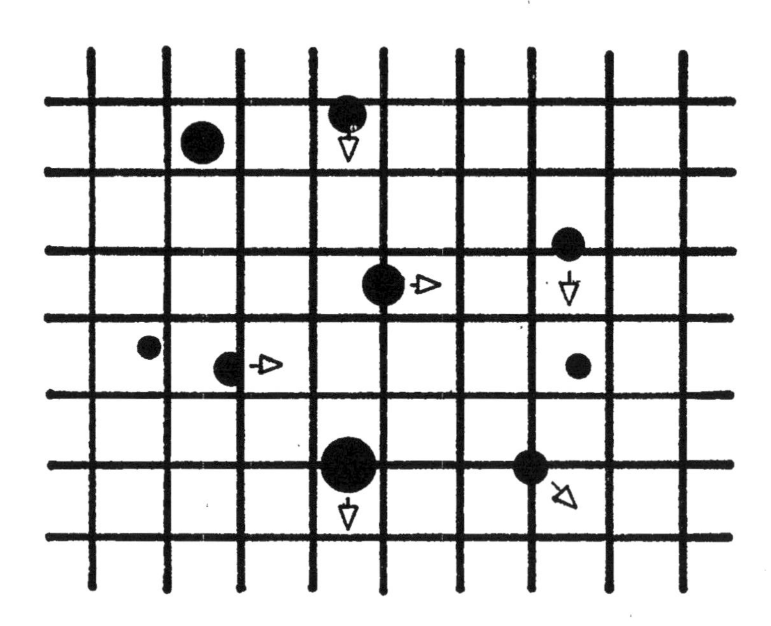

FIGURE II-C-9. Enlarged Portion of Grid-Marked Square of Filter. (Colonies are counted in squares indicated by the arrow.)

TABLE II-C-3

Acceptable Limits*

Parameters	Lower	Upper	Remarks
Total coliform bacteria	20 (0 for potable waters)	80	Limit, 200 colonies
Fecal coliform bacteria	20	60	of all types
Fecal streptococci	20	100	

* Colony counts < or > the limits cited above must be identified as outside of this range.

3.6.2 More Than One Acceptable Count

(a) If there are acceptable counts on replicate plates, carry counts independently to final reporting units, then calculate the arithmetic mean of these counts to obtain the final reported value.

For example, 1 ml volumes produce coliform counts of 26 and 36 or counts of 2600 and 3600/100 ml:

$$\frac{2600 + 3600}{2} = 3100$$

and value = 3100/100 ml

(b) If more than one dilution, independently carry counts to final reporting units, then average for final reported value.

For example, assume that volumes of 0.3, 0.1, 0.03 and 0.01 ml produced coliform colony counts of TNTC (Too Numerous To Count), 75, 30 and 8, respectively. In this example, two volumes, 0.1 and 0.03 produce colonies in the acceptable counting range.

Independently carry each MF count to a count per 100 ml:

$$\frac{75}{0.1} \times 100 = 75{,}000/100 \text{ ml}$$

$$\frac{30}{0.03} \times 100 = 100{,}000/100 \text{ ml}$$

Then calculate the arithmetic mean of these counts to obtain the final reported value:

$$\frac{75{,}000 + 100{,}000}{2} = 87{,}500$$

Report as: 88,000/100 ml.

3.6.3 If All MF Counts are Below the Lower Limit, Select the Most Nearly Acceptable Count (for non-potable waters)

For example, assume a count in which sample volumes of 1, 0.3 and .01 ml produced colony counts of 14, 3, and 0, respectively.

Here, no colony count falls within recommended limits. Calculate on the basis of the most nearly acceptable plate count, 14, and report with a qualifying remark:

$$\frac{14}{1.0} \times 100 = 1400$$

Report as: Estimated Count, 1400 per 100 ml.

3.6.4 If Counts from All Membranes are Zero, Calculate Using Count from Largest Filtration Volume

For example, sample volumes of 25, 10, and 2 ml produced colony counts of 0, 0, and 0, respectively, and no actual calculation is possible, even as an estimated report. Calculate the number of colonies per 100 ml that would have been reported if there had been one colony on the filter representing the largest filtration volume, thus:

$$\frac{1}{25} \times 100 = 4$$

Report as: < (Less than) 4 colonies per 100 ml.

3.6.5 If All Membrane Counts are Above the Upper Limit, Calculate Count with Smallest Volume Filtered

For example, assume that the volumes 1, 0.3, and 0.01 ml produced colony counts of TNTC, 150, and 110 colonies. Since all colony counts are above the recommended limit, use the colony count from the smallest sample volume filtered and estimate the count as:

$$\frac{110}{0.01} \times 100 = 1{,}100{,}000$$

Report as: Estimated Count

1,100,000 per 100 ml.

3.6.6 If Colonies are Too Numerous To Count, Use Upper Limit Count with Smallest Filtration Volume

Assume in Example 3.6.5 that the volumes 1.0, 0.3, and 0.01 ml, all produced too many colonies to show separated colonies, and that the laboratory bench record showed TNTC (Too Numerous to Count).

Use 80 colonies (Upper Limit Count for Total Coliform) as the basis of calculation with the smallest filtration volume, thus:

$$\frac{80}{0.01} \times 100 = 800{,}000$$

Report as: > (Greater Than) 8000,000 per 100 ml.

3.6.7 If there is no result because of confluency, lab accident, etc. Report as: No Result and specify reason.

3.6.8 Reporting Results: Report bacterial densities per 100 ml of sample. See Figure II-C-1 and II-C-3 for examples of forms for reporting results.

3.7 Verification: A verified membrane filter test establishes the validity of colony differentiation on a selective medium and provides support evidence of colony interpretation.

3.7.1 A percent verification can be determined for any colony validation test:

$$\frac{\text{No. of colonies meeting verification test}}{\text{No. of colonies subjected to verification}} \times 100 = \text{Percent verification}$$

3.7.2 Verification is required for all positive samples from potable waters.

3.7.3 Verification is also recommended for establishing quality control in research for use with new test waters, new procedures or new technicians, for identifying unusual colony types and as support for data used in legal actions.

3.7.4 The worker is cautioned not to apply the percentage of verification determined for one sample to other samples.

3.7.5 The careful worker may also pick non-typical colonies and follow the verification procedure to determine that false negative colonies do not occur.

3.8 Significant Figures: See 2.8.1.

3.9 Repeatability and Reproducibility of Counts: Analysts should be able to duplicate their own colony counts on the same membrane within 5% and the counts of other analysts within 10%. Failure to agree within these limits should trigger a review of procedures.

4. Most Probable Number (MPN) Method

4.1 Summary: The Most Probable Number procedure estimates the number of specific organisms in water and wastewater by the use of probability tables.

Decimal dilutions of samples are inoculated in series into liquid tube media. Positive tests are indicated by growth and/or fermentative gas production. Bacterial densities are based on combinations of positive and negative tube results read from the MPN table. The MPN procedure may be carried to three stages of completion:

4.1.1 The Presumptive Test provides a preliminary estimate of bacterial density based on enrichment in minimally-restrictive tube media. The results of this test are never used without further analyses.

4.1.2 The Confirmed Test, the second stage of the MPN, is the usual extent of testing. Growth from each positive Presumptive Test tube is inoculated into a more selective inhibitory medium. The tubes are incubated at the prescribed temperature and time, the positive reactions noted and counts calculated from the MPN table.

4.1.3 The Completed Test is the third stage of the MPN used for total coliform analyses only. Positive tubes from the Confirmed Test are submitted to additional tests to verify the identification of the isolated microorganisms. Although the Completed Test provides the greatest reliability, the amount of time and the workload restrict its use to periodic substantiation of Confirmed Test results, to other QC checks on methodology and analysts, and to research.

4.2 Scope and Application

4.2.1 Advantages: The MPN procedure has the advantages inherent in liquid nutrient media.

(a) The Presumptive and Confirmed Tests require only observing and recording of gas/no gas for coliforms and growth/no growth for fecal streptococci. The tests require minimal experience, training or interpretation by the analyst.

(b) Water samples with high turbidity or large numbers of algae have no apparent deleterious effect on the tube reactions.

(c) If a toxic substance is present in the sample, the resultant 1:10 or 1:100 dilution of that sample in the liquid broth may reduce the toxicity to the point of no effect.

(d) The MPN may be the only method applicable to problem sample materials such as bottom sludges, muds, soils and sediments (with blending).

4.2.2 Limitations: The MPN procedure has disadvantages:

(a) This method is ordinarily limited to a maximum sample volume of 10 ml per tube, but 100 ml portions are used in shellfish waters.

(b) The time required for the test may be as long as 96 hours for a Confirmed Test result.

(c) The MPN tables are probability calculations and inherently have poor precision and contain a 23% bias at the 5 tube, three dilution levels normally used.

(d) The man-hour requirements to prepare glassware and media and to perform the tests are significant.

(e) Relatively large amounts of bench space, incubator space and tube/rack storage space are required.

(f) The procedure does not lend itself to field work. As compliance monitoring of water quality and effluent standards becomes a major legal requirement, the time, precision and equipment limitations cited in (b), (c), (d) and (e) above are more serious for the large number of field analyses which will be required.

(g) Background organisms or toxic constituents in 10 ml volumes of marine water can interfere and be undetected.

4.2.3 The minimum MPN test that is acceptable for water and wastewater analyses is the Confirmed Test because of the high probability of false positive reactions in the Presumptive Test.

4.3 Apparatus and Materials

4.3.1 Water bath or air incubator set at 35 ± 0.5 C for total coliform and fecal streptococci tests. Water bath at 44.5 ± 0.2 for fecal coliform test.

4.3.2 Pipet containers of stainless steel, aluminum or pyrex glass for glass pipets.

4.3.3 Inoculation loops, 3 mm diameter, and needle of nichrome or platinum wire, 26 B&S gauge, in suitable holder.

4.3.4 Disposable applicator sticks or plastic loops as alternatives to inoculation loops in 4.3.3 above.

4.3.5 Compound microscope, oil immersion, 1000×.

4.3.6 Culture tube racks, 10 × 5 openings; each opening to accept 25 mm diameter tubes.

4.3.7 Gas burner, Bunsen/Fisher types or electric incinerator unit.

4.3.8 Sterile T.D. bacteriological or Mohr pipets, glass or plastic, of appropriate sizes.

4.3.9 Dilution bottles (milk dilution), pyrex glass, marked at 99 ml volume, screw cap with neoprene liner.

4.3.10 Pyrex culture test tubes 150 × 25 or 150 × 20 mm containing inverted fermentation vials, 75 × 10 mm and proper closures.

4.3.11 Gram stain solutions (optional). See This Section 5.3.

(a) Hucker's Crystal Violet Solution (stain).

(b) Lugol's Iodine Solution (mordant).

(c) Acetone Alcohol (decolorizer).

(d) Safranin Solution (counterstain).

4.3.12 Glass microscope slides, 2.5 × 7.6 cm (1 × 3 inches).

4.4 Media: Appropriate media dispensed in test tubes or in fermentation test tubes. See Part IIB. for specific media.

4.5 Dilution Water: Sterile 99 ± 2 ml volumes in screw-cap pyrex glass bottles. See Part II-B, 7.

4.6 Presumptive Test

4.6.1 Shake the sample or dilution container vigorously about 25 times.

4.6.2 To perform the Presumptive Test, arrange a series of three or more rows of culture tubes containing the test medium in a rack, providing for five replicates in each row. Use five rows for samples of unknown density. Inoculate each successive row with decreasing decimal dilutions of the sample. For example, in testing polluted waters for total coliforms, the initial sample inoculations might be 0.1, 0.01, 0.001, 0.0001, 0.00001 ml of original sample into successive rows each containing five replicate volumes. This series of sample volumes would yield determinate results from test waters containing up to 16,000,000 organisms per 100 ml by use of the MPN tables.

When removing sample aliquots or dilutions for further inoculations, do not insert the pipet tip more than 2.5 cm (1 inch) below the surface of the sample.

4.6.3 Incubate tubes for 24±2 hours at 35 C. A positive presumptive test is gas production for the coliforms or growth for fecal streptococci. After 24 hours incubation, examine the tubes for gas formation and/or growth. Inoculate positive tubes into Confirmed Test media. If there is no gas or growth reincubate these negative tubes for an additional 24 hours.

4.6.4 If the Presumptive tubes are negative after 48 ± 3 hours, discard tubes. If the Presumptive tubes are positive, the cultures are verified in the Confirmed Test. Record the negative and positive results.

4.7 Confirmed Test

4.7.1 The Confirmed Test is performed by verifying positive tubes from the Presumptive Test at 24 and 48 hours. If Presumptive tubes are positive at 24 hours, confirm them at that time.

4.7.2 A positive test is indicated by gas production for the coliform bacteria or growth for fecal streptococci. After 24 ± 2 hours incubation, examine the tubes for gas formation and/or growth. If there is no gas/growth, reincubate these negative tubes for a second 24 hours.

4.7.3 After 48 hours ± 3 hours examine tubes for gas and/or growth, record positive and negative results. Discard negative tubes. Retain positive tubes if the test is to be carried to completion for total coliform tests.

4.7.4 The fecal coliform MPN test is performed by inoculating EC Broth tubes with growth from all positive Presumptive tubes and incubating them at the elevated temperature of 44.5 C for 24 hours. Gas production is the positive reaction.

4.7.5 Passage of positive Presumptive cultures through the Confirmed Test completes the MPN series for fecal streptococci and fecal coliform bacteria.

4.7.6 In routine practice, most sample examinations for total coliform are terminated at the end of the Confirmed Test. However, for quality control, at least five percent of the Confirmed Test samples (and a minimum of one sample per test run) should be carried through the Completed Test.

4.8 Completed Test for Total Coliform MPN

Positive Confirmed Test cultures may be subjected to final Completed Test identification through application of further culture tests, as follows:

4.8.1 Streak Levine's EMB agar plates from each positive confirmatory tube and incubate at 35 C for 24 hours. Pick typical coliform colonies (or atypical colonies if no typical colonies are present), inoculate into lauryl tryptose fermentation tubes and incubate at 35 C for 24-48 hours. The formation of gas in any amount in the fermentation tubes constitutes a positive Completed Test for total coliforms.

Typical colonies show a golden green metallic sheen or reddish purple color with nucleation.

Atypical colonies are red, pink or colorless, unnucleated and mucoid.

4.8.2 Optional Gram Stain Procedure

The gram stain test has been used in the Completed MPN Test for demonstrating gram negative, nonsporeforming rods from isolated colonies. Although the gram stain procedure is proposed for revision of the 15th edition of *Standard Methods,* it provides a final check on results and remains useful for evaluating questionable colony types.

After incubation of the EMB agar plates for 24 hours at 35 C (in 4.8.1) pick at least two typical colonies (or atypicals if no typical colonies are present) and inoculate onto nutrient agar slants. Incubate for 24 hours at 35 C, and proceed as in 5, this Section.

4.9 Calculation of MPN Value

The calculated density of the Confirmed or Completed Test may be obtained from the MPN table based on the number of positive tubes and reactions in each dilution.

4.9.1 Table II-C-4 illustrates the MPN indices and 95% Confidence Limits for general use.

4.9.2 Table II-C-5 shows the MPN indices and limits for potable water testing.

4.9.3 Three dilutions are necessary to formulate the MPN code. For example in Table II-C-4 if five 10 ml, five 1.0 ml, and five 0.1 ml portions are used as inocula and positive results are observed in five of the 10 ml inocula, three of the 1.0 inocula, and none of the 0.1

TABLE II-C-4

Most Probable Number Index and 95% Confidence Limits for Five Tube, Three Dilution Series (8, 9)

No. of Tubes Giving Positive Reaction out of			MPN Index per 100 ml	95% Confidence Limits	
5 of 10 ml Each	5 of 1 ml Each	5 of 0.1 ml Each		Lower	Upper
0	0	0	<2		
0	0	1	2	<0.5	7
0	1	0	2	<0.5	7
0	2	0	4	<0.5	11
1	0	0	2	<0.5	7
1	0	1	4	<0.5	11
1	1	0	4	<0.5	11
1	1	1	6	<0.5	15
1	2	0	6	<0.5	15
2	0	0	5	<0.5	13
2	0	1	7	1	17
2	1	0	7	1	17
2	1	1	9	2	21
2	2	0	9	2	21
2	3	0	12	3	28
3	0	0	8	1	19
3	0	1	11	2	25
3	1	0	11	2	25
3	1	1	14	4	34
3	2	0	14	4	34
3	2	1	17	5	46
3	3	0	17	5	46
4	0	0	13	3	31
4	0	1	17	5	46
4	1	0	17	5	46
4	1	1	21	7	63
4	1	2	26	9	78
4	2	0	22	7	67
4	2	1	26	9	78
4	3	0	27	9	80
4	3	1	33	11	93
4	4	0	34	12	93
5	0	0	23	7	70
5	0	1	31	11	89
5	0	2	43	15	110
5	1	0	33	11	93
5	1	1	46	16	120
5	1	2	63	21	150
5	2	0	49	17	130
5	2	1	70	23	170
5	2	2	94	28	220
5	3	0	79	25	190
5	3	1	110	31	250
5	3	2	140	37	340
5	3	3	180	44	500
5	4	0	130	35	300
5	4	1	170	43	490
5	4	2	220	57	700
5	4	3	280	90	850
5	4	4	350	120	1,000
5	5	0	240	68	750
5	5	1	350	120	1,000
5	5	2	540	180	1,400
5	5	3	920	300	3,200
5	5	4	1600	640	5,800
5	5	5	≥2400		

TABLE II-C-5

Most Probable Number Index and 95% Confidence Limits for Testing Potable Waters

Number of Positive Tubes from five 10 ml Portions	MPN Index/100 ml	95% Confidence Limits	
		Lower	Upper
0	<2.2	0	6.0
1	2.2	0.1	12.6
2	5.1	0.5	19.2
3	9.2	1.6	29.4
4	16.	3.3	52.9
5	>16.	8.0	Infinite

ml inocula, the coded results of the test are 5-3-0. The code is located in the MPN Table, and the MPN index of 79 per 100 ml is recorded.

4.9.4 When the series of decimal dilutions is other than 10, 1.0 and 0.1 ml, select the MPN value from Table II-C-4 and calculate according to the following formula:

$$\text{MPN (From Table)} \times \frac{10}{\text{Largest Quantity Tested}} = \text{MPN/100 ml}$$

As an example, five out of five 0.01 ml portions, two out of five 0.001 ml portions, and zero out of five 0.0001 ml portions from a sample of water, gave positive reactions. From the code 5–2–0 in MPN Table (Table II-C-4), the MPN index 49 is adjusted for dilutions:

$$49 \text{ (From Table)} \times \frac{10}{0.01} = 49{,}000$$

The final corrected MPN Value = 49,000/100 ml.

4.9.5 If more than the above three sample volumes are inoculated, the three significant dilutions must be determined. The significant dilutions are selected using the following rules:

(a) Only three dilutions are used in the code for calculating an MPN value.

(b) To obtain the proper three dilutions, select the smallest sample volume giving all positive results and the two succeeding lesser sample volumes. See Table II-C-6, Test 1 and 2.

(c) If less than three dilutions show positive tubes, select the three highest sample volumes which will include the dilutions with the positive tubes. See Table II-C-6, Test 3.

(d) If there are positive tubes in the dilutions higher than these dilutions selected, positive results are moved up from these dilutions sample volume to increase the positive tubes in the highest dilution selected. See Table II-C-6, Test 4.

(e) There should be no negative results in higher sample volumes than those chosen. However, if negative tubes are present, e.g., 4/5, 5/5, 3/5 and 0/5 the highest sample volume with all positive tubes must be used along with the next two lower sample volumes. See Table II-C-6, Test 5.

(f) If all tubes are positive, choose the three highest dilutions. See Table II-C-6, Test 6.

(g) If all tubes are negative, choose the three lowest dilutions. See Table II-C-6, Test 7.

(h) If positive tubes skip a dilution, select the highest dilution with positive tubes and the two lower dilutions. See Table II-C-6, Test 8.

(i) If only the middle dilution is positive, select this dilution and one higher and lower dilution. See Table II-C-6, Test 9.

4.9.6 A number of theoretically possible combinations of positive tube results are omitted in Table II-C-4 because the probability of their occurrence is less than 1%. If such unlikely tube combinations occur in more than 1% of samples, review the laboratory procedures for errors and note sample types. Collect fresh samples for analyses.

4.9.7 The MPN can also be computed for each sample based upon the number of positive and negative Presumptive, Confirmed or Completed Tests, and the total number of milliliters tested (10). MPN/100 ml =

$$\frac{\text{No. of Positive Tubes} \times 100}{\sqrt{\text{(No. of ml in Negative Tubes)} \times \text{(No. of ml in All Tubes)}}}$$

Example: From a sample of water, five out of five 10 ml portions, two out of five 1.0 ml portions, and zero out of five 0.1 ml portions

TABLE II-C-6

Selection of Coded Results, Five Tube Series

	Positive Tubes/ml Sample Volume						
Test	10	1.0	0.1	0.01	0.001	0.0001	Code
1	5*	3	0	0	0	0	5-3-0
2	5	5	4	0	0	0	5-4-0
3		4	1	0	0	0	4-1-0
4	5	5	4	1	1	0	5-4-2
5	4	5	3	0	0		5-3-0
6		5	5	5	5	5	5-5-5
7		0	0	0	0	0	0-0-0
8		4	0	2	0	0	4-0-2
9		0	1	0	0	0	0-1-0

*Underlines indicate positive tube series selected for code.

gave positive results. Therefore, MPN/100 ml =

$$\frac{7 \times 100}{\sqrt{(3.5) \times (55.5)}} = 50.22$$

or

MPN/100 ml = 50

4.10 Reporting Results: Report the MPN value for water samples on the basis of 100 ml of sample. Report the MPN values of solid type samples on the basis of 1 gram of dry weight sample.

Examples of bench forms are shown in Figures II-C-1, 2 and 3.

4.11 Precision and Accuracy

4.11.1 The precision of the MPN value increases proportionately with the number of replicates tested.

4.11.2 Multiple-tube values are generally high because MPN tables include a 23% positive bias.

4.11.3 MPN numbers represent only a statistical estimate of the true bacterial density in the sample. The 95% Confidence Limits for each MPN value included in the Tables II-C-4 and 5 show the limited precision of these estimates.

5. Staining Procedures

5.1 Preparation of Bacterial Smears

5.1.1 Place a small drop of laboratory pure water on a clean slide.

5.1.2 Using a sterilized inoculating needle, pick a small amount of growth from the agar slant. Mix the bacteria with the drop of water on the slide and spread evenly over an area the size of a quarter. Use loop for broth cultures.

5.1.3 Air-dry the smear and fix by quickly passing the slide several times through a portion of the flame.

5.2 Gram Stain

Gram staining is a general test for characterization of bacteria and for examination of culture purity.

5.2.1 Prepare and fix a bacterial smear as in 5.1. For quality control, prepare a separate smear of known gram positive cocci and gram negative rods.

5.2.2 Flood the smear with ammonium oxalate-crystal violet stain for one minute.

5.2.3 Wash the slide in a gentle stream of tap or pure water and flood with Lugol's iodine solution (mordant). Allow it to remain for one minute.

5.2.4 Wash the slide in water and blot dry.

5.2.5 Decolorize with acetone alcohol, either by adding it dropwise on the tilted slide until the blue color stops flowing from the smear, or by gently agitating the slide up and down in a beaker containing the alcohol wash for about 30 seconds.

5.2.6 Flood the smear with the safranin counterstain for 10 seconds. Wash and air-dry.

5.2.7 Examine under the oil immersion objective. Gram positive cells retain the crystal violet stain and are blue in color. Gram negative cells are decolorized by the acetone alcohol so that they accept the safranin counterstain and appear pink to red.

5.3 Stain Solutions

Only those stains and dyes which are certified by the National Biological Stain Commission should be used.

5.3.1 Loeffler's Methylene Blue

(a) Methylene Blue-ethyl alcohol solution.

(1) Dissolve 0.3 grams of methylene blue (90% dye content) in 30 ml of 95% ethyl alcohol (Solution A).

(2) Dissolve 0.01 grams of potassium hydroxide in 100 ml of laboratory pure water (Solution B).

(3) Mix solution A and B.

5.3.2 Solutions for Gram Staining

(a) Ammonium oxalate-crystal violet solution:

(1) Dissolve 2 grams crystal violet (approximately 85% dye content) in 20 ml of 95% ethyl alcohol (Solution A).

(2) Dissolve 0.8 grams ammonium oxalate in 80 ml laboratory pure water (Solution B).

(3) Mix Solutions A and B.

(4) Filter through cheesecloth or coarse filter paper.

(5) Problems with the gram stain technique are frequently traceable to the ammonium oxalate-crystal violet solution. In the event that decolorization is insufficient, the amount of crystal violet in the solution can be reduced to as little as 10% of the recommended amount.

(b) Lugol's iodine: Dissolve 1 gram iodine crystals and 2 grams potassium iodide in about 5 ml of pure water. After crystals are in solution, add sufficient laboratory pure water to bring the final solution to a volume of 300 ml.

(c) Acetone-Alcohol: Combine fifty ml volumes of acetone and 95% ethyl alcohol.

(d) Safranin: Dissolve 2.5 grams of safranin in 100 ml of 95% ethyl alcohol. Store as a stock solution. For working solution, add 10 ml of stock to 100 ml of laboratory pure water, mix and store.

6. Shipment of Cultures

6.1 Confirmation or further identification to serotype may be required if the bacteriological data are to be used for specific needs such as enforcement cases, epidemiological studies, tracing sources of pollution or scientific publication. The selected cultures should be sent to an official typing center or state health laboratory with pertinent information for the confirmatory identification. This service is usually available if the cultures are of public health significance, but permission should be obtained from the reference laboratory before sending cultures.

Observe the following instructions:

6.1.1 Send only pure cultures.

6.1.2 Provide a culture with discernable growth on brain heart infusion agar, blood agar base or nutrient agar stab in a screw-cap tube or vial sealed with a cork soaked in hot paraffin. Triple sugar iron agar or other sugar-containing agars should not be used.

6.1.3 Complete the reference laboratory form which requires information on the source of the culture, tests completed, results obtained and identification of the originating laboratory. Include form with cultures, but on outside of secondary container.

6.2 Shipping Regulations for Cultures: Transportation (DOT) regulations apply to surface and air transportation, but shipment of cultures beyond 100–200 miles is only practical by air. Air freight service is limited in many areas, hence passenger-carrying aircraft must be used for safe and quick service. Strict shipping regulations are imposed on such passenger service shipments (11). Packaging and labeling of the cultures must conform with current federal shipping regulations for etiological agents described in: 49 CFR 173.387 (12) and 42 CFR 72.25 (c) (13). The requirements in 6.2.1–6.2.8 that follow are also shown in Figure II-C-10.

6.2.1 Place each culture in a securely closed, watertight PRIMARY CONTAINER (screw-cap test tube or vial) and seal the cap with tape.

6.2.2 Wrap the PRIMARY CONTAINER with sufficient absorbent material (paper towel, tissue, etc.) to absorb the entire contents should breakage or leakage occur.

6.2.3 Place the wrapped, sealed, PRIMARY CONTAINER in a durable, watertight SECONDARY CONTAINER (screw-cap metal mailing tube or sealed metal can). Screw-cap metal mailing tubes should be sealed with tape. Several PRIMARY CONTAINERS of cultures, each individually wrapped in absorbent material, may be placed in the SECONDARY CONTAINER provided that the total aggregate volume does not exceed 50 ml. (NOTE: Multiple secondary containers of cultures, which individually meet the packaging requirements for shipment of 50 ml or less, can be overpacked in a single outer shipping container, provided that the total aggregate volume does not exceed 4000 ml).

6.2.4 Data forms, letters and other information identifying or describing the cultures should be placed around the outside of the SECONDARY CONTAINER. DO NOT ENCLOSE WITHIN THE SECONDARY CONTAINER.

6.2.5 Place the SECONDARY CONTAINER and information form in an OUTER MAILING TUBE OR BOX.

6.2.6 Place an address label and ETIOLOGIC AGENT/MICROBIOLOGICAL CULTURES label on the outer mailing tube or box.

6.2.7 Individual primary containers of greater than 50 ml of culture material require special packaging and cannot be transported on passenger-carrying aircraft.

6.2.8 International shipments must also conform to the added regulations: US Post Office Publication 51, Air Cargo Restricted Articles Tariff 6-D and DOT Regulations 49 CFR, Section 173.

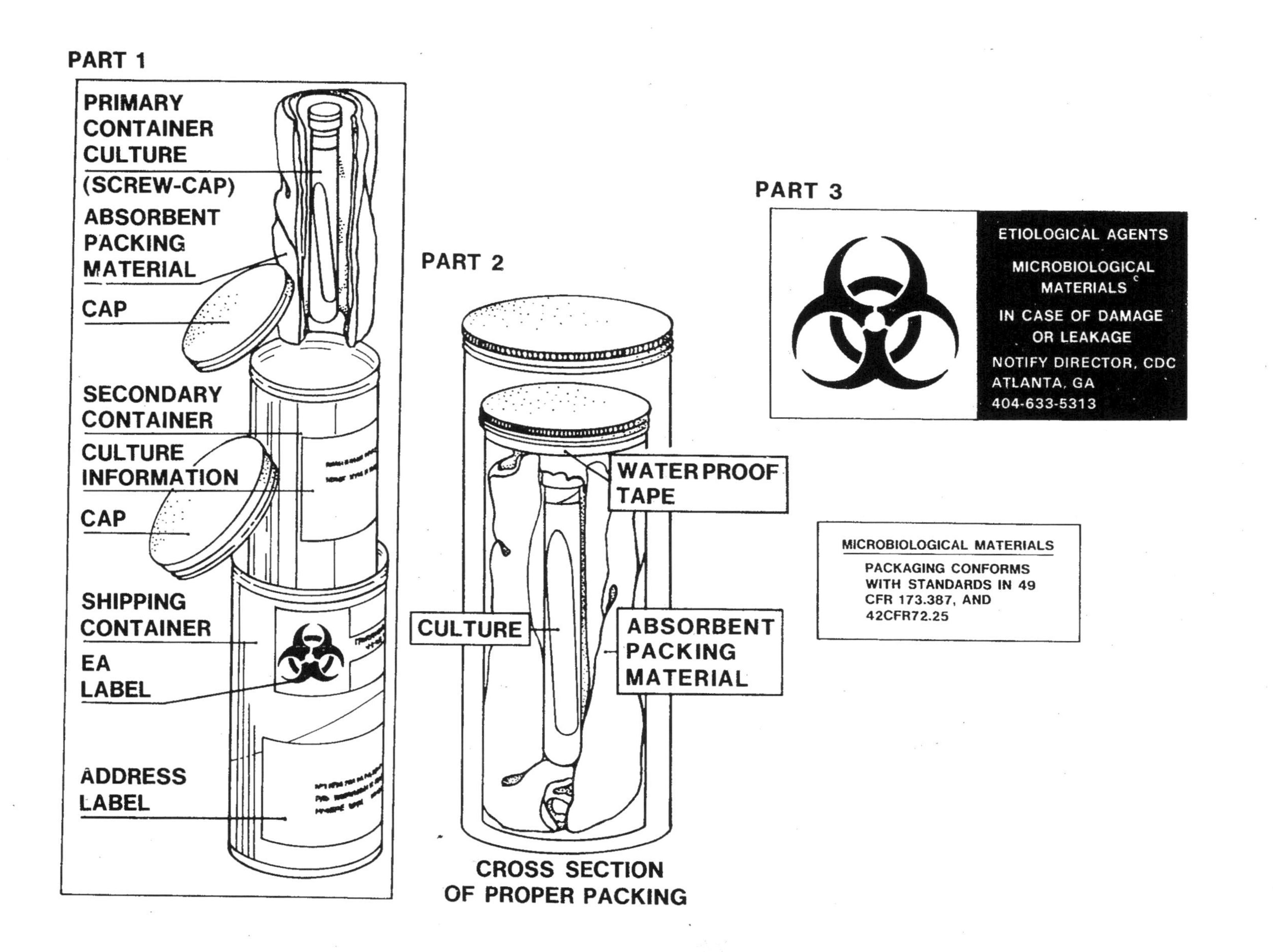

FIGURE II-C-10. Packaging and Labelling of Microbiological Cultures for Shipment.

REFERENCES

1. Clark, H. F., P. W. Kabler and E. E. Geldreich, 1957. Advantages and limitations of the membrane filter procedures. Water Sewage Works 104:385.

2. Shipe, E. L. and A. Fields, 1954. A comparison of the molecular filter technique with agar plate count for enumeration of *Escherichia coli*. Appl. Microbiol. 2:382.

3. Shipe, E. L. and G. M. Cameron, 1954. A comparison of the membrane filter with the most probable number method for coliform determinations from several waters. Appl. Microbiol. 2:85.

4. Geldreich, E. E., H. L. Jeter and J. A. Winter, 1967. Technical considerations in applying the membrane filter procedure. Health Laboratory Science 4:113.

5. Proceedings of the Symposium on the Recovery of Indicator Organisms Employing Membrane Filters. Jan 20–21, 1975. Environmental Monitoring and Support Laboratory, U.S. Environmental Protection Agency, Cincinnati, OH 45268. (1977).

6. Rhines, C. E. and W. P. Cheevers, 1965. Decontamination of membrane filter holders by ultraviolet light. J. Am. Water Works Association 57:500.

7. Manning, H., 1975. AQC Newsletter #26, July 1975. U.S. Environmental Protection Agency, Environmental Monitoring and Support Laboratory, Cincinnati, OH 45268. p. 15.

8. Swaroop, S., 1938. Numerical estimation of B. coli by dilution method. Indian J. Med. Research 26:353.

9. Swaroop, S., 1951. The range of variation of the most probable number of organisms estimated by the dilution method. Indian J. Med. Research 39:107.

10. Thomas, H. A., Jr., 1942. Bacterial densities from fermentation tubes. J. Am. Water Works Association 34:572.

11. Morbidity and Mortality Report, 1975. Center for Disease Control, Public Health Service, USDHEW, Atlanta, GA 24:49.

12. Title 49, Code of Federal Regulations (CFR), Part 173.

13. Interstate Quarantine, regulations of the shipment of etiologic agents. Title 42, Code of Federal Regulations, (CFR), Part 72, Section 25.

PART II. GENERAL OPERATIONS

Section D Selection of Analytical Methods

This Section discusses the selection of methods for monitoring water and wastewater in response to the Laws, the microbiological standards that have been established, and the criteria that have been recommended to enforce the laws. The major problems that have developed in the application of the methods are identified and solutions are given where they are available.

1. Methodology

Test procedures have been specified and published in Federal Register for drinking water, wastewater discharges (NPDES) and vessel discharges.

1.1 National Interim Primary Drinking Water Regulations

Although the National Interim Primary Drinking Water Regulations (Title 40 CFR Part 141) state that the total coliform analyses can be performed by the membrane filter or MPN procedures, the MF procedure is preferred because large volumes of samples can be analyzed in a much shorter time, a critical factor for potable water. Samples containing excessive noncoliform populations or turbidity must be analyzed by the MPN technique. These regulations specify the testing of sample sizes of 100 ml for the MF technique and the testing of five replicate 10 or 100 ml volumes for the MPN procedure. The law directs that the samples be taken at points representative of the distribution system. The minimal schedules for the frequency of sampling are based on population and the required response is given for positive test results. A detailed description of the proposed criteria for interim certification of microbiology laboratories under the Safe Drinking Water Act is given in Appendix B.

1.2 National Pollution Discharge Elimination System (NPDES) Guidelines

The NPDES established guidelines for analysis of pollutants under PL 92–500, Section 304 (g). The parameters and methods are described in 40 CFR Part 136, as amended (40 Code of Federal Regulations, Protection of the Environment, ch. 1 – Environmental Protection Agency, Part 136, Guidelines Establishing Test Procedures for the Analysis of Pollutants). The method must be specified and MPNs must be five tube, five dilution. See Table II-D-1.

1.3 Marine Sanitation Regulations

The regulations for marine sanitation devices (40 CFR Part 140) established performance standards and specified the analytical methods as those promulgated in 40 CFR Part 136, cited in 1.2 above.

1.4 Water Quality Standards

Water quality standards (limits) have been established by law for drinking water and certain sewage and industrial effluents. These standards and the reference sources are listed in Table II-D-2. A standard must be specified in the NPDES permit to be enforceable.

1.5 Water Quality Criteria

Water quality criteria have been recommended by the EPA for certain types of water classified according to use. These criteria are listed in Table II-D-3.

1.6 Alternate Test Procedures

The amendments to 304 (g) also provide procedures for approval of alternate methods. National approval for test methods is obtained by application to EPA through EMSL-Cincinnati while case by case approval is obtained by application through the EPA Regional Offices (40 CFR 136.4).

2. Problems in Application

Although the methods described in this Manual are judged the best available, there are difficulties in the application of methods in different geographical areas, in certain wastes and in some potable and surface waters. Additional problems can stem from the indiscriminate use of new and simplified equipment, supplies or media that have been proposed for use in these procedures.

2.1 Stressed Microorganisms

Some water and wastewater samples contain microorganisms which should reproduce but do not under the conditions of test. These organisms have been described as injured or stressed cells. The stress may be caused by temperature changes or chemical treatment such as chlorine or toxic wastes (1).

Stressed organisms are particularly important in environmental measurements because tests for bacterial indicators or pathogens can give negative responses, then recover later and multiply to produce dangerous conditions. Subsections 2.1.1 and 2.1.2 describe efforts to recover stressed microorganisms.

2.1.1 Ambient Temperature Effects

Extreme ambient temperatures stress microorganisms and reduce recovery of microbiological indicators. For example, in Alaska and other extremely cold areas, the severe change from cold stream temperature to 44.5 C temperature of incubation reduces recovery of fecal coliforms. The two-step MF test for fecal coliforms increases recoveries by use of a 2-hour acclimation on an enrichment medium at 35 C before normal incubation at 44.5 C.

In contrast, water samples from natural waters at high temperatures may include large numbers of non-coliform organisms which interfere with sheen production on MF's and with positive gas production in MPN analyses. An improved MF medium that provides greater selectivity is desirable but may not be possible without sacrificing recovery.

TABLE II-D-1

Approved Test Procedures for the Analysis of Pollutants (40 CFR 136)

Parameter per 100 ml	Method	Reference and Page No. SM[1]	USGS[2]	This Manual
Fecal Coliforms	MPN	922		Part III C, 5
	MF	937	45	Part III C, 2
Fecal Coliforms in presence of chlorine[3]	MPN	922		Part III C, 5
	MF	928, 937		Part III C, 2
Total Coliforms	MPN	916		Part III B, 4
	MF	928	35	Part III B, 2
Total Coliforms in presence of Chlorine	MPN	916		Part III B, 4
	MF	933		Part III B, 2
Fecal Streptococci	MPN	943		Part III D, 4
	MF	944	50	Part III D, 2
	Plate Count	947		Part III D, 5

[1] Standard Methods for the Examination of Water and Wastewater, 14th Edition, (1975).

[2] Slack, K. V., et.al. Methods for Collection and Analysis of Aquatic Biological and Microbiological Samples. USGS Techniques of Water Resources Inv., Book 5, ch. A4 (1973).

[3] Since the MF technique usually yields low and variable recovery from chlorinated wastewaters, the MPN method will be required to resolve any controversies.

TABLE II-D-2

Water Quality Standards

Water or Wastewater	Microbiological Standards Coliforms/100 ml		Reference Source
	Total	Fecal	
Potable Water	<5	—	PL 93-523
Chlorinated Effluents	—	200-400	PL 92-500
2° Treatment Wastes	—	200-400	40 CFR Part 133
Selected Industrial Wastes	—	200-400	PL 92-500
Leather and Tanning	—	400	40 CFR Part 425
Feed Lots	—	400	40 CFR Part 412
Meat Products	—	400	40 CFR Part 432
Beet Sugar	—	400	40 CFR Part 409
Canned Fruits and Vegetables	—	400	40 CFR Part 407
Textiles	—	400	40 CFR Part 410
Effluents from Marine Sanitation Devices with Discharges Type I	—	1000	40 CFR Part 140 and Amendments
Type II	—	200	40 CFR Part 140 and Amendments

TABLE II-D-3

Water Quality Criteria

Water or Wastewater	Statistical Measure	Microbiological Criteria Coliforms/100 ml		Reference Source
		Total	Fecal	
Public Water Supply	log $\overline{X}$	20000	2000	A
Recreational Water:				
Primary Contact	log $\overline{X}$/30 days		200	B
	maximum/30 days, in 10% of Samples		400	B
General Contact	log $\overline{X}$/30 days		1000	B
	maximum/30 days, in 10% of Samples		2000	B
Agricultural Water	monthly $\overline{X}$	5000	1000	B
Shellfish-Raising Waters				
	Daily Median	70	14	C & D
	Highest 10% of Daily Values	230	43	

A Water Quality Criteria, EPA. March, 1973. Superintendent of Documents, U.S. Government Printing Office, Washington, DC 20402.

B Water Quality Criteria, FWPCA, April 1,1968. Superintendent of Documents, U.S. Government Printing Office, Washington, DC 20402.

C National Shellfish Sanitation Program Manual of Operation. U.S. Dept. of HEW, 1965. Public Health Service Publ. No. 33. Superintendent of Documents, U.S. Government Printing Office, Washington, DC 20402.

D Quality Criteria for Water, July 1976, O.W.H.M., US EPA.

2.1.2 Chlorinated Effluents and Toxic Wastes

Although thiosulfate is added to all samples suspected of containing chlorine, to neutralize its toxic effects, the membrane filter procedure yields poor recovery of coliforms from chlorinated effluents as compared to MPN recovery (1-6). A recent amendment to 40 CFR 136 added Coliform bacteria (Fecal) in the presence of chlorine, as a specific parameter and recommended analysis by the MF or MPN techniques (7). A qualifying statement appended to the method in 40 CFR Part 136 requires the five tube, five dilution MPN and states: "Since the membrane filter technique usually yields low and variable recovery from chlorinated wastewaters, the MPN method will be required to resolve any controversies." Therefore, the MPN procedure should be used in analysis of chlorinated effluents where the data may be challenged by legal or enforcement actions. The MF may be used currently for self-monitoring situations. (See Table II-D-1).

Proposed changes in MF materials and procedures include new membrane filter formulations, an agar overlay technique, modified media and twostep methods (1). Present modifications of the MF method have not produced recoveries of fecal coliforms from chlorinated effluents equivalent to MPN recoveries. Thorough evaluation and approval of proposed procedures by EPA are required before changes will be acceptable.

Certain types of wastes show recovery problems for total and fecal coliforms:

1. Primary and Chlorinated-Primary Waste Effluents.

2. Chlorinated-Secondary and Chlorinated-Tertiary Waste Effluents.

3. Industrial wastes containing toxic metals or phenols.

When turbidity and low recovery prevent the application of the MF technique to coliform analyses of primary and secondary effluents or industrial wastes containing toxic materials, the MPN procedure is required. However, the two-step MF procedure for total coliforms described in this Manual and in *Standard Methods* is acceptable for toxic wastes.

If the MF procedure is applied to chlorinated or toxic samples, the laboratory should require data from at least 10 samples collected over 1 week of plant processing (but not less than 5 calendar days) to show comparability of the MF to the MPN technique. See Part IV-C, 3 for details.

2.2 Incomplete Recovery/Suppression

When coliforms are present in low numbers in drinking water, high levels of non-coliforms can suppress growth or mask detection. This problem may appear as a mass of confluent growth on a membrane filter or as spots of sheen in this confluent growth. In the MPN procedure, presumptive tubes may show heavy growth with no gas bubbles, dilution skips or unusual tube combinations. When these negative presumptive tubes are transferred to BGLB, they confirm in this more restrictive medium, indicating that the coliform gas production in the Presumptive Test was suppressed by non-coliforms.

2.3 Interference by Turbidity

The tendency of bacteria to clump and adhere to particles can produce inaccurate results in the analysis of water samples. The National Interim Primary Drinking Water Regulations (NIPDWR) specify one turbidity unit as the primary maximum allowable level but permit up to five turbidity units if this level does not interfere with disinfection or microbiological analyses. Turbidity can interfere with filtration by causing a clumping of indicators or clogging of pores. The turbidity as organic solids can also provide nutrients for bacterial growth and subsequently produce higher counts. The type of particles variably affects the filtration rate; for example, clay, silt or organic debris clog more easily than sand. Background organisms may also be imbedded in the particles and interfere with the coliform detection.

2.4 Analysis of Ground Water

Although total coliforms are a valid measure of pollution, their use as indicators in analyzing ground waters and rural community supplies may not sufficiently describe the water quality. For example, ground waters frequently contain high total counts of bacteria with no coliforms. Such waters pass Interim Drinking Water Regulations but technical judgment must conclude these are not acceptable as potable waters.

2.5 Field Problems

Assurance of data validity demands sample analyses within the shortest time interval after collection. This need requires field analyses using either a mobile laboratory or field kit equipment. Since a mobile laboratory may not be available for a survey, it is likely that at least a part of the analyses will need to be completed in an onsite facility. If the analyses can be done using membrane filtration techniques, field kits such as Millipore's Water Laboratory and MF Portable Incubator (heat sink) are particularly helpful for rapid set-up and analyses of limited samples. However, if large numbers of samples are tested per day or the survey covers more than a few days, the heat-sink incubator is impractical because of limited capacity and high cost. In such surveys, a mobile laboratory utilizing water-jacketed incubators is more practical.

2.6 Method Modifications and Kits

Commercial manufacturers continue to offer proprietory kits and method modifications to speed or simplify the procedures used in coliform and fecal coliform analyses, primarily for field use. Most of these units have not been demonstrated to produce results comparable to the official procedures. If not tested to the satisfaction of EPA, such method modifications and kits cannot be used for establishing total or fecal coliform numbers for permits under NPDES or for total coliform numbers under the Safe Drinking Water Act. The procedure required for acceptance of an alternative procedure is described in 40 CFR Parts 136.4 and 136.5, as amended.

2.7 Changes in Membrane Filters and Methodology

There is an expected pattern of changes in materials and methodology used in the manufacture of membrane filters. The changes may or may not be announced by the manufacturer. Therefore, it is important for the laboratory to monitor membrane performance as described in Section A of Quality Control in this Manual.

These changes include modification of formulations and the replacement of the 0.45 μm pore MF by a 0.7 μm retention pore MF for improved recovery. Tests by independent investigators show that several MF's give comparable recovery (5, 6, 8, 9), however, enrichment or two-temperature incubations are needed before recoveries approach the MPN values (See 2.1.2 in this Section).

This discussion of problems with new methodology and membrane materials should not be interpreted as indicating that EPA discourages new developments. Rather EPA encourages the MF supply industry to test and examine procedures, to innovate and to research. The membrane filter manufacturers should be commended and encouraged to continue their efforts toward solving problems and improving materials and techniques in water microbiology.

2.8 *Klebsiella* in Industrial Wastes

Klebsiella bacteria (part of the coliform group) multiply in certain industrial wastes, are not differentiated from fecal coliforms by MF and MPN procedures and consequently are included in the results. These recoveries have been reported in textile, paper and pulp mills and other wastes. Objections have been raised to the application of fecal coliform standards

TABLE II-D-4

Selection of Methods for Problem Samples

Problem Area	Parameter Chosen	Method of Choice
Shellfish-harvesting waters	Total coliform	MPN*
	Fecal Coliform	MPN*
Marine & Estuarine Waters	Total Coliform	MF/MPN
	Fecal Coliform	MF/MPN
Treated Industrial Wastes (non-chlorinated, non-toxic) Low Solids Wastes	Fecal Coliform	MF
Toxic Industrial Wastes (metals, phenolics) and High Solids Wastes	Fecal Coliform	MPN or alternate procedure, tested and approved**
Primary and Chlorinated-Primary Municipal/Industrial Effluent	Total Coliform	MPN or alternate procedure, tested and approved**
	Fecal Coliform	MPN or alternate procedure, tested and approved**
Chlorinated-Secondary Effluent	Total Coliform	Two-Step MF
	Fecal Coliform	MPN or alternate procedure, tested and approved**

*MPN recommended to conform with the MPN method specified for examination of shellfish.

**Requires proof of comparability under EPA's specified test regime that the alternate procedure (MF, streak plate, etc.) is valid. See This Manual, IV-C, 3.

to these wastes because *Klebsiella* originate from other than sanitary sources. However, EPA does consider large numbers of *Klebsiella, Aeromonas* and other noncoliforms as indicators of organic pollution. Further, these organisms do occur in low densities in human and animal wastes.

3. Recommendations for Methods in Waters and Wastewaters

The amended Federal Water Pollution Control Act, the Marine Protection, Research and Sanctuaries Act and the Safe Drinking Water Act require recommendations on analytical methodology. Generally, the membrane filter methods are preferred over MPN and other techniques, where proven applicable.

In Table II-D-4, problem samples are identified and the analytical method recommended for parameters of choice.

REFERENCES

1. Bordner, R. H., C. F. Frith and J. A. Winter, eds., 1977. Proceedings of the Symposium on Recovery of Indicator Organisms Employing Membrane Filters, U.S. Environmental Protection Agency, EPA-600-19-77-024, EMSL-Cincinnati, Cincinnati, OH 45268.

2. Lin, S. D., 1973. Evaluation of coliform tests for chlorinated secondary effluents. JWPCF, 45:3:498.

3. Greene, R. A., R. H. Bordner and P. V. Scarpino, 1974. Applicability of the membrane filter and most probable number coliform procedures to chlorinated wastewaters. Paper G87 given at 74th Annual Meeting of the American Society for Microbiology, May 12-17, 1974, Chicago, IL.

4. Rose, R. E., E. E. Geldreich and W. Litsky, 1975. Improved membrane filter method for fecal coliform analysis. Appl. Microbiol. 29:4:532.

5. Lin, S. D., 1976. Evaluation of Millipore HA and HC membrane filters for the enumeration of indicator bacteria. Appl. Environ. Microbiol. 32:300.

6. Green, B. L., E. Clausen and W. Litsky, 1975. Comparison of the new Millipore HC with conventional membrane filters for the enumeration of fecal coliform bacteria. Appl. Microbiol. 30:697.

7. Guidelines for Establishing Test Procedures, 40 Code of Federal Regulations (CFR) Part 136, Published in Federal Register, 40, 52780, Dec. 1, 1976.

8. Tobin, R. S. and B. J. Dutka, 1977. Comparison of the surface structure, metal binding, and fecal coliform recoveries of nine membrane filters. Appl. Environ. Microbiol. 34:69.

9. Lin, S. D., 1977. Comparison of membranes for fecal coliform recovery in chlorinated effluents. JWPCF, 49:2255.

PART III. ANALYTICAL METHODOLOGY

Part III of the manual describes the specific analytical procedures selected in response to the parameters required under PL 92–500 and 93–523 (see Part V-D, Legal Considerations) and to related parameters for indicators and pathogens which supplement the required information. New parameters and new methodology will be added as proven in actual usage. The methods are presented in Sections as follows:

Section A **Standard Plate Count**

Section B **Total Coliforms**

Section C **Fecal Coliforms**

Section D **Fecal Streptococci**

Section E ***Salmonella***

Section F **Actinomycetes**

PART III. ANALYTICAL METHODOLOGY

Section A Standard Plate Count

1. Summary of Method

The Standard Plate Count (SPC) Method is a direct quantitative measurement of the viable aerobic and facultative anaerobic bacteria in a water environment, capable of growth on the selected plating medium. An aliquot of the water sample or its dilution is pipetted into a sterile glass or plastic petri dish and a liquified, tempered agar medium added. The plate is rotated to evenly distribute the bacteria. Each colony that develops on or in the agar medium originates theoretically from one bacterial cell. Although no one set of plate count conditions can enumerate all organisms present, the Standard Plate Count Method provides the uniform technique required for comparative testing and for monitoring water quality in selected situations.

2. Scope and Application (1–6)

This simple technique is a useful tool for determining the bacterial density of potable waters and for quality control studies of water treatment processes. The Standard Plate Count provides a method for monitoring changes in the bacteriological quality of finished water throughout a distribution system, thus giving an indication of the effectiveness of chlorine in the system as well as the possible existence of cross-connections, sediment accumulations and other problems within the distribution lines. Total bacterial densities greater than 500–1000 organisms per ml may indicate coliform suppression or desensitization of quantitative tests for coliforms (1–3). The procedure may also be used to monitor quality changes in bottled water or emergency water supplies.

2.1 Theoretically, each bacterium present in a sample multiplies into a visible colony of millions of bacteria. However, no standard plate count or any other total count procedure yields the true number because not all viable bacterial cells in the water sample can reproduce under a single set of cultural conditions imposed in the test. The number and types of bacteria that develop are influenced by the time and temperature of incubation, the pH of the medium, the level of oxygen, the presence of specific nutrients in the growth medium, competition among cells for nutrients, antibiosis, predation, etc.

2.2 This procedure does not allow the more fastidious aerobes or obligate anaerobes to develop. Also, bacteria of possible importance in water such as *Crenothrix*, *Sphaerotilus*, and the actinomycetes will not develop within the incubation period specified for potable water analysis.

2.3 Clumps of organisms in the water sample which are not broken up by shaking result in underestimates of bacterial density, since an aggregate of cells will appear as one colony on the growth medium.

3. Apparatus and Materials

3.1 Incubator that maintains a stable 35 ±0.5 C. Temperature is checked against an NBS certified thermometer or one of equivalent accuracy.

3.2 Water bath for tempering agar set at 44–46 C.

3.3 Colony Counter, Quebec darkfield model or equivalent.

3.4 Hand tally or electronic counting device (optional).

3.5 Pipet containers of stainless steel, aluminum or pyrex glass for glass pipets.

3.6 Petri dish containers of stainless steel or aluminum for glass petri dishes.

3.7 Thermometer certified by National Bureau of Standards or one of equivalent accuracy, with calibration chart.

3.8 Sterile TD (To Deliver) bacteriological or Mohr pipets, glass or plastic of appropriate volumes, see Part II-B, 1.8.1.

3.9 Sterile 100 mm × 15 mm petri dishes, glass or plastic.

3.10 Dilution bottles (milk dilution), pyrex glass, marked at 99 ml volume, screw cap with neoprene rubber liner.

3.11 Bunsen/Fisher gas burner or electric incinerator.

4. Media

4.1 Sterile Plate Count Agar (Tryptone Glucose Yeast Agar) dispensed in tubes (15 to 20 ml per tube) or in bulk quantities in screw cap flasks or dilution bottles. See Part II-B, 5.1.5.

4.2 Sterile buffered dilution water, 99 ± 2 ml volumes, in screwcapped dilution bottles. See Part II-B, 7.

5. Procedure

5.1 Dilution of Sample (See Part II-C, 1.4 for details)

5.1.1 The sample is diluted to obtain final plate counts of 30–300 colonies. In this range, the plate counts are the most accurate and precise possible. Since the microbial population in the original water sample is not known beforehand, a series of dilutions must be prepared and plated to obtain a plate count within this range.

5.1.2 For most potable water samples, countable plates can be obtained by plating 1 and 0.1 ml of the undiluted sample, and 1 ml of the 1:100 sample dilution (see Figure III-A-1). Higher dilutions may be necessary with some potable waters.

5.1.3 Shake the sample vigorously about 25 times.

5.1.4 Prepare an initial 1:100 dilution by pipetting 1 ml of the sample into a 99 ml dilution water blank using a sterile 1 ml pipet (see Figure III-A-1).

5.1.5 The 1:100 dilution bottle is vigorously shaken and further dilutions made by pipetting aliquots (usually 1 ml) into additional dilution blanks. A new sterile pipet must be used for each transfer and each dilution must be thoroughly shaken before removing an aliquot for subsequent dilution.

5.1.6 When an aliquot is removed, the pipet tip should not be inserted more than 2.5 cm (1 inch) below the surface of the liquid.

5.2 Preparation of Agar

5.2.1 Melt prepared plate count agar (tryptone glucose yeast agar) by heating in boiling water or by flowing steam in an autoclave at 100 C. Do not allow the medium to remain at these high temperatures beyond the time necessary to melt it. Prepared agar should be melted once only.

5.2.2 Place melted agar in a tempering water bath maintained at a temperature of 44–46 C. Do not hold agar at this temperature

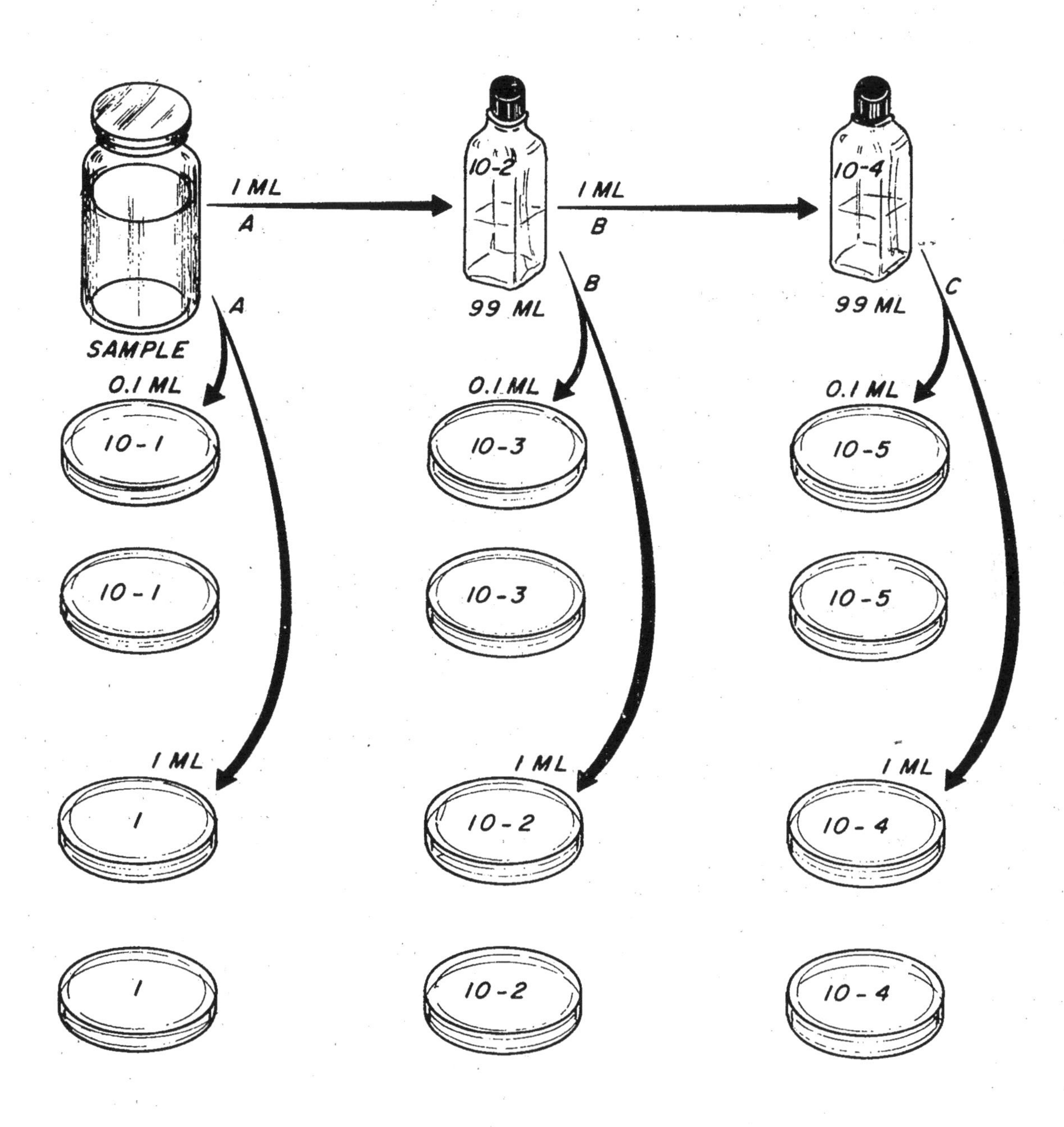

FIGURE III-A-1. Typical Dilution Series for Standard Plate Count

longer than three hours because precipitates may form which confuse the counting of colonies. Maintain a thermometer immersed in a separate bottle or flask in the water bath to monitor the temperature.

5.3 Preparation for Plating

5.3.1 Prepare at least duplicate plates for each sample or dilution tested. Mark and arrange plates in a reasonable order for use. Prepare a bench sheet or card, including sample identity, dilutions, date and other relevant information.

5.3.2 Aseptically pipet an aliquot from the appropriate dilution into the bottom of each petri dish. Use a separate sterile pipet to transfer an aliquot to each set of petri dishes for each sample or sample dilution used. Vigorously shake the undiluted sample and dilution containers before each transfer is made.

5.3.3 Pipet sample or sample dilution into marked petri dish. After delivery, touch the tip once to a dry spot in the dish.

5.3.4 To minimize bacterial density changes in the samples, do not prepare any more samples than can be diluted and plated within 20–25 minutes.

5.4 Pouring Agar Plates

5.4.1 Use the thermometer in the control bottle in the tempering bath to check the temperature of the plating medium before pouring.

5.4.2 Add not less than 12 ml (usually 12–15 ml) of the melted and cooled (44–46 C) agar medium to each petri dish containing an aliquot of the sample or its dilution. Mix the inoculated medium carefully to prevent spilling. Avoid splashing the inside of the cover. One recommended technique rotates plate five times to left, five times to the right and five times in a back and forth motion.

5.4.3 Pipet a one ml volume of sterile dilution water into a petri dish, add agar, mix and incubate with test plates. This control plate will check the sterility of pipets, agar, dilution water and petri dishes. See Part IV-C, 1.3.

5.5 Incubation of Plated Samples

5.5.1 After agar plates have hardened on a level surface (usually within 10 minutes), invert the plates and immediately incubate at 35 C.

5.5.2 Incubate tests on all water samples except bottled water at 35 $\pm$ 0.5 C for 48 $\pm$ 3 hours. Incubate the tests on bottled water at 35 $\pm$ 0.5 C for 72 $\pm$ 4 hours. The longer incubation is required to recover organisms in bottled water with longer generation times.

5.5.3 Stacks of plates should be at least 2.5 cm from adjacent stacks, the top or sides of the incubator. Do not stack plates more than four high. These precautions allow proper circulation of air to maintain uniform temperature throughout the incubator and speed equilibration.

5.6 Counting and Recording Colonies:

After the required incubation period, examine plates and select those with 30–300 colonies. Count these plates immediately. A Quebec-type colony counter equipped with a guide plate, appropriate magnification and light is recommended for use with a hand tally.

5.6.1 Electronic-assist devices are available which register colony counts with a sensing probe and automatically tabulate the total plate count.

Fully-automatic colony counters are available which count all colonies (particles) larger than a preset threshold-size. These counters scan and provide digital register and a visual image of the plate for further examination and recounting with different threshold if so desired.

Because the accuracy of automatic conters varies with the size and number of colonies per plate, the analyst should periodically compare its results with manual counts.

5.6.2 The following rules should be used to report the Standard Plate Count:

(a) Plates with 30 to 300 Colonies: Count all colonies and divide by the volume tested (in ml). If replicate plates from one dilution are countable (30–300), sum the counts of colonies on all plates and divide by the volumes tested (in ml) as follows:

$$\frac{\text{Sum of Colonies}}{\text{Sum of Volumes Tested, ml}} = \text{S.P. Count/ml}$$

Record the dilutions used, the number of colonies on each plate and report as the Standard Plate Count per milliliter.

If two or more consecutive dilutions are countable, independently carry each calculation of plate count to a final count per ml, then calculate the mean of these counts/ml for the reported value.

For example, if 280 and 34 colonies are counted in the 1:100 and 1:1000 dilutions of a water sample, the calculation is:

$$\frac{280}{.01} = 28{,}000/\text{ml}$$

$$\frac{34}{.001} = 34{,}000/\text{ml}$$

$$\text{Reporting Value} = \frac{28000 + 34000}{2}$$

$$= 31000 \quad \text{SPC/ml}$$

(b) All Plates with Fewer than 30 Colonies: If there are less than 30 colonies on all plates, record the actual number of colonies on the lowest dilution plated and report the count as: Estimated Standard Plate Count per milliliter. For example, if volumes of 0.1, 0.01 and 0.001 ml were plated and produced counts of 22, 2 and 0 colonies respectively, the colony count of 22 from the largest sample volume (0.1 ml) would be selected, calculated and reported as follows:

$$\frac{\text{Plate Count}}{\text{Volume Plate}} = \frac{22}{0.1} = 220$$

Count reported: Estimated Standard Plate Count, 220/ml.

(c) If 1 ml volumes of original sample produce counts < 30, actual counts are reported.

(d) Plate with No Colonies: If all plates from dilutions tested show no colonies, report the count as < 1 times the lowest dilution plated. For example, if 0.1, 0.01 and 0.001 ml volumes of sample were tested with no visible colonies developing, the lowest dilution, 0.1 ml would be used to calculate a less than (<) count as follows:

$$\frac{1}{\text{Volume Tested}} = \frac{1}{0.1} = < 10$$

Count reported: Standard Plate Count, > 10/ml.

(e) All Plates Greater than 300 Colonies: When counts per plate in the highest dilution exceed 300 colonies, compute the count by multiplying the mean count by the dilution used and report as a greater than (>), Standard Plate Count per milliliter. For example, if duplicate 1.0, 0.1 and 0.01 volumes of sample were tested with average counts of > 500, > 500 and 340 developing in the dilutions, the count would be calculated as follows:

$$\frac{\text{Plate Count}}{\text{Volume Tested}} = \frac{340}{0.01} = 34{,}000$$

or count reported as: Standard Plate Count, > 34,000/ml.

5.6.3 Count Estimations on Crowded Plates: The square divisions of the grid on the Quebec or similar colony counter can be used to estimate the numbers of bacteria per plate. With less than 10 colonies per sq cm count the colonies in 13 squares with representative distribution of colonies. Select 7 consecutive horizontal squares and 6 consecutive vertical squares for counting. Sum the colonies in these 13 sq cm, and multiply by 5 to estimate the colonies per plate for glass plates (area of 65 sq cm) or multiply by 4.32 for plastic plates (area of 57 sq cm). With more than 10 colonies per sq cm, count 4 representative squares, average the count per sq cm, multiply by the number of sq cm/plate (usually 65 for glass plates and 57 for plastic plates) to estimate the colonies per plate. Then multiply by the reciprocal of the dilution to determine the count/ml. When bacterial counts on crowded plates are greater than 100 colonies per sq cm, report the result as Estimated Standard Plate Count greater than (>) 6,500 times the highest dilution plated.

5.6.4 Spreaders: Plates containing spreading colonies must be so reported on the data sheet. If spreaders exceed one-half of the total plate area, the plate is not used. Report as: No results, spreaders.

Colonies can be counted on representative portions of plates if spreading colonies constitute less than one-half of the total plate area, and the colonies are well-distributed in the remaining portion of the plate.

(a) Count each chain of colonies as a single colony.

(b) Count each spreader colony that develops as a film of growth between the agar and the petri dish bottom as one colony.

(c) Count the growth that develops in a film of water at the edge or over the surface of the agar as one colony.

(d) Adjust count for entire plate and report as: Estimated Standard Plate Count/ml.

5.6.5 Remarks on Data Sheet: Any unusual occurrences such as missed dilutions, loss of plates through breakage, contamination of equipment, materials, media, or the laboratory environment, as shown by sterility control plates, must be noted on the data sheet. Report as: Lab Accident, etc.

6. Reporting Results

Report Standard Plate Count or Estimated Standard Plate Count as colonies per ml, not per 100 ml.

Standard Plate Counts should be rounded to the number of significant figures (S.F.) obtainable in the procedure: 1 S.F. for 0–9 actual plate counts, 2 S.F. for 10–99 actual plate counts and 3 S.F. for 100–300 actual plate counts. See Part II-C, 2.8.1 of this manual.

7. Precision and Accuracy

7.1 Prescott et al (7) reported that the standard deviation of individual counts from 30–300 will vary from 0–30 percent. This plating error was 10% for plate counts within the 100–300 range. A dilution error of about 3% for each dilution stage is incurred in addition to the plating error. Large variations can be expected from high density samples such as sewage for which several dilutions are necessary.

7.2 Laboratory personnel should be able to duplicate their plate count values on the same plate within 5%, and the counts of others within 10%. If analysts' counts do not agree, review counting procedures for analyst error.

REFERENCES

1. Geldreich, E. E., H. D. Nash, D. J. Reasoner and R. H. Taylor, 1972. The necessity of controlling bacterial populations in potable waters: Community Water Supply. J. Amer. Water Works Assoc. 64:596.

2. Geldreich, E. E., H. D. Nash, D. J. Reasoner and R. H. Taylor, 1975. The necessity of controlling bacterial populations in potable waters: Bottled Water and Emergency Water Supplies. J. Amer. Water Works Assoc. 67:117.

3. Geldreich, E. E., 1973. Is the total count necessary? Ist AWWA Technology Conference Proceedings, Amer. Water Works Assoc. VII-1, Cincinnati, Ohio.

4. Clark, D. S., 1971. Studies on the surface plate method of counting bacteria. Can. J. Microbiol. 17:943

5. Klein, D. A. and S. Wu, 1974. Stress: a factor to be considered in heterotrophic microorganism enumeration from aquatic environments. Appl. Microbiol. 27:429.

6. Van Soestbergen, A. A. and C. H. Lee, 1969. Pour plates or streak plates? Appl. Microbiol. 18:1092.

7. Prescott, S. C., C-E. A. Winslow, and M. H. McCrady, 1946. Water Bacteriology. (6th ed.) John Wiley and Sons, Inc., p. 46–50.

PART III. ANALYTICAL METHODOLOGY

Section B Total Coliform Methods

This section describes the enumerative techniques for total coliform bacteria in water and wastewater. The method chosen depends upon the characteristics of the sample. The Section is divided as follows:

1. **Definition of the Coliform Group**
2. **Single-Step, Two-Step and Delayed-Incubation Membrane Filter (MF) Methods**
3. **Verification**
4. **Most Probable Number (MPN) Method**
5. **Differentiation of the Coliform Group by Further Biochemical Tests**

1. Definition of the Coliform Group

The coliform or total coliform group includes all of the aerobic and facultative anaerobic, gram-negative, nonspore-forming, rod-shaped bacteria that ferment lactose in 24-48 hours at 35 C. The definition includes the genera: *Escherichia, Citrobacter, Enterobacter, and Klebsiella.*

2. Single-Step, Two-Step and Delayed-Incubation Membrane Filter Methods

2.1 Summary: An appropriate volume of a water sample or its dilution is passed through a membrane filter that retains the bacteria present in the sample.

In the single-step procedure the filter retaining the microorganisms is placed on M-Endo agar, LES M-Endo agar or on an absorbent pad saturated with M-Endo broth in a petri dish. The test is incubated at 35 C for 24 hours.

In the two-step enrichment procedure the filter retaining the microorganisms is placed on an absorbent pad saturated with lauryl tryptose (lauryl sulfate) broth. After incubation for 2 hours at 35 C, the filter is transferred to an absorbent pad saturated with M-Endo broth, M-Endo agar, or LES M-Endo agar, and incubated for an additional 20–22 hours at 35 C. The sheen colonies are counted under low magnification and the numbers of total coliforms are reported per 100 ml of original sample.

In the delayed-incubation procedure, the filter retaining the microorganisms is placed on an absorbent pad saturated with M-Endo preservative medium in a tight-lidded petri dish and transported from field site to the laboratory. In the laboratory, the filter is transferred to M-Endo growth medium and incubated at 35 C for 24 hours. Sheen colonies are counted as total coliforms per 100 ml.

2.2 Scope and Application: The total coliform test can be used for any type of water or wastewater, but since the development of the fecal coliform procedure there has been increasing use of this more specific test as an indicator of fecal pollution. However, the total coliform test remains the primary indicator of bacteriological quality for potable water, distribution system waters, and public water supplies because a broader measure of pollution is desired for these waters. It is also a useful measure in shellfish-raising waters.

Although the majority of water and wastewater samples can be examined for total coliforms by the single-step MF procedure, coliforms may be suppressed by high background organisms, and potable water samples may require the two-step method.

If the membrane filtration method is used to measure total coliforms in chlorinated secondary or tertiary sewage effluents the two-step enrichment procedure is required. However, it may be necessary to use the MPN method because of high solids in the wastes or toxicity from an industrial waste (see Part II-D, this Manual).

The delayed-incubation MF method is useful in survey monitoring or emergency situations when the single step coliform test cannot be performed at the sample site, or when time and temperature limits for sample storage cannot be met. The method eliminates field processing and equipment needs. Also, examination at a central laboratory permits confirmation and biochemical identification of the organisms as necessary. Consistent results have been obtained with this method using water samples from a variety of sources (1, 2). The applicability of this method for a specific water source must be determined in preliminary studies by comparison with the standard MF method.

2.3 Apparatus and Materials

2.3.1 Water jacket, air, or heat sink incubator that maintains 35 ± 0.5 C. Temperature is checked against an NBS certified thermometer or equivalent. Incubator must have humidity control if loose-lidded pertri dishes are used. See Part II-B, 1.2.

2.3.2 A binocular (dissection) microscope, with magnification of 10 or 15×, and a daylight type fluorescent lamp angled to give maximum sheen appearance.

2.3.3 Hand tally.

2.3.4 Pipet container of stainless steel, aluminum or pyrex glass for glass pipets.

2.3.5 Sterile 50–100 ml graduated cylinders covered with aluminum foil or kraft paper,

2.3.6 Sterile, unassembled membrane filtration units (filter base and funnel), glass, plastic or stainless steel, wrapped with aluminum foil or kraft paper. Portable field filtration units are available.

2.3.7 Vacuum source.

2.3.8 Vacuum filter flask with appropriate tubing. Filter manifolds which hold a number of filter bases can also be used.

2.3.9 Ultraviolet sterilizer for MF filtration units (optional).

2.3.10 Safety trap flask between the filter flask and the vacuum source.

2.3.11 Forceps with smooth tips.

2.3.12 Methanol or ethanol, 95%, in small vial, for flaming forceps.

2.3.13 Bunsen/Fisher burner or electric incinerator.

2.3.14 Sterile TD bacteriological or Mohr pipets, glass or plastic, of appropriate size.

2.3.15 Sterile petri dishes with tight-fitting lids, 50 × 12 mm or loose-fitting lids 60 × 15 mm, glass or plastic.

2.3.16 Dilution bottles (milk dilution), pyrex, marked at 99 ml volume, screw cap with neoprene rubber liner.

2.3.17 Membrane filters, white, grid-marked, 47 mm diameter, with 0.45 μm ± 0.02 μm pore size, or other pore size, as recommended by manufacturer for water analyses.

2.3.18 Absorbent pads.

2.3.19 Inoculation loops, at least 3 mm diameter, or needles, nichrome or platinum wire, 26 B&S gauge, in suitable holder.

2.3.20 Disposable applicator sticks or plastic loops as alternatives to inoculation loops.

2.3.21 Shipping tubes, labels, and packing materials for mailing delayed incubation plates.

2.4 Media: Media are prepared in pre-sterilized erlenmeyer flasks with metal caps, aluminum foil covers, or screw caps.

2.4.1 M-Endo broth or agar (See Part II-B, 5.2.2).

2.4.2 LES M-Endo agar (See Part II-B, 5.2.4).

2.4.3 Lauryl tryptose broth (See Part II-B, 5.3.1).

2.4.4 Brilliant green lactose bile broth (See Part II-B, 5.3.2).

2.4.5 M-Endo holding medium (See Part II-B, 5.2.3).

2.4.6 Sodium benzoate, U.S.P., for use in the delayed incubation procedure (See Part II-B, 5.2.3).

2.4.7 Cycloheximide (Actidione – Upjohn, Kalamazoo, MI) for use as antifungal agent in delayed incubation procedure (See Part II-B, 5.2.3).

2.5 Dilution Water (See Part II-B, 7 for preparation).

2.5.1 Sterile dilution water dispensed in 99 ± 2 ml amounts in screw-capped dilution bottles.

2.5.2 Sterile dilution water prepared in 1 liter or larger volumes for wetting membranes before addition of small sample volumes and for rinsing the funnel after sample filtration.

2.6 Procedure: Refer to the general procedure in Part II-C for more complete details.

2.6.1 Single-Step Procedure

(a) Prepare the M-Endo broth, M-Endo agar or LES M-Endo agar as directed in Part II-B.

(b) Place one sterile absorbent pad in the bottom half of each petri dish. Pipet 1.8–2.0 ml M-Endo broth onto the pad to saturate it. Pour off excess broth. Alternatively, pipet 5–6 ml of melted agar into each dish (2–3 mm) and allow to harden before use. Mark dishes and bench forms with sample identities and volumes.

(c) Place a sterile membrane filter on the filter base, grid-side up and attach the funnel to the base of the filter unit; the membrane filter is now held between the funnel and the base.

(d) Shake the sample bottle vigorously about 25 times and measure the desired volume of sample into the funnel. Select sample volumes based on previous knowledge to produce membrane filters with 20–80 coliform colonies. See Table II-C-1. If sample volume is < 10 ml, add 10 ml of sterile dilution water to the filter before adding sample.

It is desirable to filter the largest possible sample volumes for greatest accuracy. However, if past analyses of specific samples have resulted in confluent growth, "too numerous to count" membranes, or lack of sheen from excessive turbidity, additional samples should be collected and filtration volumes adjusted to provide isolated colonies from smaller volumes. See 2.7.2 in this Section for details on adjusting sample volumes for potable waters.

The suggested method for measuring sample volumes is described in Part II-C, 3.4.6.

(e) Filter sample and rinse the sides of the funnel at least twice with 20–30 ml of sterile dilution water. Turn off the vacuum and remove the funnel from the filter base. Aseptically remove the membrane filter from the filter base and place grid-side up on the agar or pad.

(f) Filter samples in order of increasing sample volume, filter potable waters first.

(g) If M-Endo broth is used, place the filter on an absorbent pad saturated with the broth. Reseat the membrane, if air bubbles occur, as evidenced by non-wetted areas on the membrane. Invert dish and incubate for 24 ± 2 hours at 35 ± 0.5 C in an atmosphere with near saturated humidity.

(h) If M-Endo agar or LES M-Endo agar is used, place the inoculated filter directly on the agar surface. Reseat the membrane if bubbles occur. Invert the dish and incubate for 24 ± 2 hours at 35 ± 0.5 C in an atmosphere with near saturated humidity.

(i) If tight-lidded dishes are used, there is no requirement for near-saturated humidity.

(j) After incubation remove the dishes from the incubator and examine for sheen colonies.

(k) Proceed to 2.7 for Counting and Recording Colonies.

2.6.2 Two-Step Enrichment Procedure

(a) Place a sterile absorbent pad in the top of each petri dish.

(b) Prepare lauryl tryptose broth as directed in Part II-B. Pipet 1.8–2.0 ml lauryl tryptose broth onto the pad to saturate it. Pour off excess broth.

(c) Place a sterile membrane filter on the filter holder, grid-side up and attach the funnel to the base of the filter unit; the membrane filter is now held between the funnel and the base.

(d) Shake the sample bottle vigorously about 25 times to obtain uniform distribution of bacteria. Select sample volumes based on previous knowledge to produce membrane filters with 20–80 coliform colonies. See Table II-C-1. If sample volume is < 10 ml, add 10 ml of sterile dilution water to filter before adding sample.

(e) Filter samples in order of increasing sample volume, rinsing with sterile buffered dilution water between filtrations. The methods of measurement and dispensation of the sample into the funnel are given in Part II-C, 3.4.6.

(f) Turn on the vacuum to filter the sample through the membrane, rinse the sides of the funnel at least twice with 20–30 ml of sterile dilution water. Turn off vacuum and remove funnel from base.

(g) Remove the membrane filter aseptically from the filter base and place grid-side up on the pad in the top of the petri dish. Reseat MF if air bubbles are observed.

(h) Incubate the filter in the petri dish without inverting for 1 1/2 – 2 hours at 35 ± 0.5 C in an atmosphere of near saturated humidity. This completes the first step in the Two-Step Enrichment Procedure.

(i) Prepare M-Endo broth, M-Endo agar, or LES M-Endo agar as directed in Part II-B.

If M-Endo broth is used, place a new sterile absorbent pad in the bottom half of the dish and saturate with 1.8-2.0 ml of the M-Endo broth. Transfer the filter to the new pad. Reseat MF if air bubbles are observed. Remove the used pad and discard.

If M-Endo or LES M-Endo agar is used, pour 5-6 ml of agar into the bottom of each petri dish and allow to solidfy. The agar medium can be refrigerated for up to two weeks.

(j) Transfer the filter from the lauryl tryptose broth onto the Endo medium. Reseat if air bubbles are observed.

(k) Incubate dishes in an inverted position for an additional 20–22 hours at 35 $\pm$ 0.5 C. This completes the second step in the Two-Step Enrichment Procedure.

(l) Proceed to 2.7 Counting and Recording.

2.6.3 Delayed Incubation Procedure

(a) Prepare the M-Endo Holding Medium or LES Holding Medium as outlined in Part II-B, 5.2.3 or 5.2.5. Saturate the sterile absorbent pads with about 2.0 ml of holding broth. Pour off excess broth. Mark dishes and bench forms with sample identity and volumes.

(b) Using sterile forceps place a membrane filter on the filter base grid side up.

(c) Attach the funnel to the base of the filter unit; the membrane filter is now held between the funnel and base.

(d) Shake the sample vigorously about 25 times and measure into the funnel with the vacuum off. If the sample is $<$ 10 ml, add 10 ml of sterile dilution water to the membrane filter before adding the sample.

(1) Select sample volumes based on previous knowledge to produce counts of 20–80 coliform colonies. See Table II-C-1.

(2) Follow the methods for sample measurement and dispensation given in Part II-C, 3.4.6

(e) Filter the sample through the membrane and rinse the sides of the funnel walls at least twice with 20–30 ml of sterile dilution water.

(f) Turn off the vacuum and remove the funnel from the base of the filter unit.

(g) Aseptically remove the membrane filter from the filter base and place grid side up on an absorbent pad saturated with M-Endo Holding Medium or LES Holding Medium.

(h) Place the culture dish in shipping container and send to the examining laboratory. Coliform bacteria can be held on the holding medium for up to 72 hours with little effect on the final counts. The holding period should be kept to a minimum.

(i) At the examining laboratory remove the membrane from the holding medium, place it in another dish containing M-Endo broth or agar medium, and complete testing for coliforms as described above under 2.6.1.

2.7 Counting and Recording Colonies: After incubation, count colonies on those membrane filters containing 20–80 golden-green metallic surface sheen colonies and less than 200 total bacterial colonies. A binocular (dissection) microscope with a magnification of 10 or 15$\times$ is recommended. Count the colonies according to the general directions given in Part II-C, 3.5.

2.7.1 The following general rules are used in calculating the total coliform count per 100 ml of sample. Specific rules for analysis and counting of water supply samples are given in 2.7.2.

(a) Countable Membranes with 20–80 Sheen Colonies, and Less Than 200 Total Bacterial Colonies: Select the plate counts to be used according to the rules given in Part II-C, 3.6, and calculate the final value using the formula.

Total Coliforms/100 ml:

$$\frac{\text{No. of Total Coliform Colonies Counted}}{\text{Volume in ml of Sample Filtered}} \times 100$$

(b) Counts Greater Than the Upper Limit of 80 Colonies: All colony counts are above the recommended limits. For example, sample volumes of 1, 0.3, and 0.01 ml are filtered to produce total coliform colony counts of TNC, 150, and 110 colonies.

Use the count from the smallest filtration volume and report as a greater than count/100 ml. In the example above:

$$\frac{110}{0.01} \times 100 = 1{,}100{,}000$$

or > 1,100,000 coliforms/100 ml.

(c) Membranes with More Than 200 Total Colonies (Coliforms plus Non-coliforms).

(1) Estimate sheen colonies if possible, calculate total coliform density as in (a) above.

Report as: Estimated Count/100 ml.

(2) If estimate of sheen colonies is not possible, report count as Too Numerous to Count (TNTC).

(d) Membranes with Confluent Growth

Report as: Confluent Growth and specify the presence or absence of sheen.

2.7.2 Special Rules for Potable Waters

(a) Countable Membranes with 0-80 Sheen Colonies, and Less than 200 Total Colonies

Count the sheen colonies per volume filtered. Calculate and report the number of Total Coliforms/100 ml.

(b) Uncountable Membranes for Potable Water Samples

If 100 ml portions of potable water samples cannot be tested because of high background counts or confluency, multiple volumes of less than 100 ml can be filtered. For example, if 60 colonies appear on the surface of one membrane through which a 50 ml portion of the sample was passed, and 50 colonies on a second membrane through which a second 50 ml portion of the sample was passed, the colonies are totaled and reported as 110 total coliforms per 100 ml.

If filtration of multiple volumes of less than 100 ml still results in confluency or high background count, the coliforms may be present but suppressed. These samples should be analyzed by the MPN Test. This MPN check should be made on at least one sample for each problem water once every three months.

(c) Membranes with Confluent Growth

For potable water samples, confluence requires resampling and retesting.

(d) Verification. Because unsatisfactory samples from public water supplies containing 5 or more coliform colonies must be verified, at least 5 colonies need to be verified for each positive sample. Reported counts are adjusted based on verification.

(e) Quality control procedures are specified by EPA under the law, and described in Appendix C in this Manual.

2.7.3 Reporting Results: Report total coliform densities per 100 ml of sample. See Figure II-C-3 for an example of a bench form for reporting results. A discussion on significant figures is given in Part II-C, 2.8.

2.8 Precision and Accuracy: There are no established precision and accuracy data available at this time.

3. Verification

Verification of total coliform colonies from M-Endo type media validates sheen as evidence of coliforms. Verification of representative numbers of colonies may be required in evidence gathering or for quality control procedures. The verification procedure follows:

3.1 Using a sterile inoculating needle, pick growth from the centers of at least 10 well-isolated sheen colonies (5 sheen colonies per plate for potable waters). Inoculate each into a tube of lauryl tryptose broth and incubate 24-48 hours at 35 C ± 0.5 C. Do not transfer exclusively into brilliant green bile lactose broth. However, colonies may be transferred to LTB and BGLB simultaneously.

3.2 At the 24 and 48 hour readings, confirm gas-positive lauryl tryptose broth tubes by inoculating a loopful of growth into brilliant green lactose bile broth and incubate for 24–48 hours at 35 ± 0.5 C. Cultures that are positive in BGLB are interpreted as verified coliform colonies (see Figure III-B-1).

3.3 If questionable sheen occurs, the worker should also verify these colonies.

4. Most Probable Number (MPN) Method

4.1 Summary: This method detects and estimates the total coliforms in water samples by the multiple fermentation tube technique. The method has three stages: the Presumptive, the Confirmed, and the Completed Tests. In the Presumptive Test, a series of lauryl tryptose broth fermentation tubes are inoculated with decimal dilutions of the sample. The formation of gas at 35 C within 48 hours constitutes a positive Presumptive Test for members of the total coliform group. However, the MPN must be carried through the Confirmed Test for valid results. In this test, inocula from positive Presumptive tubes are transferred to tubes of brilliant green lactose bile (BGLB) broth. The BGLB medium contains selective and inhibitive agents to suppress the growth of all noncoliform organisms. Gas production after incubation for 24 or 48 hours at 35 C constitutes a positive Confirmed Test and is the point at which most MPN tests are terminated. The Completed Test begins with streaking inoculum from the positive BGLB tubes onto EMB plates and incubating the plates for 24 hours at 35 C. Typical and atypical colonies are transferred into lauryl tryptose broth fermentation tubes and onto nutrient agar slants. Gas formation in the fermentation tubes and presence of gram-negative rods constitute a positive Completed Test for total coliforms. See Figure III-B-2. The MPN per 100 ml is calculated from the MPN table based upon the Confirmed or Completed test results.

4.2 Scope and Application

4.2.1 Advantages: The MPN procedure is a tube-dilution method using a nutrient-rich medium, which is less sensitive to toxicity and supports the growth of environmentally-stressed organisms. The method is applicable to the examination of total coliforms in chlorinated primary effluents and under other stressed conditions. The multiple-tube procedure is also better suited for the examination of turbid samples, muds, sediments, or sludges because particulates do not interfere visibly with the test.

4.2.2 Limitations: Certain non-coliform bacteria may suppress coliforms or act synergistically to ferment lauryl tryptose broth and yield false positive results. A significant number of false positive results can also occur in the brilliant green bile broth when chlorinated primary effluents are tested, especially when stormwater is mixed with the sewage (3). False negatives may occur with waters containing nitrates (4). False positives are more common in sediments.

4.3 Apparatus and Materials

4.3.1 Water bath or air incubator set at 35 ± 0.5 C.

4.3.2 Pipet containers of stainless steel, aluminum, or pyrex glass for glass pipets.

4.3.3 Inoculation loops, at least 3 mm diameter and needles of nichrome or platinum wire, 26 B & S gauge, in suitable holders.

4.3.4 Disposable sterile applicator sticks or plastic loops as alternatives to inoculating loops.

4.3.5 Compound microscope, oil immersion.

4.3.6 Bunsen/Fisher burner or electric incinerator unit.

4.3.7 Sterile TD Mohr or bacteriological pipets, glass or plastic, of appropriate size.

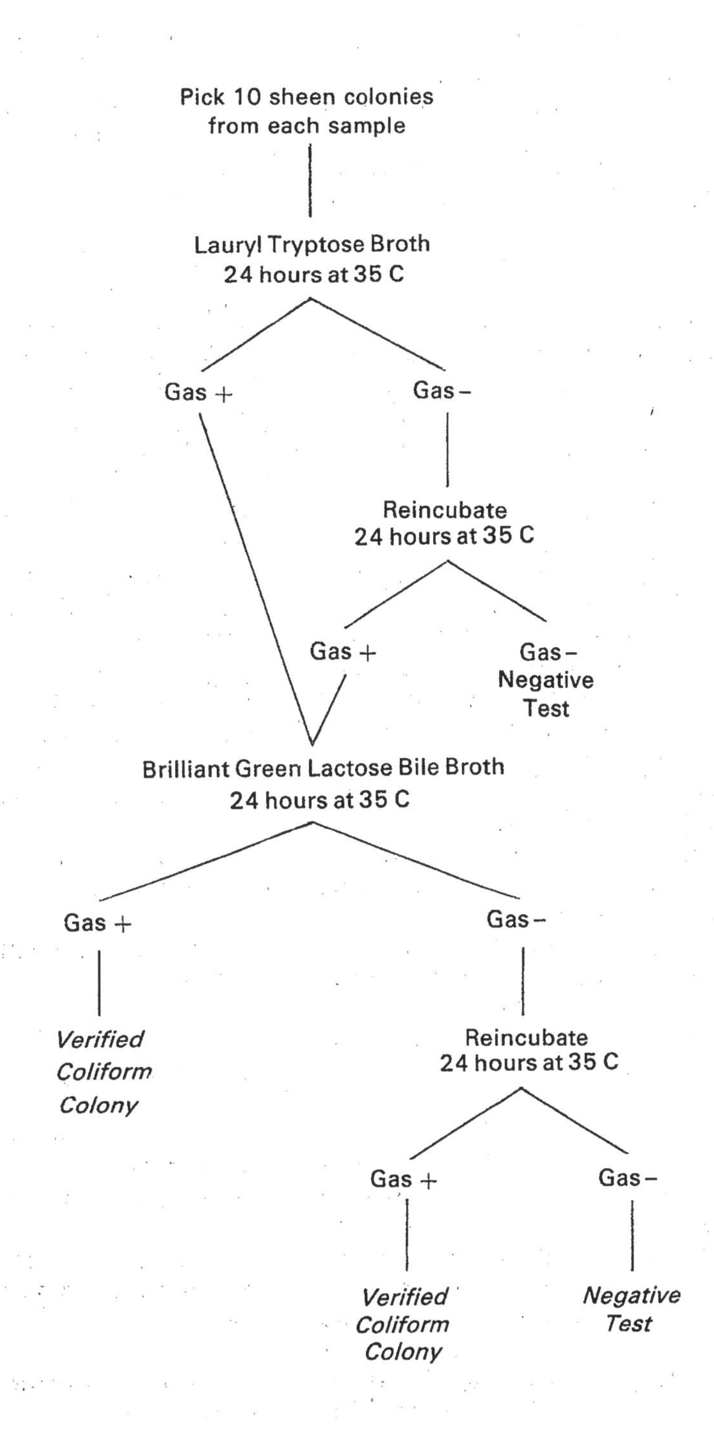

FIGURE III-B-1. Verification of Total Coliform Colonies on the Membrane Filter

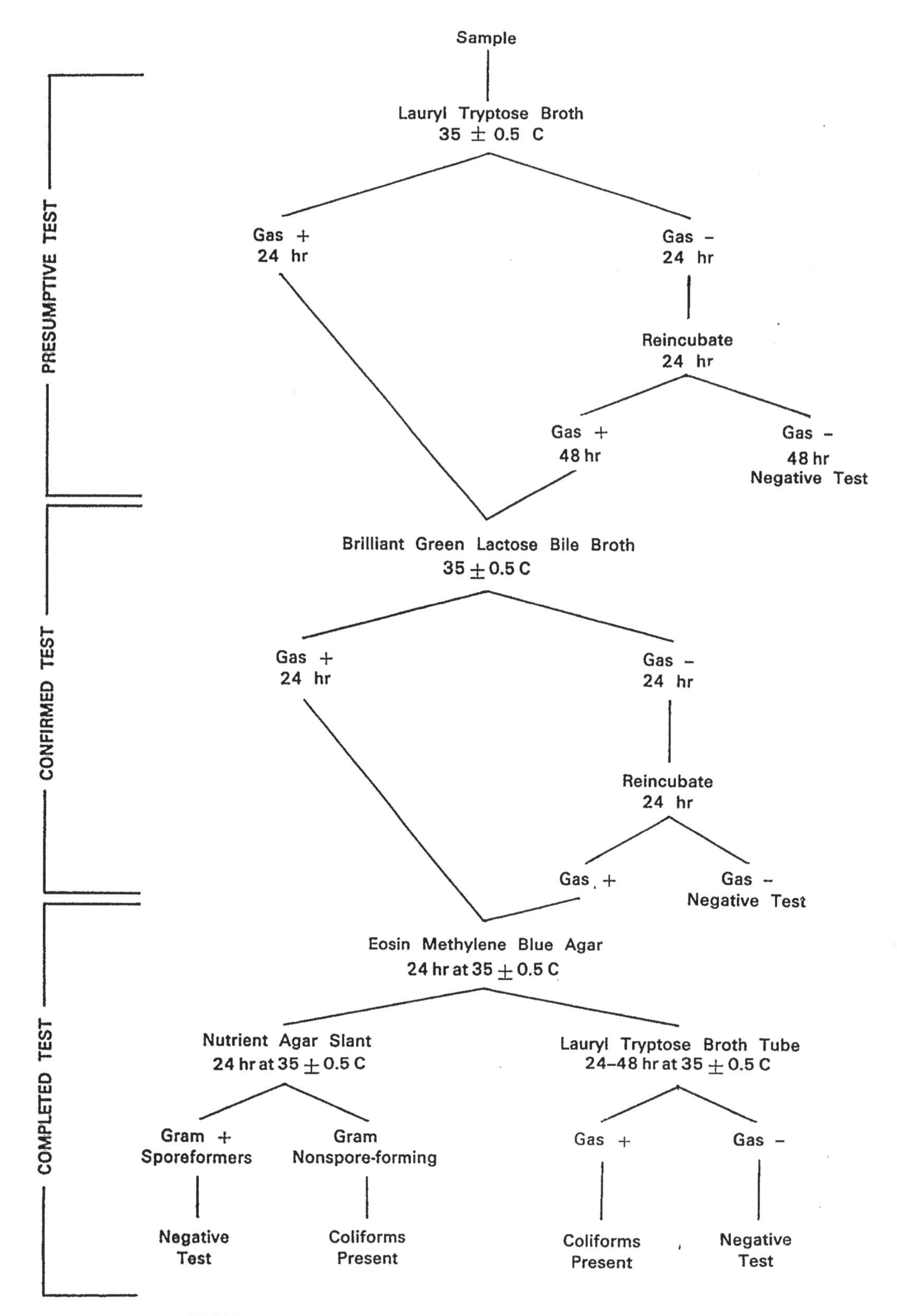

FIGURE III-B-2. Flow Chart for the Total Coliform MPN Test

4.3.8 Pyrex culture tubes, 150 × 25 mm or 150 × 20 mm, containing inverted fermentation vials, 75 × 10 mm with caps.

4.3.9 Culture tube racks to hold fifty, 25 mm diameter tubes.

4.3.10 Dilution bottles (milk dilution) pyrex glass, 99 ml volume, screw cap with neoprene rubber liners.

4.4 Media

4.4.1 Presumptive Test: Lauryl tryptose broth. See Part II-B, 5.3.1 . Lactose broth is not used because of false positive reactions.

4.4.2 Confirmed Test: Brilliant green bile broth. (See Part II-B, 5.3.2).

4.4.3 Completed Test:

(a) Eosin methylene blue agar (see Part II-B, 5.3.3).

(b) Nutrient agar or plate count agar slants (see Part II-B, 5.1.1 and 5.1.5).

4.5 Dilution Water: Sterile dilution water dispensed in 99 ± 2 ml amounts preferably in screw-capped bottles. (See Part II-B, 7).

4.6 Procedure: Part II-C describes the general MPN procedure in detail.

4.6.1 Prepare the media for Presumptive, Confirmed or Completed Tests selected. (See Part II-B, 5.3).

4.6.2 Presumptive Test (See Figure III-B-2): To begin the Presumptive Test, arrange fermentation tubes of lauryl tryptose broth in rows of 5 tubes each in the tube rack. Select sample volumes and clearly label each bank of tubes to identify the sample and volume inoculated.

(a) For potable waters, five portions of 10 ml each or five portions of 100 ml each are used.

(b) For relatively-unpolluted waters the sample volumes for the five rows might be 100, 10, 1, 0.1 and 0.01 ml, respectively; the latter two volumes delivered as dilutions of original sample.

(c) For known polluted waters the initial sample inoculations might be 0.1, 0.01, 0.001, 0.0001, and 0.00001 ml of original sample delivered as dilutions into successive rows each containing five replicate volumes. This series of sample volumes will yield determinate results from a low of 200 to a high of 16,000,000 organisms per 100 ml.

(d) Shake the sample and dilutions vigorously about 25 times. Inoculate each 5-tube row with replicate sample volumes in increasing decimal dilutions and incubate at 35 C ± 0.5 C.

(e) After 24 ± 2 hours incubation at 35 C, gently agitate the tubes in the rack and examine the tubes for gas. Any amount of gas constitutes a positive test. If there is no gas production in the tubes, reincubate for an additional 24 hours and reexamine for gas. Positive Presumption tubes are submitted directly to the Confirmed Test. Results are recorded on laboratory bench forms.

(f) If a laboratory using the MPN test on water supplies finds frequent numbers of Presumptive test tubes with heavy growth but no gas, these negative tubes should be submitted to the Confirmed Test to check for suppression of coliforms.

(g) If The Presumptive Test tubes are gas-negative after 48 ± 3 hours, they are discarded and the results recorded as negative Presumptive Tests. Positive Presumptive tubes are verified by the Confirmed Test.

(h) If the fecal colifrom test is to be run, (Part III-C), the analyst can inoculate growth from positive Presumptive Test tubes into EC medium at the same time as he inoculates the Confirmed Test Medium.

4.6.3 Confirmed Test (See Figure III-B-2)

(a) Carefully shake each positive Presumptive tube. With a sterile 3 mm loop or a sterile applicator stick, transfer growth from each tube to BGLB. Gently agitate the tubes to mix the inoculum and incubate at 35 ± 0.5 C.

(b) After 24 ± 2 hours incubation at 35 C examine the tubes for gas. Any amount of gas in BGLB constitutes a positive Confirmed Test. If there is no gas production in the tubes (negative test) reincubate tubes for an additional 24 hours. Record the gas-positive and gas-negative tubes. Hold the positive tubes for the Completed Test if required for quality control or for checks on questionable reactions.

(c) After 48 ± 3 hours reexamine the Confirmed Test Tubes. Record the positive and negative tube results. Discard the negative tubes and hold the positive tubes for the Completed Test if required as in (b) above.

(d) In routine practice most sample analyses are terminated at the end of the Confirmed Test. However, the Confirmed Test data should be verified by carrying 5% of Confirmed Tests with a minimum of one sample per test run through the Completed Test.

(e) For certification of water supply laboratories, the MPN test is carried to completion (except for gram stain) on 10 percent of positive confirmed samples and at least one sample quarterly.

4.6.4 Completed Test (See Figure III-B-2)

Positive Confirmed Test cultures may be subjected to final Completed Test identification through application of further biochemical and culture tests, as follows:

(a) Streak one or more EMB agar plates from each positive BGLB tube. Incubate the plates at 35 ± 0.5 C for 24 ± 2 hours.

(b) Transfer one or more well-isolated typical colonies (nucleated with or without a metallic sheen) to lauryl tryptose broth fermentation tubes and to nutrient or plate count agar slants. Incubate the slants for 24 ± 2 or 48 ± 3 hours at 35 ± 0.5 C. If no typical colonies are present, pick and inoculate at least two atypical (pink, mucoid and unnucleated) colonies into lauryl tryptose fermentation tubes and incubate tubes for up to 48 ± 3 hours.

(c) The formation of gas in any amount in the fermentation tubes and presence of gram negative rods constitute a positive Completed Test for total coliforms.

4.6.5 Special Considerations for Potable Waters

Sample Size – For potable waters the standard sample shall be five times the standard portion which is either 10 milliliters or 100 milliliters as described in 40 CFR 141 (5).

Confirmation – If a laboratory using the MPN test on water supplies finds frequent numbers of Presumptive test tubes with heavy growth but no gas, these negative tubes should be submitted to the Confirmed Test to check for suppression of coliforms.

Completion — In water supply laboratories, 10% of all samples and at least one sample quarterly must be carried to completion but no gram stain of cultures is required.

4.7 Calculations: The results of the Confirmed or Completed Test may be obtained from the MPN table based on the number of positive tubes in each dilution. See Part II-C, 4.9 for details on calculation of MPN results.

4.7.1 Table II-C-4 illustrates the MPN index and 95% Confidence Limits for combinations of positive and negative results when five 10 ml, five 1.0 ml, and five 0.1 ml volumes of sample are tested.

4.7.2 Table II-C-5 provides the MPN indices and limits for the five tube, single volumes used for potable water supplies.

4.7.3 When the series of decimal dilutions is other than those in the tables select the MPN value from Table II-C-4 and calculate according to the following formula:

$$\text{MPN (From Table)} \times \frac{10}{\text{Largest Volume Tested}} = \text{MPN/100 ml}$$

4.8 Reporting Results: Report the MPN values per 100 ml of sample. See an example of a report form in Figures III-D-2 and III-D-3.

4.9 Precision and Accuracy: The precision of the MPN value increases with increased numbers of replicates tested. A five tube, five dilution MPN is recommended for natural and waste waters. Only a five tube, single volume series is required for potable waters.

5. Differentiation of the Coliform Group by by Further Biochemical Tests

5.1 Summary: The differentiation of the members of the coliform group into genera and species is based on additional biochemical and cultural tests (see Table III-B-1). These tests require specific training for valid results.

5.2 Apparatus and Materials

5.2.1 Incubator set at 35 ± 0.5 C.

5.2.2 Pipet containers of stainless steel, aluminum or pyrex glass for glass pipets.

5.2.3 Inoculation loop, 3 mm diameter and needle.

5.2.4 Bunsen/Fisher type burner or electric incinerator.

5.2.5 Sterile TD Mohr and bacteriological pipets, glass or plastic, of appropriate volumes.

5.2.6 Graduates, 25 - 500 ml.

5.2.7 Test tubes, 100 × 13 mm or 150 × 20 mm with caps, in racks.

5.2.8 Reagents

(a) Indole Test Reagent: Dissolve 5 grams para-dimethylamino benzaldehyde in 75 ml isoamyl (or normal amyl) alcohol, ACS grade, and slowly add 25 ml conc HCl. The reagent should be yellow and have a pH below 6.0. If the final reagent is dark in color it should be discarded. Some brands are not satisfactory and others become unsatisfactory after aging. Both amyl alcohol and benzaldehyde compound should be purchased in as small amounts as will be consistent with the volume of work anticipated. Store the reagent in the dark in a brown bottle with a glass stopper.

(b) Methyl Red Test Reagent: Dissolve 0.1 gram methyl red in 300 ml of 95% ethyl alcohol and dilute to 500 ml with distilled water.

(c) Voges-Proskauer Test Reagents

(1) Naphthol solution: Dissolve 5 grams purified alphanaphthol (melting point 92.5 C or higher) in 100 ml absolute ethyl alcohol. This solution must be freshly prepared each day.

(2) Potassium hydroxide solution: Dissolve 40 grams KOH in 100 ml distilled water.

(d) Oxidase Test Reagents

(1) Reagent A: Weigh out 1 gram alphanapthol and dissolve in 100 ml of 95% ethanol.

(2) Reagent B: Weigh out 1 gram para-aminodimethylaniline HCl (or oxylate) and dissolve in 100 ml of distilled water. Prepare frequently and store in refrigerator.

5.3 Media

5.3.1 Tryptophane broth for demonstrating indole production in the Indole Test. (See Part II-B, 5.1.9 (a) for preparation).

5.3.2 MR-VP broth (buffered glucose) to demonstrate acid production by methyl red color change in the Methyl Red Test and to demonstrate acetyl methyl carbinol production in the Voges-Proskauer test. (See Part II-B, 5.1.9 (b) for preparation).

5.3.3 Simmon's Citrate Agar to demonstrate utilization of citrate as a sole source of carbon. (See Part II-B, 5.1.9 (c) for preparation).

TABLE III-B-1

Differentiation of the Coliform and Related Organisms Based Upon Biochemical Reactions

Bacterium	Tests								
	Indole	Methyl Red	Voges-Proskauer	Citrate	Cytochrome Oxidase	Ornithine Decarboxylase	Lysine Decarboxylase	Arginine Dehydrolase	Motility
Escherichia coli	+	+	–	–	–	V	V	V	±
Citrobacter freundii	–	+	–	+	–	V	V	V	+
Enterobacter aerogenes	–	–	+	+	–	+	+	–	+
Klebsiella	±	–	+	+	–	–	+	–	–
Pseudomonas	–	–	–	+	+	–	–	–(+)[1]	+
Aeromonas	+	–	+	+	+	+	–	+	+

V = variable
( )[1] = reaction of *P. aeruginosa*

5.3.4 Nutrient agar slant for oxidase test. (See Part II-B, 5.1.1 for preparation).

5.3.5 Decarboxylase medium base containing lysine HCl, arginine HCl or ornithine HCl to demonstrate utilization of the specific amino acids. (See Part II-B, 5.5.14 for preparation).

5.3.6 Motility test medium (Edwards and Ewing). (See Section II-B, 5.1.10 for preparation).

5.3.7 Multitest Systems (optional to Single Test Series)

(a) API Enteric 20 (Analytab Products, Inc.).

(b) Enterotube (Roche Diagnostics).

(c) Inolex (Inolex Biomedical Division of Wilson Pharmaceutical and Chemical Corp.).

(d) Minitek (Baltimore Biological Laboratories, Bioquest).

(e) Pathotec Test Strips (General Diagnostics Division of Warner-Lambert Company).

(f) r/b Enteric Differential System (Diagnostic Research, Inc.).

5.4 Procedure

5.4.1 Biochemical tests should always be performed along with positive and negative controls. See Table IV-A-5.

5.4.2 Indole Test

(a) Inoculate a pure culture into 5 ml of tryptophane broth.

(b) Incubate the tryptophane broth at 35 ± 0.5 C for 24 ± 2 hours and mix well.

(c) Add 0.2–0.3 ml test reagent to the 24 hour culture, shake and allow the mixture to stand for 10 minutes. Observe and record the results.

(d) A dark red color in the amyl alcohol layer on top of the culture is a positive indole test; the original color of the reagent, a negative test. An orange color may indicate the presence of skatole and is reported as a $\pm$ reaction.

5.4.3 Methyl Red Test

(a) Inoculate a pure culture into 10 ml of buffered glucose broth.

(b) Incubate for 5 days at 35 C.

(c) To 5 ml of the five day culture, add 5 drops of methyl red indicator.

(d) A distinct red color is positive and distinct yellow, negative. Orange color is dubious, may indicate a mixed culture and should be repeated.

5.4.4 Voges Proskauer Test: This procedure detects the production of acetyl methyl carbinol which in the presence of alphanapthol and potassium hydroxide develops a reddish color.

(a) Use a pure culture to inoculate 10 ml of buffered glucose broth or 5 ml of salt peptone glucose broth or use the previously inoculated buffered glucose broth from the Methyl Red Test.

(b) Incubate the inoculated salt peptone glucose broth or the buffered glucose broth at 35 ± 0.5 C for 48 hours.

(c) Add 0.6 ml naphthol solution and 0.2 ml KOH solution to 1 ml of the 48 hour salt peptone or buffered glucose broth culture in a separate clean test tube. Shake vigorously for 10 seconds and allow the mixture to stand for 2–4 hours.

(d) Observe the results and record. A pink to crimson color is a positive test. Do not read after 4 hours. A negative test may develop a copper or faint brown color.

5.4.5 Citrate Test

(a) Lightly inoculate a pure culture into a tube of Simmon's Citrate Agar, using a needle to stab, then streak the medium. Be careful not to carry over any nutrient material.

(b) Incubate at 35 C for 48 hours.

(c) Examine agar tube for growth and color change. A distinct Prussian blue color in the presence of growth indicates a positive test; no color change is a negative test.

5.4.6 Cytochrome Oxidase Test (Indophenol): The cytochrome oxidase test can be done with commercially-prepared paper strips or on a nutrient agar slant as follows:

(a) Inoculate nutrient agar slant and incubate at 35 C for 18–24 hours. Older cultures should not be used.

(b) Add 2–3 drops of reagent A and reagent B to the slant, tilt to mix and read reaction within 2 minutes.

(c) Strong positive reaction (blue color slant or paper strip) occurs in 30 seconds. Ignore weak reactions that occur after 2 minutes.

5.4.7 Decarboxylase Tests (lysine, arginine and ornithine)

(a) The complete decarboxylase test series requires tubes of each of the amino acids and a control tube containing no amino acids.

(b) Inoculate each tube lightly.

(c) Add sufficient sterile mineral oil to the broths to make 3–4 mm layers on the surface and tighten the screw caps.

(d) Incubate for 18–24 hours at 35 C and read. Negative reactions should be reincubated up to 4 days.

(e) Positive reactions are purple and negative reactions are yellow. Read the control tube without amino acid first; it must be yellow for the reactions of the other tubes to be valid. Positive purple tubes must have growth as evidenced by turbidity because uninoculated tubes are also purple; nonfermenters may remain alkaline throughout incubation.

5.4.8 Motility Test

(a) Stab-inoculate the center of the tube of Motility Test Medium to at least half depth.

(b) Incubate tubes 24–48 hours at 35 C.

(c) Examine tubes for growth. If negative, reincubate at room temperature for 5 more days.

(d) Non-motile organisms grow only along the line of inoculation. Motile organisms grow outward from the line of inoculation and spread throughout the medium producing a cloudy appearance.

(e) Addition of 2, 3, 5 triphenyl tetrazolium chloride (TTC) will aid recognition of motility. Growth of microorganisms reduces TTC and produces red color along the line of growth.

5.4.9 Additional Biochemical Tests: If other biochemical tests are necessary to further identify enteric bacteria, for example specific carbohydrate fermentation, see the Table III-E-5, Biochemical Characteristics of Enterobacteriaceae.

5.4.10 Multitest Systems: Multitest systems are available which use tubes containing agar media that provide numerous biochemical tests, plastic units containing a series of dehydrated media, media-impregnated discs and reagent-impregnated paper strips. Some of the systems use numerical codes to aid identification. Others provide computerized identification of bacteria. A number of independent investigators have compared one or more multitest systems with conventional or traditional biochemical tests. Some of the earlier systems have been improved. Most of the recent studies report the correct identification of high percentages of isolates. The systems are described in Part III-E, 5.6.

REFERENCES

1. Geldreich, E. E., P. W. Kabler, H. L. Jeter and H. F. Clark, 1955. A delayed incubation membrane filter test for coliform bacteria in water. Amer. Jour. Public Health 45:1462.

2. Brezenski, F. T. and J. A. Winter, 1969. Use of the delayed incubation membrane filter test for determining coliform bacteria in sea water. Water Res. 3:583.

3. Geldreich, E. E., 1975. Handbook for Evaluating Water Bacteriological Laboratories (2nd ed.), EPA-670/9–75–006. U.S. Environmental Protection Agency, Cincinnati, Ohio.

4. Tubiash, H., 1951. The Anaerogenic Effect of Nitrates and Nitrites on Gram-negative Enteric Bacteria. Amer. Jour. Public Health 41:833.

5. National Interim Primary Drinking Water Regulations, 40 Code of Federal Regulations (CFR) Part 141.14 (b) and (c), Published in Federal Register, 40, 59566, December 24, 1975.

PART III. ANALYTICAL METHODOLOGY

Section C Fecal Coliform Methods

The direct membrane filter (MF), the delayed-incubation MF and the multiple-tube, most probable number (MPN) methods can be used to enumerate fecal coliforms in water and wastewater. For a general description of the fundamental laboratory techniques refer to Part II-C. The method chosen depends upon the characteristics of the sample. The Section is divided as follows:

1. Definition of the Fecal Coliform Group
2. Direct Membrane Filter (MF) Method
3. Delayed-Incubation Membrane Filter Method
4. Verification
5. Most Probable Number (MPN) Method

1. Definition of the Fecal Coliform Group

1.1 The fecal coliforms are part of the total coliform group. They are defined as gram-negative nonspore-forming rods that ferment lactose in 24 $\pm$ 2 hours at 44.5 $\pm$ 0.2 C with the production of gas in a multiple-tube procedure or produce acidity with blue colonies in a membrane filter procedure.

1.2 The major species in the fecal coliform group is *Escherichia coli*, a species indicative of fecal pollution and the possible presence of enteric pathogens.

2. Direct Membrane Filter (MF) Method

2.1 Summary: An appropriate volume of a water sample or its dilution is passed through a membrane filter that retains the bacteria present in the sample. The filter containing the microorganisms is placed on an absorbent pad saturated with M-FC broth or on M-FC agar in a petri dish. The dish is incubated at 44.5 C for 24 hours. After incubation, the typical blue colonies are counted under low magnification and the number of fecal coliforms is reported per 100 ml of original sample.

2.2 Scope and Application

2.2.1 Advantages: The results of the MF test are obtained in 24 hours. Up to 72 hours are required for the multiple-tube fermentation method. The M-FC method provides direct enumeration of the fecal coliform group without enrichment or subsequent testing. Over 93% of the blue colonies that develop in this test using M-FC medium at the elevated temperature of 44.5 C $\pm$ 0.2 C are reported to be fecal coliforms (1). The test is applicable to the examination of lakes and reservoirs, wells and springs, public water supplies, natural bathing waters, secondary non-chlorinated effluents from sewage treatment plants, farm ponds, stormwater runoff, raw municipal sewage, and feedlot runoff. The MF test has been used with varied success in marine waters.

2.2.2 Limitations: Recent data (2, 3) indicate that the single-step MF fecal coliform procedure may produce lower results than those obtained with the fecal coliform multiple-tube procedure, particularly for chlorinated effluents. Since chlorination stresses fecal coliforms and significantly reduces recovery, this method should not be used with chlorinated wastewater. Disinfection and toxic materials such as metals, phenols, acids or caustics also affect recovery of fecal coliforms on the membrane filter. Any decision to use this test for stressed microorganisms requires parallel MF/MPN evaluation based on the procedure described in Part IV-C, 3.

Recently-proposed solutions to problems of lower recovery (2, 4, 5, 6) include the use of two-step incubation, two-step incubation overlay and/or enrichment techniques and modification of membrane filter structures.

2.3 Apparatus and Materials

2.3.1 Water bath, aluminum heat sink, or other incubator that maintains a stable 44.5 ± 0.2 C. Temperature is checked against an NBS certified thermometer or one of equivalent accuracy.

2.3.2 Binocular (dissecting type) microscope, with magnification of 10–15× and daylight-type fluorescent lamp.

2.3.3 Hand tally.

2.3.4 Pipet containers of stainless steel, aluminum or pyrex glass for glass pipets.

2.3.5 Graduated cylinders, covered with aluminum foil or kraft paper before sterilization.

2.3.6 Sterile, unassembled membrane filtration units (filter base and funnel), glass, plastic or stainless steel, wrapped with aluminum foil and kraft paper.

2.3.7 Vacuum source.

2.3.8 Vacuum filter flask, with appropriate tubing. Filter manifolds which hold a number of filter bases can also be used.

2.3.9 Safety trap flask between the filter flask and the vacuum source.

2.3.10 Forceps with smooth tips.

2.3.11 Ethanol, 95% or methanol, in small vial, for sterilizing forceps.

2.3.12 Bunsen/Fisher burner or electric incinerator.

2.3.13 Sterile TD bacteriological or Mohr pipets, glass or plastic, of appropriate size.

2.3.14 Sterile petri dishes, 50 × 12 mm plastic with tight-fitting lids.

2.3.15 Dilution bottles (milk dilution), pyrex glass, marked at 99 ml volume, screw-cap with neoprene rubber liner.

2.3.16 Membrane filters, white, grid marked, 47 mm diameter, 0.45±0.02 μm pore size or other pore size recommended by manufacturer for water analyses. The Millipore HC MF, not the HA, is recommended.

2.3.17 Absorbent pads.

2.3.18 Water-proof plastic bags.

2.3.19 Inoculation loops, 3 mm diameter, or needle of nichrome or platinum wire, 26 B&S gauge, in suitable holder.

2.3.20 Disposable applicator sticks or plastic loops as alternatives to inoculation loops.

2.3.21 Ultraviolet sterilizer for MF filtration units (optional).

2.4 Media

2.4.1 M-FC broth or agar prepared in presterilized erlenmeyer flasks (See Part II-B, 5.2.1).

2.4.2 Lauryl tryptose broth prepared in 10 ml volumes in fermentation tubes (see Part II-B, 5.3.1) for verification.

2.4.3 EC medium prepared in 10 ml volumes in fermentation tubes (see Part II-B, 5.3.4) for verification.

2.5 Dilution Water (See Part II,B, 7 for preparation).

2.5.1 Sterile buffered dilution water or peptone water dispensed in 99$\pm$2 ml amounts in screw-capped dilution bottles.

2.5.2 Sterile buffered water or peptone water prepared in 500 ml or larger volumes for wetting membranes before addition of the sample, and for rinsing the funnel after sample filtration.

2.6 Procedure: The general membrane filter procedure is described in detail in Part II-C.

2.6.1 Prepare the M-FC broth or agar medium as outlined in Part II-B, 5.2.1. Saturate the sterile absorbent pads with about 2.0 ml of broth or add 5–6 ml of M-FC agar to the bottom of each 50 $\times$ 12 mm petri dish (to a depth of 2–3 mm). Pour off excess liquid from broth-saturated pads. Mark dishes and bench forms with sample identity and sample volumes.

2.6.2 Using a sterile forceps place a sterile membrane filter on the filter base, grid side up.

Attach the funnel to the base of the filter unit; the membrane filter is now held between the funnel and base.

2.6.3 Shake the sample vigorously about 25 times and measure the sample into the funnel with the vacuum off. If sample volume is $<$ 10 ml, add 10 ml of sterile dilution water to the filter before adding the sample.

2.6.4 Sample volumes for fecal coliform enumeration in different waters and wastewaters are suggested in Table III-C-1. These volumes should provide the recommended count of 20–60 colonies on a membrane filter. Fecal coliform levels are generally lower than total coliform densities in the same sample; therefore larger volumes are sampled.

2.6.5 Do not filter less than 1.0 ml of undiluted sample.

2.6.6 Filter the sample and rinse the sides of the funnel walls at least twice with 20–30 ml of sterile dilution water.

2.6.7 Turn off the vacuum and remove the funnel from the filter base.

2.6.8 Aseptically remove the membrane filter from the filter base. Place the filter, grid side up, on the absorbent pad saturated with M-FC Broth or on M-FC agar, using a rolling motion to prevent air bubbles.

2.6.9 Incubate the petri dishes for 24 $\pm$ 2 hours at 44.5 $\pm$ 0.2 C in sealed waterproof plastic bags submerged (with the petri dishes inverted) in a waterbath, or without plastic bag in a heat-sink incubator. MF cultures should be placed in incubator within 30 minutes of filtration.

2.6.10 After 24 hours remove dishes from the incubator and examine for blue colonies.

2.7 Counting and Recording Colonies: Select those plates with 20-60 blue (sometimes greenish-blue) colonies. Non-fecal colonies are gray, buff or colorless and are not counted. Pinpoint blue colonies should be counted and confirmed. The colonies are counted using a microscope of 10-15$\times$ and a fluorescent lamp. Use of hand lens or other simple optical devices of lower magnification make difficult the identification and differentiation of typical and atypical blue colonies.

2.7.1 The general counting rules are given in Part II-C, 3.5. The following rules are used in calculating the fecal coliform count per 100 ml of sample:

(a) Countable Membranes with 20–60 Blue Colonies. Count all blue colonies using the formula:

$$\frac{\text{No. of Fecal Coliform Colonies Counted}}{\text{Volume in ml of Sample Filtered}} \times 100 = \text{fecal coliform count/100 ml}$$

TABLE III-C-1

Suggested Range of Sample Volumes for Fecal Coliform Tests Using the Membrane Filter Method

Sample Source	100	30	10	3	1	0.3	0.1	0.03	0.01	0.003	0.001	0.0003	0.0001
Swimming Pools	X												
Wells, Springs	X	X											
Lakes, Reservoirs	X	X	X										
Water Supply Intakes			X	X	X	X	X						
Bathing Beaches			X	X	X	X	X						
River Water					X	X	X	X	X				
Chlorinated Sewage Effluent					X	X	X	X	X				
Raw Sewage									X	X	X	X	X

For example, if 40 colonies are counted after the filtration of 50 ml of sample, the calculation is:

$$\frac{40}{50} \times 100 = 80 \text{ fecal coliforms/100 ml.}$$

(b) Countable Membranes With Less Than 20 Blue Colonies. Report as: Estimated Count/100 ml and specify the reason.

(c) Membranes With No Colonies. Report the count as: Less than (calculated value)/100 ml, based upon the largest single volume filtered.

For example, if 10, 3 and 1 ml are filtered and all plates show zero counts, select the largest volume, apply the general formula and report the count as a <(less than) value:

$$\frac{1}{10} \times 100 = 10$$

or < 10 fecal coliforms/100 ml.

(d) Countable Membranes With More Than 60 Blue Colonies. Calculate count from highest dilution and report as a > value.

(e) Uncountable Membranes With More Than 60 Colonies. Use 60 colonies as the basis of calculation with the smallest filtration volume, e.g., 0.01 ml:

$$\frac{60}{0.01} \times 100 = 600{,}000$$

Report as: > 600,000 fecal coliforms/100 ml.

2.7.2 Reporting Results. Report fecal coliform densities per 100 ml. See discussion on significant figures in Part II-C, 2.8.1.

2.8 Precision and Accuracy

2.8.1 Ninety-three percent of the blue colonies that develop on M-FC medium at the elevated temperature of 44.5 ± 0.2 C were verified as fecal coliform (1).

2.8.2 Laboratory personnel should be able to duplicate their own colony counts on the same plate within 5%, and the counts of other analysts on the same plate within 10%.

3. Delayed-Incubation Membrane Filter (MF) Method

3.1 Summary: Bacteria are retained on 0.45 µm filters after passage of selected sample volumes through the filters. The filters are placed on M-VFC broth (a minimum growth medium) and transported from field sites to the laboratory. In the laboratory, the filters are transferred to the M-FC medium and incubated at 44.5 C for 24 hours. Blue colonies are counted as fecal coliforms.

3.2 Scope and Application

3.2.1 Advantages: The delayed incubation MF method is useful in survey monitoring or emergency situations when the standard fecal coliform test cannot be performed at the sample site, or when time and temperature limits for sample storage cannot be met. The method eliminates field processing and equipment needs. Also, examination at a central laboratory permits confirmation and biochemical identification of the organisms as necessary. Consistent results have been obtained with this method using water samples from a variety of sources (7).

3.2.2 Limitations: The applicability of this method for a specific water source must be determined in preliminary studies by comparison with the standard MF method. For example, limited testing has indicated that the delayed-incubation method is not as effective in saline waters (7).

3.3 Apparatus and Materials

3.3.1 Water bath, aluminum heat sink, or equivalent incubator that maintains a 44.5 ± 0.2 C temperature.

3.3.2 Binocular (dissection) microscope, with magnification 10 or 15×, binocular, wide-field type. A microscope lamp producing diffuse daylight from cool white fluorescent lamps.

3.3.3 Hand tally.

3.3.4 Pipet containers of stainless steel, aluminum or pyrex glass for glass pipets.

3.3.5 Graduated cylinders, covered with aluminum foil or kraft paper before sterilization.

3.3.6 Sterile unassembled membrane filtration units (filter base and funnel), glass, plastic or stainless steel wrapped with aluminum foil or kraft paper.

3.3.7 Vacuum source.

3.3.8 Filter flask to hold filter base, with appropriate tubing. Filter manifold to hold a number of filter bases can also be used. In the field, portable field kits are also used.

3.3.9 Safety trap flask between the filtering flask and the vacuum source.

3.3.10 Forceps with smooth tip.

3.3.11 Ethanol, 95% or methanol, in small vial, for sterilizing forceps.

3.3.12 Bunsen/Fisher type burner.

3.3.13 Sterile TD bacteriological or Mohr pipets, glass or plastic, in appropriate volumes.

3.3.14 Sterile petri dishes, 50 $\times$ 12 mm plastic with tight-fitting lids.

3.3.15 Dilution bottles (milk dilution), pyrex glass, 99 ml volume, screw-caps with neoprene rubber liners.

3.3.16 Membrane filters, white, grid marked, 47 mm in diameter, 0.45$\pm$0.02 μm pore size, or other pore size recommended by the manufacturer for water analyses. The Millipore HC MF, not the HA is recommended.

3.3.17 Shipping tubes, labels, and packing materials for mailing delayed incubation plates.

3.3.18 Ultraviolet sterilizer for MF filtration units (optional).

3.4 Media: The following media are prepared in pre-sterilized erlenmeyer flasks with metal caps, aluminum foil covers, or screw-caps:

3.4.1 M-VFC holding media (see Part II-B, 5.2.6).

3.4.2 M-FC broth or agar (see Part II-B, 5.2.1).

3.5 Dilution Water

3.5.1 Sterile dilution water dispensed in 99 $\pm$ 2 ml volumes in screw-capped bottles.

3.5.2 Sterile dilution water prepared in large volumes for wetting membranes before the addition of the sample, and for rinsing the funnel after sample filtration.

3.6 Procedure: The general membrane filter procedure is described in detail in Part II-C.

3.6.1 Prepare the M-VFC holding medium as outlined in Part II-B, 5.2.6. Saturate the sterile absorbent pads with about 2.0 ml of M-VFC broth. Pour off excess broth. Mark dishes and bench forms with sample identity and volumes.

3.6.2 Using sterile forceps place a membrane filter on the filter base grid side up.

3.6.3 Attach the funnel to the base of the filter unit; the membrane filter is now held between the funnel and base.

3.6.4 Shake the sample vigorously about 25 times and measure into the funnel with the vacuum off. If the sample is $<$ 10 ml, add 10 ml of sterile dilution water to the membrane filter before adding the sample.

(a) Sample volumes for fecal coliform enumeration in different waters and wastewaters are suggested in Table III-C-1. These volumes should produce membrane filters with a recommended count of 20–60 colonies.

(b) Follow the methods for sample measurement and dispensation given in Part II-C, 3.4.6.

3.6.5 Filter the sample through the membrane and rinse the sides of the funnel walls at least twice with 20–30 ml of sterile dilution water.

3.6.6 Turn off the vacuum and remove the funnel from the base of the filter unit.

3.6.7 Aseptically remove the membrane filter from the filter base and place grid side up on an absorbent pad saturated with VFC medium.

3.6.8 Place the culture dish in shipping container and send to the examining laboratory. Fecal coliform bacteria can be held on the VFC holding medium for up to 72 hours with little effect on the final counts. The holding period should be kept to a minimum.

3.6.9 At the examining laboratory remove the membrane from the holding medium, place it in another dish containing M-FC broth or agar medium, and complete testing for fecal coliforms as described above under 2.6.

3.7 Counting and Recording Colonies: After the required incubation select those plates with 20-60 blue (sometimes greenish-blue) colonies. Gray to cream colored colonies are not counted. Pin-point blue colonies are not counted unless confirmed. The colonies are enumerated using a binocular microscope with a magnification of 10 or 15×.

Refer to 2.7.1, for rules used in reporting the fecal coliform MF counts.

3.8 Reporting Results: Record densities as fecal coliforms per 100 ml. Refer to Part II-C, 2.8, for discussions on the use of significant figures and rounding off values.

3.9 Precision and Accuracy: As reported in 2.8, this Section.

4. Verification

Verification of the membrane filter test for fecal coliforms establishes the validity of colony differentiation by blue color and provides supporting evidence of colony interpretation. The verification procedure corresponds to the fecal coliform MPN (EC Medium) test.

4.1 Pick from the centers of at least 10 well-isolated blue colonies. Inoculate into lauryl tryptose broth and incubate 24-48 hours at 35 ± 0.5 C.

4.2 Confirm gas-positive lauryl tryptose broth tubes at 24 and 48 hours by inoculating a loopful of growth into EC tubes and incubating for 24 hours at 44.5 ± 0.2 C. Cultures that produce gas in EC tubes are interpreted as verified fecal coliform colonies (see Figure III-C-1).

4.3 A percent verification can be determined for any colony-validation test:

$$\frac{\text{No. of colonies meeting verification test}}{\text{No. of colonies subjected to verification}} \times 100 = \text{Percent verification}$$

Example: Twenty blue colonies on M-FC medium were subjected to verification studies shown in Figure III-C-1. Eighteen of these colonies proved to be fecal coliforms according to provisions of the test:

$$\text{Percent verification} = \frac{18}{20} \times 100 = 90\%$$

4.4 A percent verification figure can be applied to the direct test results to determine the verified fecal coliform count per 100 ml:

$$\frac{\text{Percent verification}}{100} \times \text{count per 100 ml} = \text{Verified fecal coliform count}$$

Example: For a given sample, by the M-FC test, the fecal coliform count was found to be

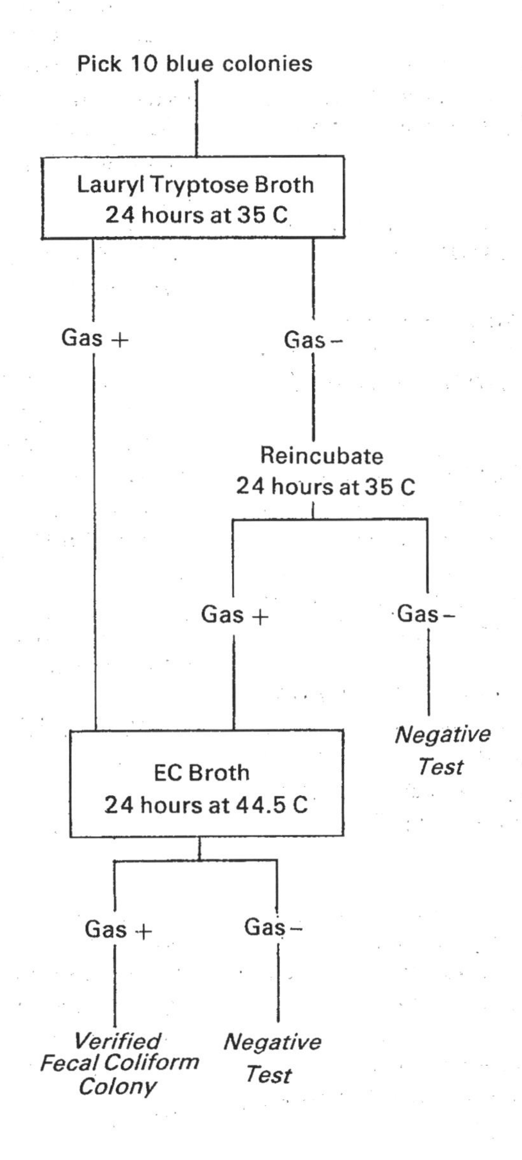

FIGURE III-C-1. Verification of Fecal Coliform Colonies on the Membrane Filter

42,000 organisms per 100 ml. Supplemental studies on selected colonies showed 92% verification.

$$\text{Verified fecal coliform count} = \frac{92}{100} \times 42{,}000 = 38{,}640$$

Rounding off = 39,000 fecal coliforms per 100 ml

The worker is cautioned not to apply percentage of verification determined on one sample to other samples.

5. Most Probable Number (MPN) Method

5.1 Summary: Culture from positive tubes of the lauryl tryptose broth (same as presumptive MPN Method, Part III-B) is inoculated into EC Broth and incubated at 44.5 C for 24 hours (see Figure III-C-2). Formation of gas in any quantity in the inverted vial is a positive reaction confirming fecal coliforms. Fecal coliform densities are calculated from the MPN table on the basis of the positive EC tubes (8).

5.2 Apparatus and Materials

5.2.1 Incubator that maintains 35 ± 0.5 C.

5.2.2 Water bath or equivalent incubator that maintains a 44.5 ± 0.2 C temperature.

5.2.3 Pipet containers of stainless steel, aluminum or pyrex glass for glass pipets.

5.2.4 Inoculation loop, 3 mm diameter and needle of nichrome or platinum wire, 26 B & S gauge, in suitable holder. Sterile applicator sticks are a suitable alternative.

5.2.5 Sterile pipets T.D., Mohr or bacteriological, glass or plastic, of appropriate size.

5.2.6 Dilution bottles (milk dilution), pyrex, 99 ml volume, screw-cap with neoprene liners.

5.2.7 Bunsen or Fisher-type burner or electric incinerator unit.

5.2.8 Pyrex test tubes, 150 × 20 mm, containing inverted fermentation vials, 75 × 10 mm, with caps.

5.2.9 Culture tube racks to hold fifty, 25 mm diameter tubes.

5.3 Media

5.3.1 Lauryl tryptose broth (same as total coliform Presumptive Test medium) prepared in 10 ml volumes in appropriate concentration for sample volumes used. (Part II-B, 5.3.1).

5.3.2 EC medium prepared in 10 ml volumes in fermentation tubes (Part II-B, 5.3.4).

5.4 Dilution Water: Sterile buffered or peptone dilution water dispensed in 99 ± 2 ml volumes in screw-capped bottles.

5.5 Procedure: Part II-C describes in detail the general MPN procedure. See Figure III-C-2.

5.5.1 Prepare the total coliform Presumptive Test medium, (lauryl tryptose broth) and EC medium. Clearly mark each bank of tubes, identifying the sample and the volume inoculated.

5.5.2 Inoculate the Presumptive Test medium with appropriate quantities of sample following the Presumptive Test total coliform procedure, (Part III-B).

5.5.3 Gently shake the Presumptive tube. Using a sterile inoculating loop or a sterile wooden applicator, transfer inocula from positive Presumptive Test tubes at 24 and 48 hours to EC confirmatory tubes. Gently shake the rack of inoculated EC tubes to insure mixing of inoculum with medium.

5.5.4 Incubate inoculated EC tubes at 44.5 ± 0.2 C for 24 ± 2 hours. Tubes must be placed in the incubator within 30 minutes after inoculation. The water depth in the water bath incubator must come to the top level of the culture medium in the tubes.

5.5.5 The presence of gas in any quantity in the EC confirmatory fermentation tubes af-

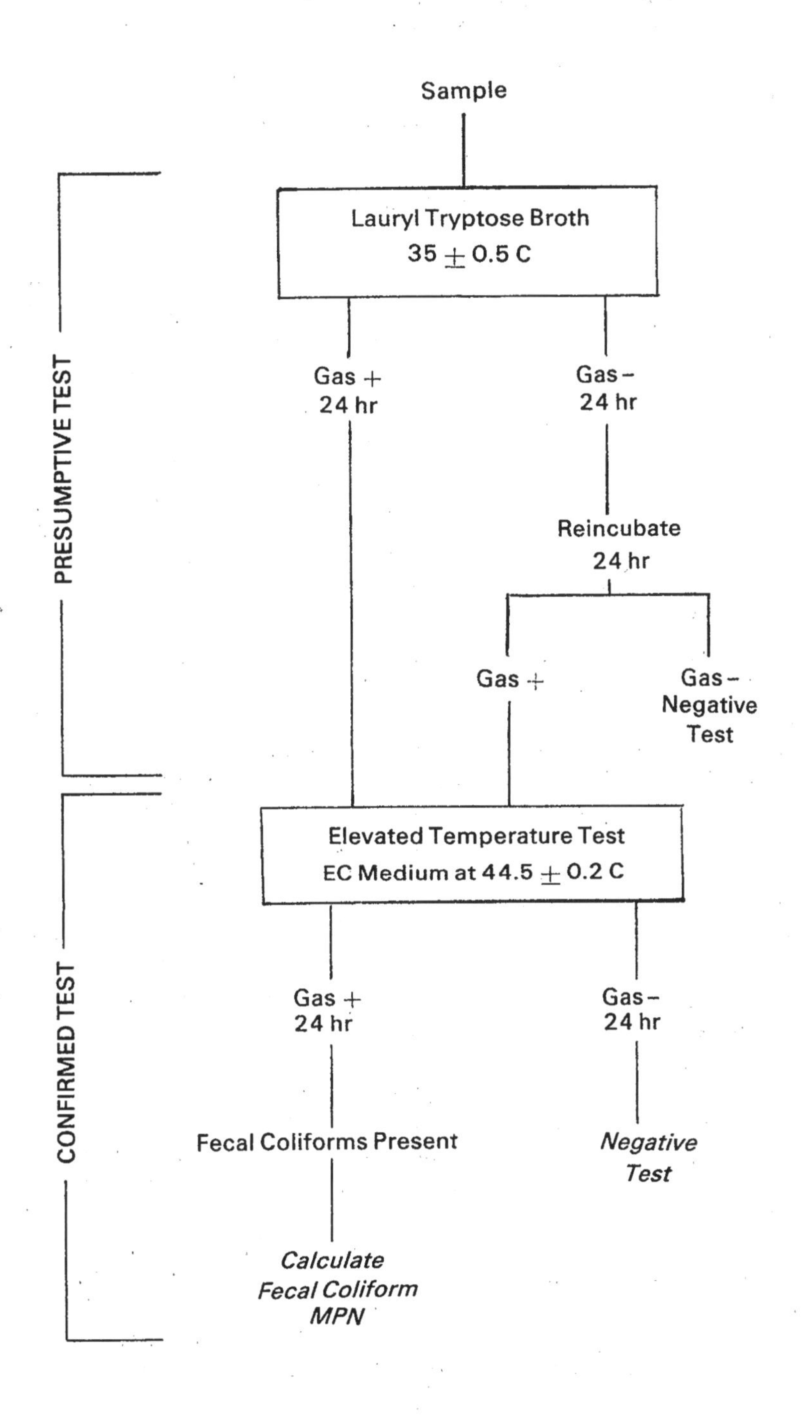

FIGURE III-C-2. Flow Chart for the Fecal Coliform MPN Tests.

ter 24 ± 2 hours constitutes a positive test for fecal coliforms.

5.6 Calculations

5.6.1 Calculate fecal coliform densities on the basis of the number of positive EC fermentation tubes, using the table of most probable numbers (MPN).

5.6.2 The MPN results are computed from three dilutions that include the highest dilution with all positive tubes and the next two higher dilutions. For example, if five 10 ml, five 1.0 ml, and five 0.1 ml sample portions are inoculated initially into Presumptive Test medium, and positive EC confirmatory results are obtained from five of the 10 ml portions, three of the 1.0 ml portions, and none of the 0.1 ml portions, the coded result of the test is 5-3-0. The code is located in the MPN Table II-C-4, and the MPN per 100 ml is recorded. See Part II-C, 4.9 for rules on selection of significant dilutions.

5.7 Reporting Results: Report the fecal coliform MPN values per 100 ml of sample.

5.8 Precision and Accuracy: The precisions of the MPN counts are given as confidence limits in the MPN tables. Note that the precision of the MPN value increases with increased numbers of replicates per sample tested.

REFERENCES

1. Geldreich, E. E., H. F. Clark, C. B. Huff, and L. C. Best, 1965. Fecal-coliform-organism medium for the membrane filter technique. J. Amer. Water Works Assoc. 57:208.

2. Bordner, R. H., C. F. Frith and J. A. Winter, (Editors), 1977. Proceedings of the Symposium on the Recovery of Indicator Organisms Employing Membrane Filters. U.S. Environmental Protection Agency, Environmental Monitoring and Support Laboratory, Cincinnati, OH. (EPA-600/9-77-024).

3. Lin, S. D., 1973. Evaluation of coliform tests for chlorinated secondary effluents. J. Water Pollution Control Federation 45:498.

4. Lin, S. D., 1976. Evaluation of Millipore HA and HC membrane filters for the enumeration of indicator bacteria. Appl. Environ. Microbiol. 32:300.

5. Lin, S. D., 1976. Membrane filter method for recovery of fecal coliforms in chlorinated sewage effluents. Appl. Environ. Microbiol. 32:547.

6. Green, B. L., E. M. Clausen, and W. Litsky, 1975. Two-temperature membrane filter method for enumeration of fecal coliform bacteria from chlorinated effluents. Appl. Environ. Microbiol. 33: 1259.

7. Taylor, R. H., R. H. Bordner, and P. V. Scarpino, 1973. Delayed-incubation membrane-filter test for for fecal coliforms. Appl. Microbiol. 25:363.

8. Geldreich, E. E., 1966. Sanitary Significance of Fecal Coliforms in the Environment, U.S. Dept. of the Interior, FWPCA, WP-20-3, 122 pp.

PART III. ANALYTICAL METHODOLOGY

Section D Fecal Streptococci

The membrane filter (MF), most probable number (MPN), and direct pour plate procedures can be used to enumerate and identify fecal streptococci in water and wastewater. For a general description of these fundamental techniques refer to Part II-C. The method selected depends upon the characteristics of the sample. The Section is divided as follows:

1. **The Fecal Streptococcus Group**
2. **Membrane Filter Method**
3. **Verification**
4. **Most Probable Number Method**
5. **Pour Plate Method**
6. **Determination of FC/FS Ratios**
7. **Identification to Species**

1. The Fecal Streptococcus Group

1.1 Fecal Streptococci and Lancefield's Group D Streptococcus: The terms "Fecal Streptococcus" and "Lancefield's Group D Streptococcus" have been use synonymously. When used as indicators of fecal contamination the following species and varieties are implied: *S. faecalis, S. faecalis* subsp. *liquefaciens, S. faecalis* subsp. *zymogenes, S. faecium, S. bovis* and *S. equinus*. For sanitary analyses, media and methodology for quantification are selective for these organisms.

1.2 Fecal Streptococcal Intermediates and Biotypes: Current information indicates that other streptococci belonging to Lancefield's serological Group Q occur in the feces of humans and other warm-blooded animals, especially chickens. *S. avium* is characteristically found in the feces of chickens and occasionally in the feces of man, dogs and pigs. The Group Q antigen is found in the cell wall of these organisms, and in addition, the Group D antigen is located between the cell wall and the cytoplasmic membrane where it occurs naturally in the established Group D species. These common antigens indicate a relationship between Group D and Group Q organisms. The Group Q Streptococcci may account for the occurrence of the "intermediate strains or biotypes" of Group D Streptococci (1). Group Q organisms grow on media commonly used for the isolation and enumeration of enterococci. Kenner et al. (2) observed that enterococcus biotypes occur more commonly in the feces of fowl than of pigs, sheep, cows and humans. Forty percent of the streptococci isolated from the feces of fowl by Kenner et al. were enterococcus biotypes with the rest being the enterococcus group. It is probable that some of the biotypes they described were Group Q Streptococci. The Group Q Streptococci, with a type species, should be considered in the Fecal Streptococcus Group.

1.3 Definition of the Fecal Streptococcus Group: The term, fecal streptococci, will be used to describe the streptococci which indicate the sanitary quality of water and wastewater. The fecal streptococci group includes the serological groups D and Q.

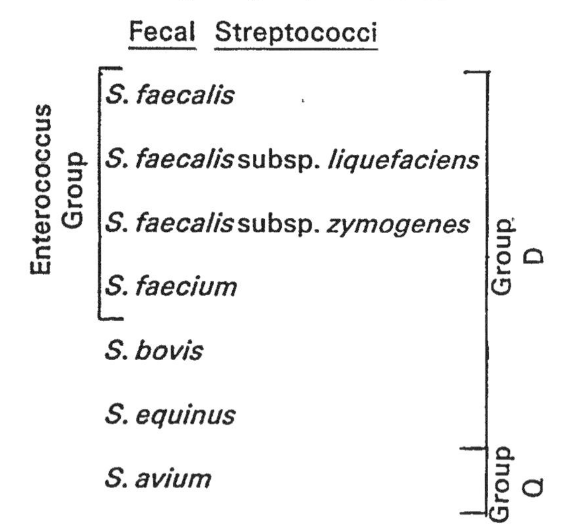

1.4 Viridans Streptococci: The viridans streptococci, primarily *S. salivarius* and *S. mitis* are not considered as part of the fecal streptococci as defined in 1.2 and 1.3. These inhabitants of the nasopharyngeal tract have been reported by a few workers in feces and do grow on some fecal streptococci media. However, their low numbers when present, the low frequency of occurrence and the limited data available at this time concerning their presence, have resulted in their exclusion from the classification of fecal streptococci.

1.5 Scope and Application: Fecal streptococci data verify fecal pollution and may provide additional information concerning the recency and probable origin of pollution. In combination with data on coliform bacteria, fecal streptococci are used in sanitary evaluation as a supplement to fecal coliforms when a more precise determination of sources of contamination is necessary. The occurrence of fecal streptococci in water indicates fecal contamination by warm-blooded animals. They are not known to multiply in the environment. Further identification of streptococcal types present in the sample may be obtained by biochemical characterization. (See Figure III-D-2 "Isolation and Identification of Fecal Streptococci"). Such information is useful for source investigations. For example, *S. bovis* and *S. equinus* are host specific and are associated with the fecal excrement of non-human warm-blooded animals. High numbers of these organisms are associated with pollution from meat processing plants, dairy wastes, and run-off from feedlots and farmlands. Because of limited survival time outside the animal intestinal tract their presence indicates very recent contamination from farm animals.

2. Membrane Filter (MF) Method

2.1 Summary: A suitable volume of sample is passed through the 0.45 µm membrane filter which retains the bacteria. The filter is placed on KF Streptococcus agar and incubated at 35 C for 48 hours. Red and pink colonies are counted as streptococci (1, 2).

2.2 Scope and Application: The membrane filter technique is recommended as the standard method for assaying fecal streptococci in fresh and marine waters and in nonchlorinated sewage. Wastewaters from food processing plants, slaughter houses, canneries, sugar processing plants, dairy plants, feedlot and farmland run-off may be analyzed by this procedure. Colonies on a membrane filter can be transferred to biochemical media for identification and speciation to provide information on the source of contamination. The general advantages and limitations of the MF method are given in Part II-C.

2.3 Apparatus and Materials

2.3.1 Water bath, aluminum heat sink or air incubator set at 35 $\pm$0.5 C. Temperature checked with an NBS thermometer or one of equivalent accuracy.

2.3.2 Stereoscopic (dissection) microscope, with magnification of 10 to 15$\times$, preferably wide field type. A microscope lamp with diffuse light from cool, white fluorescent tubes is recommended.

2.3.3 Hand tally.

2.3.4 Pipet containers of stainless steel, aluminum or pyrex glass for glass pipets.

2.3.5 Sterile graduated cylinders, covered with aluminum foil or kraft paper.

2.3.6 Sterile, unassembled membrane filtration units (filter base and funnel), glass, plastic or stainless steel, wrapped with aluminum foil or kraft paper.

2.3.7 Vacuum source.

2.3.8 Filter suction flask to hold filter base, with appropriate tubing. Filter manifolds that hold a number of filter bases can also be used.

2.3.9 Safety vacuum flask.

2.3.10 Forceps, with smooth tips.

2.3.11 Ethanol, 95%, or methanol, in small vial for sterilizing forceps.

2.3.12 Bunsen/Fisher burner or electric incinerator.

2.3.13 Sterile T.D. bacteriological or Mohr pipets, glass or plastic, of appropriate size.

2.3.14 Sterile petri dishes, 50 × 12 mm plastic or 60 × 15 mm glass or plastic.

2.3.15 Dilution bottles (milk dilution), pyrex glass, marked at 99 ml volumes, screw-cap with neoprene rubber liner.

2.3.16 Membrane filters, manufactured from cellulose ester materials, white, grid marked, 47 mm in diameter, 0.45 ± 0.02 µm pore size or other pore size as recommended by the manufacturer for fecal streptococci analyses.

2.3.17 Ultraviolet sterilizer for MF filtration units (optional).

2.4 Media: KF Streptococcus agar prepared as described in Part II-B, 5.4.1.

2.5 Dilution Water

2.5.1 Sterile buffered dilution water or peptone water dispensed in 99 ± 2 ml volumes in screw-capped dilution bottles.

2.5.2 Sterile buffered water or peptone water prepared as described in Part II-B, 7, in large volumes for wetting membranes before the addition of the sample, and for rinsing the funnel after filtration.

2.6 Immediate MF Procedure: The general membrane filter procedure is described in detail in Part II-C, 3.

2.6.1 Clearly mark each petri dish and aseptically add 5–6 ml of the liquified agar medium (to each dish to a depth of 2–3 mm).

2.6.2 Place a sterile membrane filter on the filter base, grid-side up and attach the funnel to the base of the filter unit; the membrane filter is now held securely between the funnel and base.

2.6.3 Shake the sample vigorously about 25 times and measure the sample into the funnel with the vacuum off.

2.6.4 Filter appropriate volumes of water sample through the sterile membrane to obtain 20-100 colonies on the membrane surface.

At least 3 sample increments should be filtered in order of increasing volumes. Where no background information is available, more may be necessary. The methods of measurement and dispensation of the sample into the funnel are given in Part II-C, 3.4.6.

2.6.5 Filter the sample and rinse the sides of the funnel at least twice with 20–30 ml of sterile buffered dilution water. Turn off the vacuum and remove the funnel from the filter base.

2.6.6 Aseptically remove the membrane from the filter base and place grid-side up on the agar.

2.6.7 Incubate the petri dishes in the inverted position at 35 ± 0.5 C for 48 hours.

2.6.8 After incubation, remove dishes and examine for red to pink colonies and count.

2.7 Counting and Recording Colonies: Select those plates with 20-100 pink to dark red colonies. These may range in size from barely visible to about 2 mm in diameter. Colonies of other colors are not counted. Count the colonies as described in Part II-C, 3.5, using low-power (10–15×) microscope equipped with overhead illumination.

Fecal streptococcal density is reported as organisms per 100 ml. Use the general formula to calculate fecal streptococci densities:

$$\text{Fecal Streptococci/100 ml} = \frac{\text{No. of Fecal Streptococcus Colonies Counted}}{\text{Volume of Sample Filtered, ml}} \times 100$$

For example, if 40 colonies are counted after the filtration of 50 ml of sample the calculation is:

$$\text{Fecal Streptococci/100 ml} = \frac{40}{50} \times 100 = 80.$$

See Part II-C, 3.6 for calculation for results.

Reporting Results: Report fecal streptococcal densities per 100 ml of sample. See discussion on significant figures in Part II-C, 2.8.

2.8 Precision and Accuracy

2.8.1 Extensive precision and accuracy data are not available, however, KF Streptococcus agar has been reported to be highly efficient in the recovery of fecal streptococci (2, 3, 4). In the analyses of feces, sewage and foods, KF yielded a high recovery of fecal streptococci with a low percent (18.6) of non-fecal streptococci.

2.8.2 Laboratory personnel should be able to duplicate their colony counts on the same plates within 5%, and the counts of others within 10%.

2.9 Delayed MF Procedure: Because of the stability of the KF agar and its extreme selectivity for fecal streptococci, it is possible to filter water samples at a field site, place membranes on the KF agar medium in tight-lidded petri dishes and hold these plates for up to 3 days. After the holding period, plates are incubated for 48 hours at 35 C and counted in the normal manner. This 72 hour holding time can be used to air mail the membranes on KF agar to a central laboratory for incubation and counting. (National Pollution Surveillance System FWPCA, data collected from geographical locations around the Nation; and Kenner et al., Kansas City data.)

3. Verification

Periodically, typical colonies growing on the membrane filter should be verified. When a survey is initiated or a new body of water is being sampled, it is recommended that at least 10 typical colonies from the membrane or agar plate used in computing the final density be picked and transferred into BHI broth or onto BHI agar slants. After 24-48 hours incubation, subject the cultures to a catalase test. Catalase activity indicates the nonstreptococci. Atypical colonies should also be verified to determine false negative reactions on the membrane filter. Final confirmation of fecal streptococci is achieved by determining growth of catalase negative isolates in BHI broth at 45 C and in 40% bile within two days (see Figure III-D-1).

3.1 Apparatus and Materials

3.1.1 Incubators set at 35 ± 0.5 C and 45 ± 0.5 C.

3.1.2 Inoculating needle and loop.

3.1.3 Bunsen/Fisher burner or electric incinerator.

3.1.4 Solution of 3% hydrogen peroxide.

3.1.5 Glass microscope slides, 2.5 × 7.6 cm (1 × 3 inches).

3.1.6 Media

(a) Brain heart infusion (BHI) agar. (See Part II-B, 5.4.6).

(b) Brain heart infusion (BHI) broth. (See Part II-B, 5.4.5).

3.2 Procedure

3.2.1 Plates for verification should contain 20-100 colonies. Pick at least 10 typical colonies from the selected membrane or agar plate and inoculate into a BHI agar slant and into a BHI broth tube.

3.2.2 After 24-48 hours incubation at 35 ±0.5 C, transfer a loopful of growth from the BHI slant to a clean glass slide and add a few drops of freshly tested 3% hydrogen peroxide (H_2O_2) to the smear. If the catalase enzyme is present, it cleaves the H_2O_2 to water and visible oxygen gas. Bubbles constitute a positive catalase test and indicate non-streptococcal species. Confirmation need not be continued. Use a platinum loop, not nichrome, to avoid false positive reactions.

3.2.3 If a negative catalase reaction occurs, transfer a loopful of growth from the BHI broth to fresh BHI broth and BHI broth + 40% bile and incubate at 45 C and at 35 C. Growth within two days indicates fecal streptococcal species (see Figure III-D-1 and III-D-2).

3.2.4 Further identification of streptococcal types present in the sample may be obtained by biochemical characterization. (See Figure III-D-2 to III-D-4 for identification of fecal streptococci. Such information is useful for investigating sources of pollution. See Part II-B for preparation of media used in the schematic outlines).

4. Most Probable Number Method

4.1 Summary: The multiple-tube procedure estimates the number of fecal streptococci by inoculating decimal dilutions of the sample into broth tube media. Positive tubes in the Presumptive Test are indicated by growth (turbidity) in azide dextrose broth after incubation at 35 C for 24–48 hours. To confirm the presence of fecal streptococci, a portion of the growth from each positive azide dextrose broth tube is streaked onto PSE or equivalent esculin-azide agar and incubated at 35 C for 24 hours (5). The presence of brownish-black colonies with brown halos confirms fecal streptococci. The MPN is computed on the basis of the Confirmed Test results read from and MPN table.

4.2 Scope and Application: This method can be used for detection of fecal streptococci in water, sewage or feces, but is more time-consuming, less convenient and less direct than the other procedures. The MPN must be used for samples which cannot be examined by the MF or direct plating techniques because of turbidity, high numbers of background bacteria, metallic compounds, the presence of coagulants, the chlorination of sewage effluents or sample volume limitations of the plating technique.

4.3 Apparatus and Materials

4.3.1 Water bath or air incubator set at 35 ± 0.5 C. Temperature checked with an NBS thermometer or one of equivalent accuracy.

4.3.2 Pipet containers of stainless steel, aluminum or pyrex glass for pipets.

4.3.3 Culture tube racks to hold fifty 25 mm diameter tubes.

4.3.4 Sterile T.D. bacteriological or Mohr pipets, of appropriate sizes.

4.3.5 Dilution bottles (milk dilution), pyrex glass, 99 ml volume, screw-capped, with neoprene rubber liner.

4.3.6 Test tubes, pyrex, culture 150 × 25 or 150 × 20 mm, with caps.

4.3.7 Inoculating loop, 3 mm diameter, in holder or disposable applicator sticks or loops.

4.3.8 Bunsen/Fisher burners or electric incinerator.

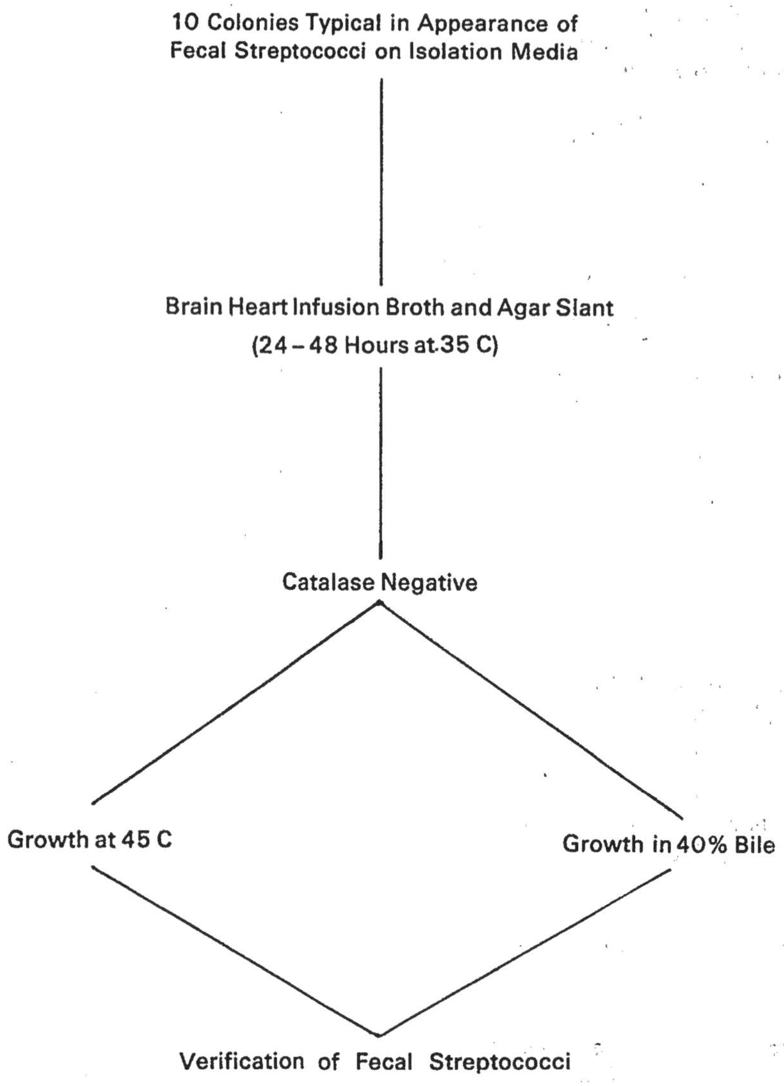

10 Colonies Typical in Appearance of
Fecal Streptococci on Isolation Media

Brain Heart Infusion Broth and Agar Slant
(24 – 48 Hours at 35 C)

Catalase Negative

Growth at 45 C

Growth in 40% Bile

Verification of Fecal Streptococci

FIGURE III-D-1. Verification Procedure for Fecal Streptococci.

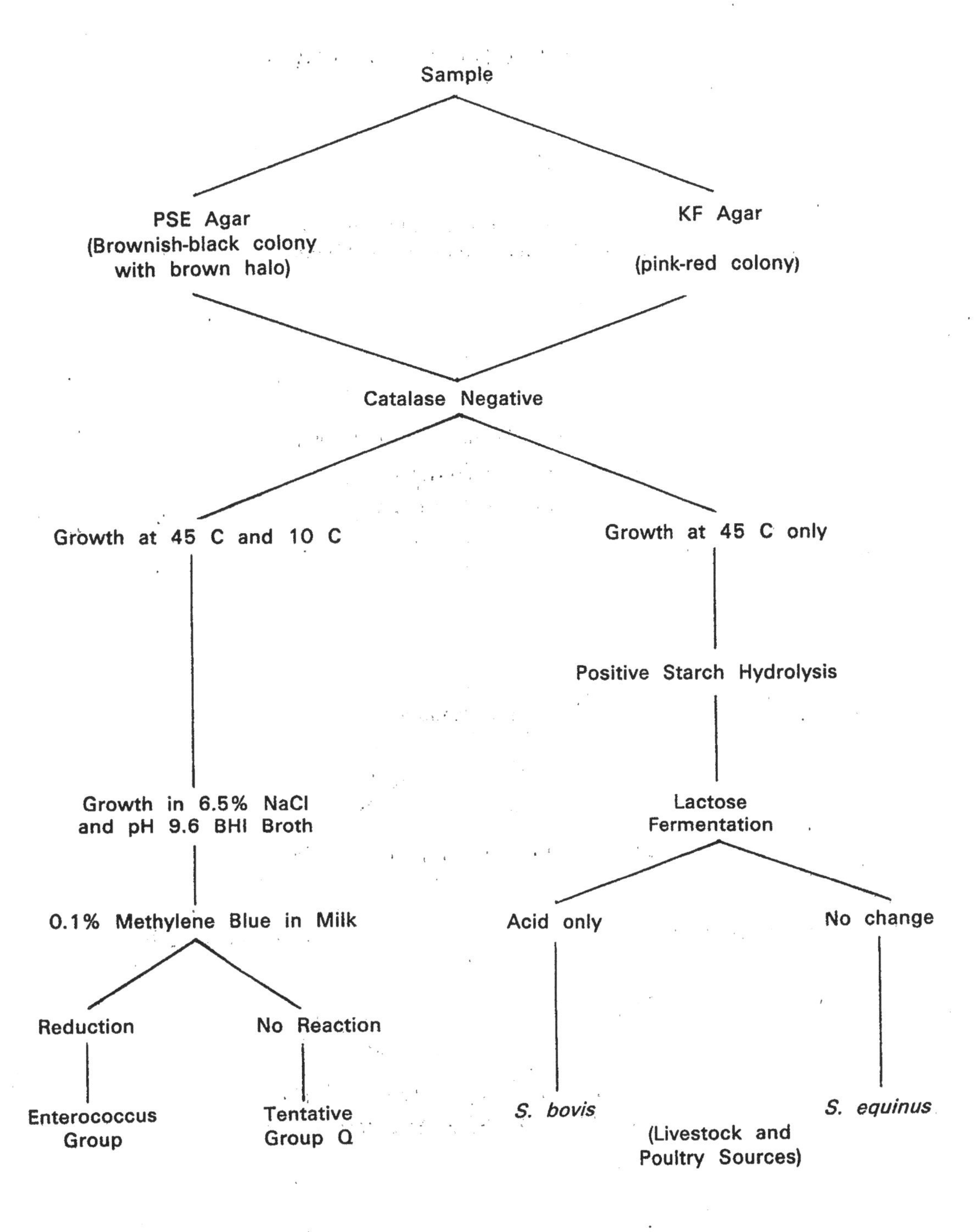

FIGURE III-D-2. Isolation and Identification of Fecal Streptococci, General Scheme

4.4 Media

4.4.1 Azide dextrose broth prepared in 10 ml volumes in test tubes without fermentation vial (see Part II-B, 5.4.2 for preparation). Ethyl violet azide broth is not used because of false positive reactions.

4.4.1 Azide dextrose broth prepared in 10 ml volumes in test tubes without fermentation vial (see Part II-B, 5.4.2 for preparation).

4.4.2 Pfizer Selective Enterococcus (PSE) or equivalent esculin-azide agar in pour plates (see Part II-B, 5.4.4 for preparation).

4.5 Dilution Water: Sterile dilution water dispensed in 99 ml $\pm$ 2 ml amounts in screw-capped bottles.

4.6 Procedure: The general MPN procedure is described in detail in Part II-C, 4.

4.6.1 Prepare the media for the Presumptive Test, (azide dextrose broth) and the Confirmed Test, (PSE Agar plates). (see Part II-B, 5.4.2 and 5.4.4 respectively).

4.6.2 Mark culture tubes to identify samples and sample volumes.

4.6.3 Shake the sample vigorously about 25 times.

4.6.4 Inoculate the azide dextrose broth with appropriate sample volumes for the Presumptive Test. The number of fecal streptococci in a water polluted with municipal wastes is generally lower than the number of coliforms. Therefore, larger sample volumes must be used to inoculate the MPN tubes for fecal streptococci than for coliforms. For example, if sample volumes of 1.0, 0.1, 0.01, and 0.001 ml are used for the coliform test, a series of 10, 1.0, 0.1, and 0.01 ml volumes are inoculated for the fecal streptococci test. Use single-strength broth, 10 ml tubes for inocula of 1.0 ml or less, and double-strength broth, 10 ml tubes for inocula of 10 ml. Sample volumes from feedlots, meat packing plants, and stormwater run-off with more fecal streptococci than coliforms must be adjusted accordingly.

4.6.5 Shake the rack of inoculated culture tubes to mix well and incubate them at 35 $\pm$ 0.5 C for 48 hours $\pm$ 3 hours. Examine tubes for turbidity after 24 $\pm$ 2 hours and 48 $\pm$ 3 hours.

4.6.6 Read and record the results from each tube. A positive Presumptive Test shows growth consisting of turbidity in the medium or a button of sediment at the bottom of the culture tube, or both.

4.6.7 For the Confirmed Test, streak growth from each positive azide dextrose broth tube onto PSE Agar plates, making certain that the label on the plate corresponds to the positive azide dextrose tube used.

4.6.8 Incubate the PSE Agar plates at 35 $\pm$ 0.5 C for 24 hours.

4.6.9 Read and record the results of each plate corresponding to the positive azide dextrose tube. A positive Confirmed Test is evidenced by the presence of brownish-black colonies with brown halos. The number of positive confirmed azide tubes in each dilution is used to compute the density from an MPN table.

4.7 Calculations

4.7.1 Calculate fecal streptococci densities on the basis of the number of positive Confirmed Tests from the PSE agar plates, using the Table of Most Probable Numbers (MPN) in Table II-C-1.

4.7.2 The MPN results are computed from 3 usable Confirmed Test dilutions. For example, if positive Confirmatory Test results are obtained from 5 of the 10 ml portions, three of the 1.0 ml portions, and none of the 0.1 ml portions, the coded results of the test is 5–3–0. The code is located in the MPN Table II-C-4, and the MPN per 100 ml is recorded. See Part II-C, 4 for details.

4.8 Precision and Accuracy: The precision of the MPN value increases with increased numbers of replicates tested. Five tubes are recommended for each dilution.

5. Pour Plate Method

5.1 Summary: Aliquots of the water sample or diluted sample are delivered to the bottom of a petri dish, and liquified Pfizer Selective Enterococcus (PSE) agar, equivalent esculin-azide agar or KF agar is added and thoroughly mixed with the water sample. Fecal streptococci on PSE agar are 1 mm in diameter and brownish-black with brown halos after 18–24 hours at 35 C. On KF agar fecal streptococci are red or pink after 48 hours at 35 C.

5.2 Scope and Application: The pour plate method is recommended as an alternate procedure to the MF technique when chlorinated sewage effluent and water samples with high turbidity are encountered. PSE agar, the medium of choice, has several advantages: (1) it requires only 24 hours incubation compared to 48 hours for other media, and (2) it exhibits consistent recovery, regardless of sources.

With the pour plate technique, only small volumes of sample may be analyzed. This is a disadvantage when the fecal streptococcal density is low and a large volume of sample would be required for an accurate density determination. Consequently, the MF technique should be used unless the water is so turbid that filtration is impossible.

5.3 Apparatus and Materials

5.3.1 Air incubator set at 35 $\pm$ 0.5 C. Temperature is checked against a National Bureau of Standards thermometer or one of equivalent accuracy.

5.3.2 Water bath for tempering agars set at 44–46 C.

5.3.3 Colony Counter, Quebec darkfield model or equivalent.

5.3.4 Pipet containers of stainless steel, aluminum or pyrex glass for glass pipets.

5.3.5 Petri dish containers for glass or plastic petri dishes.

5.3.6 Sterile T.D. bacteriological or Mohr pipets of appropriate sizes.

5.3.7 Sterile 100 mm $\times$ 15 mm petri dishes, glass or plastic.

5.3.8 Dilution bottles (milk dilution), pyrex, 99 ml volume, screw-capped, with neoprene rubber liners.

5.3.9 Bunsen/Fisher burner or electric incinerator.

5.3.10 Hand tally.

5.4 Media: Sterile Pfizer Selective Enterococcus agar (PSE), equivalent esculin-azide agar, (Part II-B, 5.4.4) or KF Streptococcus agar (Part II-B, 5.4.1) are prepared in pre-sterilized erlenmeyer flasks or bottles with metal foil covers, or screw-caps.

5.5 Dilution Water: Sterile dilution water dispensed in 99 $\pm$ 2 ml amounts preferably in screw-capped dilution bottles (see Part II-B, 7).

5.6 Procedure

5.6.1 Shake the sample bottle vigorously about 25 times to disperse the bacteria. Take care that the closure is tight to prevent leakage of sample during shaking.

5.6.2 Dilute the sample to obtain final plate counts between 30–300 colonies. The number of colonies within this range gives the most accurate estimation of the microbial population. Because the magnitude of the microbial population in the original water sample is not known beforehand, a range of dilutions must be prepared and plated to obtain a plate within this range of colony counts.

5.6.3 Transfer 0.1 and 1.0 ml from the undiluted sample to each of 2 separate petri dishes.

5.6.4 Prepare the initial 1:100 or 10^{-2} dilution by pipetting 1 ml of the sample into a 99 ml dilution water blank using a sterile 1.1 ml pipet (see Part II-C, 1.4 "Preparation of Dilutions," and Figure II-C-1).

5.6.5 Vigorously shake the 1:100 dilution bottle to obtain uniform distribution of bacteria.

5.6.6 Pipet 0.1 and 1.0 ml of this 1:100 dilution into each of 2 separate petri dishes using another sterile pipet.

5.6.7 Make additional dilutions as required for raw wastes or stormwater run-off and prepare pour plates containing the dilution aliquots.

5.6.8 Prepare duplicate petri dishes for each sample increment (Figure III-A-1). Mark each petri dish with the number of the sample, the dilution, the date, and any other necessary information. Deliver the liquid into the dish, and touch the tip once against a dry area in the petri dish bottom while holding the pipet vertically.

5.6.9 Pour 12–15 ml of liquified cooled agar medium into each petri dish containing the sample or its dilution. Mix the medium and the sample thoroughly by gently rotating and tilting the petri dish. Not more than 20 minutes should elapse between dilution, plating, and addition of the medium. Refer to Part II-C, 2.6, for further information.

5.6.10 Allow agar to solidify as rapidly as possible after pouring, and place the inverted PSE plates at 35 ± 0.5 C for 18-24 hours and KF plates at 48 ± 3 hours.

5.7 Counting and Recording Colonies

5.7.1 After the specified incubation period, select those plates with 30-300 fecal streptococcal colonies. Fecal streptococci on PSE agar are brownish-black colonies, about 1 mm in diameter with brown halos. On KF agar, fecal streptococci are pink to red and of varying sizes.

5.7.2 Count colonies in the plates with the aid of a colony counter (10-15× magnification) equipped with a grid.

5.7.3 Observe the following rules for reporting the fecal streptococcal plate counts.

(a) Plates with 30–300 Fecal Streptococcal Colonies: Count all colonies for each plate within the 30–300 range. Calculate the average count for these plates correcting for the dilution as follows:

$$\frac{\text{Sum of Colonies}}{\text{Sum of Volumes tested, ml}} \times 100 = \text{FS/100 ml}$$

(b) All Plates Greater than 300 Colonies: When counts for all dilutions contain more than 300 colonies, e.g., >500 for 1.0 ml, >500 for 0.1 ml, and 340 for 0.01 ml; compute the density by counting the plate having nearest to 300 colonies. In this case use the 0:01 ml plate.

$$\frac{340}{0.01} \times 100 = 3{,}400{,}000$$

Report as: Estimated Fecal Streptococcal Count, 3,400,000/100 ml

(c) All Plates with Fewer than 30 Colonies: If all plates are less than 30 colonies, record the actual number of colonies on the lowest dilution plated and report the count as the Estimated Fecal Streptococcal Plate Count per 100 ml.

(d) Plate with No Colonies: If plates from all dilutions show no colonies, assume a count of one (1) colony; then divide 1 by the largest volume filtered and report the value as a less than (<) count. For example, if 0.1, 0.01 and 0.001 ml were filtered with no reported colonies, the count would be:

$$\frac{1}{0.1} \times 100 = < 1000$$

Report the count as: < 10/100 ml

(e) When all plates are crowded, it is possible to use the square divisions of the grid on the Quebec or similar counter to estimate the numbers of colonies on the plate. See Part III-A, 5.6.3 for details.

5.8 Precision and Accuracy: Replicate plate counts from the same sample deviate because of errors introduced from a variety of sources.

Prescott et al. (6) reported for the Standard Plate Count (but applicable here) that the standard deviation of individual counts from 30–300 will vary from 0–20%. This plating error was 10% for higher plate counts within the 100–300 range. The authors pointed out that a dilution error of about 3% for each dilution stage is incurred in addition to the plating error. Therefore, large variations can be expected from high density samples such as sewage from which several dilutions are made.

Laboratory personnel should be able to duplicate their plate count values for the same plate within 5%, and the counts of others within 10%.

6. Determination of Fecal Coliform/Fecal Streptococcus Ratios (FC/FS)

The relationship of fecal coliform to fecal streptococcus density may provide information on the potential source(s) of contamination. Estimated per capita contributions of indicator bacteria for animals were used to develop FC/FS ratios (7, 8). These ratios are as follows:

FC/FS	Ratios
Man	4.4
Duck	0.6
Sheep	0.4
Chicken	0.4
Pig	0.4
Cow	0.2
Turkey	0.1

From the data, it was reasoned that ratios greater than 4:1 were indicative of pollution derived from domestic wastes composed of man's body wastes. Ratios of less than 0.7 suggested that contamination originated from livestock and poultry wastes, milk and food processing wastes or from stormwater run-off (non-human source). Further speciation of the fecal streptococci provides more specific source information. There are several precautions to be observed when ratios are being used.

(a) Bacterial densities can be altered drastically when the pH of the sample is below 4.0 or above 9.0.

(b) Due to limited survival capability of some of the fecal streptococci, it is essential to sample close to the pollution source to obtain reliable ratios. This is especially true for the highly sensitive *S. bovis* and *S. equinus* species.

(c) It is difficult to use ratios effectively when mixed pollution sources are present.

(d) In marine waters, bays, estuaries, and irrigation returns, FC/FS ratios have been of limited value in accurately defining major pollutional sources.

(e) If fecal streptococcal counts are $\leq$ 100/100 ml, ratios should not be applied.

7. Identification of Fecal Streptococci to Species

7.1 Summary: Although the fecal streptococci are enumerated as described in the previous sections and are verified with simple biochemical tests in Part III-D, 3 above, it is important at times to identify the fecal streptococci to species to further verify animal and human sources of pollution and to determine the sanitary significance of isolates. This identification to species is performed using the additional biochemical tests described to differentiate and confirm the Group Q streptococci, the *bovis-equinus* Group and the enterococci. The enterococci can be separated as *S. faecium* and *S. faecalis* varieties or into groups according to original source.

7.2 Scope and Application: The initial biochemical test confirms that the isolates are fecal streptococci by negative catalase reaction. Group Q streptococci and enterococci are separated from the bovis-equinus group by positive growth at 10 and 45 C and are then verified by growth in 6.5% NaCl and at pH 9.6. The enterococci are séparated from the Group Q streptococci by reduction of methylene blue milk. Subsequently, the enterococci are either speciated or separated by origin using additional biochemical tests. The *Streptococcus bovis* and *equinus* are verified by hydrolysis of starch and separated by lactose fermentation. These tests require specific training for valid results.

7.3 Apparatus and Materials

7.3.1 Incubators set at 10 ± 0.5 C, 35 ± 0.5 C, and 45 ± 0.5 C (water baths recommended for 10 and 45 C). Temperatures checked with a National Bureau of Standards thermometer or one of equivalent accuracy.

7.3.2 Pipet containers of stainless steel, aluminum or pyrex glass for glass pipets.

7.3.3 Sterile T.D. bacteriological or Mohr pipets, of appropriate sizes.

7.3.4 Sterile petri dishes 100 × 15 mm or 60 × 15 mm, glass or plastic.

7.3.5 Dilution bottles (milk dilution), pyrex, 99 ml volume, screw cap, with neoprene rubber liners.

7.3.6 Inoculation loop, 3 mm diameter, or needle.

7.3.7 Bunsen/Fisher burner or electric incinerator.

7.3.8 Media

(a) Brain heart infusion (BHI) broth. (Part II-B, 5.4.5).

(b) Brain heart infusion (BHI) agar. (Part II-B, 5.4.6).

(c) Brain heart infusion (BHI) broth with 6.5% NaCl. (Part II-B, 5.4.7).

(d) Brain heart infusion (BHI) broth, pH 9.6. (Part II-B, 5.4.8).

(e) Brain heart infusion (BHI) broth with 40% bile. (Part II-B, 5.4.9).

(f) Starch agar plates. (Part II-B, 5.4.10).

(g) Starch liquid medium. (Part II-B, 5.4.11).

(h) Nutrient gelatin. (Part II-B, 5.4.12).

(i) Litmus milk. (Part II-B, 5.4.13).

(j) Skim milk with 0.1% methylene blue. (Part II-B, 5.4.14).

(k) Potassium tellurite in brain heart infusion. (Part II-B, 5.4.15).

(1) Potassium tellurite in blood agar (optional). (Part II-B, 5.4.16).

(m) Tetrazolium glucose (TG) agar or 2, 3, 5-triphenyl tetrazolium chloride (TTC) agar. (Part II-B, 5.4.17).

(n) Blood agar with 10% blood. (Part II-B, 5.4.18).

(o) 1% D-sorbitol solution in purple broth base. (Part II-B, 5.1.7).

(p) 1% glycerol in purple broth base. (Part II-B, 5.1.7).

(q) 1% L-arabinose solution in purple broth base. (Part II-B, 5.1.7).

(r) 1% lactose solution in purple broth base. (Part II-B, 5.1.8).

(s) 1% sorbose solution in purple broth base. (Part II-B, 5.1.8).

(t) 1% sorbose solution in purple broth base at pH 10. (Part II-B, 5.1.8).

7.4 Procedure: Follow the schematic outlines in Figures III-D-2 to 4 for the identification of fecal streptococcal species.

7.4.1 Isolation and Confirmation of Fecal Streptococci (Figure III-D-2)

(a) Pick colonies typical of fecal streptococci from the membranes or the pour plates, and inoculate them onto BHI agar slants and into BHI broth tubes.

(b) After 24–48 hours incubation at 35 ± 0.5 C, transfer a loopful of growth from the BHI slant to a clean glass slide, and add a few drops of freshly-tested 3% hydrogen peroxide to the smear. A positive control such as staphylococcus and a negative control such as *S. faecalis* should be used for testing the 3% H_2O_2.

(c) If the catalase enzyme is present, it cleaves the H_2O_2 to produce water and visible oxygen gas bubbles. The presence of bubbles constitutes a positive catalase test that indicates non-streptococcal species. Verification need not be continued.

(d) If the catalase test is negative, a separation of the enterococcus and Q group organisms from *S. bovis*, and *S. equinus* can be made by testing for growth at 10 and 45 C.

7.4.2 Separation of Enterococci and Group Q Streptococci (Figure III-D-2)

(a) Transfer 1 drop of the growth from the BHI broth tube from 7.4.1 to each of 2 BHI broth tubes.

(b) Place 1 tube in a 45±0.5 C water bath and observe for growth (as evidenced by turbidity) within 2 days. Place the other tube in a 10 ± 0.5 C water bath and check for growth within 5 days. Growth at 10 and 45 C indicates that the culture is a potential member of the enterococcus or Q groups. On the other hand, *S. equinus* and *S. bovis* exhibit growth at 45 C but not at 10 C. (See 7.4.7 for these speciations).

7.4.3 Confirmation of Enterococcus Group (Figure III-D-2)

This is done by testing for growth in 6.5% NaCl and at pH 9.6 in BHI broth and observing for reduction of 0.1% methylene blue in milk. Positive reactions in all cases confirm the presence of the enterococcus group. Positive reactions in 6.5% NaCl and pH 9.6 BHI broths and no reaction in 0.1% methylene blue indicate the tentative identification of group Q.

(a) Growth Test in 6.5% NaCl-BHI Broth: Transfer 1 drop of 24 hour BHI broth culture to a tube of BHI broth containing 6.5% sodium chloride. Incubate at 35 ± 0.5 C, and check for growth as evidenced by turbidity within a 3–7 day period. Growth is a positive test.

(b) Growth Test in BHI Broth of pH 9.6: Transfer 1 drop of the 24 hour BHI broth culture to a tube of BHI broth adjusted to pH 9.6. Incubate at 35 ± 0.5 C, and check for growth as evidenced by turbidity at 1, 2, 3, and 7 days. Growth is a positive test.

(c) Reduction of 0.1% Methylene Blue in Skim Milk: Transfer 1 drop of the 24 hour BHI broth culture to a tube of sterile skim milk containing 0.1% methylene blue. Positive reduction of methylene blue is evidenced by the color change from blue to white.

7.4.4 Separation of Enterococcus Group by Species (Figure III-D-3)

The enterococci can be separated into species as described in 7.4.4 or into groups by original source as described in 7.4.5. The enterococcus group can be separated into species by observing the reduction of potassium tellurite and 2, 3, 5-triphenyl tetrazolium chloride and the fermentation of D-sorbitol and glycerol.

(a) Streak the 24 hour BHI broth culture onto an agar plate containing a final concentration of 0.4% potassium tellurite. Invert the plates, incubate at 35 ± 0.5 C, and observe the plates each day for 7 days. Colonies reducing the potassium tellurite will appear black on this medium.

(b) Streak the 24 hour BHI broth culture onto tetrazolium glucose agar (TG). Reduction of tetrazolium to formazin is observed after 48 hours at 35 ± 0.5 C. The degrees of reduction are indicated as follows:

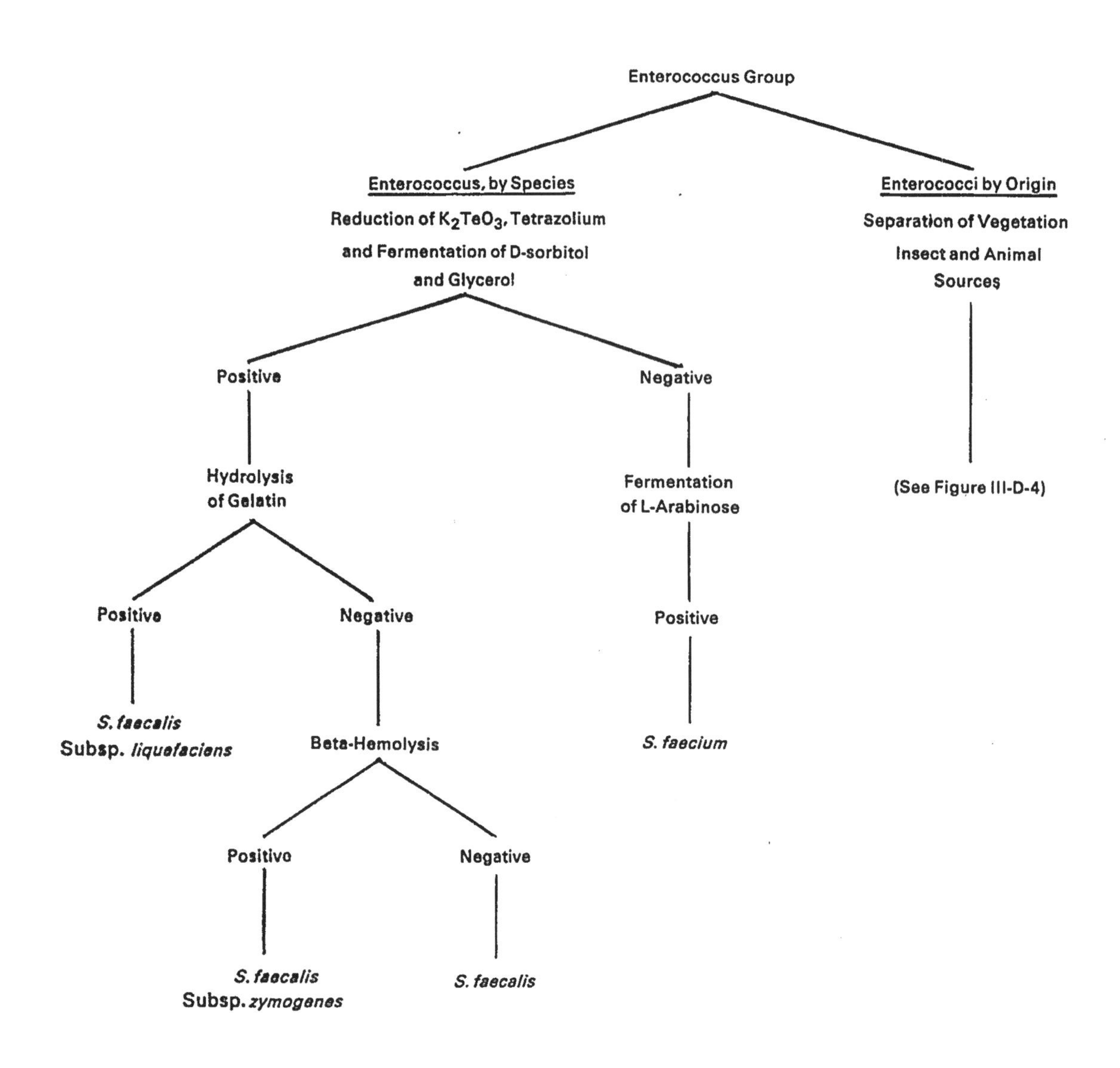

FIGURE III-D-3. Identification of Fecal Streptococci, Separation of Enterococcus Group by Species and by Original Source of Culture.

Strong Reduction (+4 to +3) = Red centered colony.
Moderate Reduction (+2) = Pink centered colony.
Weak Reduction (+1) = Pale pink centered colony.
No reduction (0) = White colony.

(c) Transfer a small amount of growth from each 24 hour BHI agar slant culture to separate purple broth base tubes containing 1% D-sorbitol and 1% glycerol. Be careful not to penetrate the agar slant and carry over agar to the carbohydrate medium. Incubate the inoculated carbohydrate media at 35 ± 0.5 C, and note acid production over a 4 day incubation period. A negative reaction in D-sorbitol or glycerol shows that *S. faecium* is present. The indicator will be unchanged (purple in color).

(d) Reduction of potassium tellurite (black colonies), reduction of tetrazolium to formazan on TTC agar and the fermentation of glycerol and D-sorbitol indicate *S. faecalis* and its varieties. The determination of *S. faecalis* subspecies is performed as follows:

(1) Stab-inoculate gelatin with a small amount of growth from a 24 hour BHI agar slant. Incubate the culture at 35 ± 0.5 C for 2–14 days, according to the rate of growth. After incubation, place the tube in a cold water bath or refrigerator 15–30 minutes to determine whether or not the gelatin will still solidify. Uninoculated controls must be run in parallel, especially when prolonged incubation periods are encountered. Liquefaction dictates *S. faecalis* subsp. *liquefaciens*, whereas solidification indicates *S. faecalis* or *S. faecalis* subsp. *zymogenes*. The latter 2 strains are separated by their hemolysis reactions.

(2) The hemolytic properties of the fecal streptococci are determined by streak or pour plates. Melt blood agar base, cool and add 10% sheep blood. Inoculate streak plates or prepare pour plates and incubate for 48 hours. After incubation, read plates. Overnight refrigeration may enhance the hemolytic reactions. Hemolysis is classified as 3 types:

Alpha-Hemolysis – Some streptococci partially lyse red blood cells and reduce hemoglobin to methemoglobin producing a discoloration of the red blood cells. This appears as greenish zones around the colonies.

Beta Hemolysis – Enzymes of fecal streptococci completely lyse red blood cells producing yellowish hue, or clear, colorless zone in the blood agar surrounding the colony. *S. faecalis* subsp. *zymogenes* demonstrates beta hemolysis of the blood.

Gamma-Hemolysis – Some streptococci produce no hemolysis which is designated as gamma hemolysis, *S. faecalis* is alpha or gamma hemolytic.

It is important to note that upon serial transfer of the culture in the laboratory, hemolysis may be lost. This is especially true for *S. faecalis* subsp. *zymogenes* where beta hemolysis may not occur after serial transfer.

(e) Negative reactions in potassium tellurite and tetrazolium media and failure to ferment D-sorbitol and glycerol indicate *S. faecium.*

(f) Transfer a small amount of growth from the 24 hour BHI agar slant to a purple broth base tube containing 1% L-arabinose. Be careful not to penetrate the agar slant surface and carry over agar to the carbohydrate medium. Incubate the inoculated carbohydrate medium at 35 ± 0.5 C, and note acid production over a 4 day incubation period. If *S. faecium* is present, the L-arabinose is fermented, acid will be produced and the medium will turn yellow.

7.4.5 Separating Enterococci by Origin (Vegetation, Insect and Animal Sources)

In contrast to the separation of enterococci by species described in 7.4.4, the members of the enterococcus group can be separated according to original source of culture. The starch hydrolysis tests separate the enterococci originating on vegetation from those typically found in insects and animals; the peptonization of litmus milk test separates the insect-origin from warm-blooded animal-source enterococci, as shown in Figure III-D-4.

S. faecalis from vegetation hydrolyzes starch while *S. faecalis* subsp. *liquefaciens* from insects and enterococci from warm-blooded animals do not. The starch hydrolysis test can be performed satisfactorily with starch agar plates or in a liquid starch tube medium (9).

(a) Starch Hydrolysis Plate Test

(1) Pour the starch agar medium into petri dishes and allow to solidify. With a wax pencil divide the bottoms of the petri dishes into 6 individual areas to allow the testing of 6 isolates.

(2) Streak each isolate from a 24 hour BHI agar slant onto one of the areas and incubate at 35 ± 0.5 C for 24 hours.

(3) After incubation, flood the agar medium with Lugol's iodine solution (Part II-C, 5.3). Streaks showing clear hydrolytic zones with absence of the usual blue-black color of the starch-iodine complex in the zones surrounding the colonies are considered positive.

(b) Starch Hydrolysis Tube Test

(1) The starch test may also be performed using a liquid medium instead of starch agar plates. Inoculate tubes of sterile liquid medium with the test organisms. After 18 hours of incubation at 35 ± 0.5 C, test the tubes for starch hydrolysis using a modification of the iodine test (8). For this modification, 0.2 ml of 2% FeCl solution and 0.2 ml of Lugol's iodine solution are added to 5 ml of the inoculated medium and to 5 ml of the uninoculated medium (control).

(2) Hold the tubes for 3 hours at room temperature. Compare the inoculated tubes to the control tube. The control tube (negative test) maintains a violet color, but positive test cultures hydrolyze the starch and decolorize the medium to a reddish-violet hue.

(c) Peptonization of Milk: Peptonization in litmus milk is used to separate *S. faecalis* subsp. *liquefaciens* (from insect sources) from enterococci derived from warm-blooded animal sources.

(1) Transfer 1 drop of the isolate growing in a 24 hour BHI broth culture to a tube of sterile litmus milk.

(2) Incubate the tube at 35 ± 0.5 C, and observe at 1, 2, 3 and 7 days.

(3) A positive peptonization includes liquifaction and clearing of the milk with the development of a brownish color. Peptonization indicates *S. faecalis* subsp. *liquefaciens.* Color changes without clearing, or coagulation are negative reactions and indicate enterococci from warm-blooded animal sources are present.

7.4.6 Identification of Group Q Streptococci

The group Q Streptococci are initially separated from the Enterococcus Group and tentatively identified by growth in 6.5% NaCl in BHI and in BHI broth at pH 9.6, but no growth in 0.1% methylene blue in milk. See Figure III-D-2.

Other pertinent physiological characteristics of Group Q Streptococci are: fermentation of sorbose, growth in sorbose medium at pH 10, but no growth in tellurite medium and negative hydrolysis of starch and gelatin. The tests may be carried out as follows:

(a) Inoculate a sorbose fermentation tube with a small amount of growth from an isolate on a 24 hour BHI slant. Be careful not to penetrate the agar slant surface and carry over agar to the carbohydrate medium. Incubate the carbohydrate medium at 35 ± 0.5 C and note acid production over a 4 day incubation period. Sorbose fermentation is indicated by acid production and change in the medium color from purple to yellow.

(b) Transfer a small amount of growth from the 24 hour BHI slant to the 1% sorbose me-

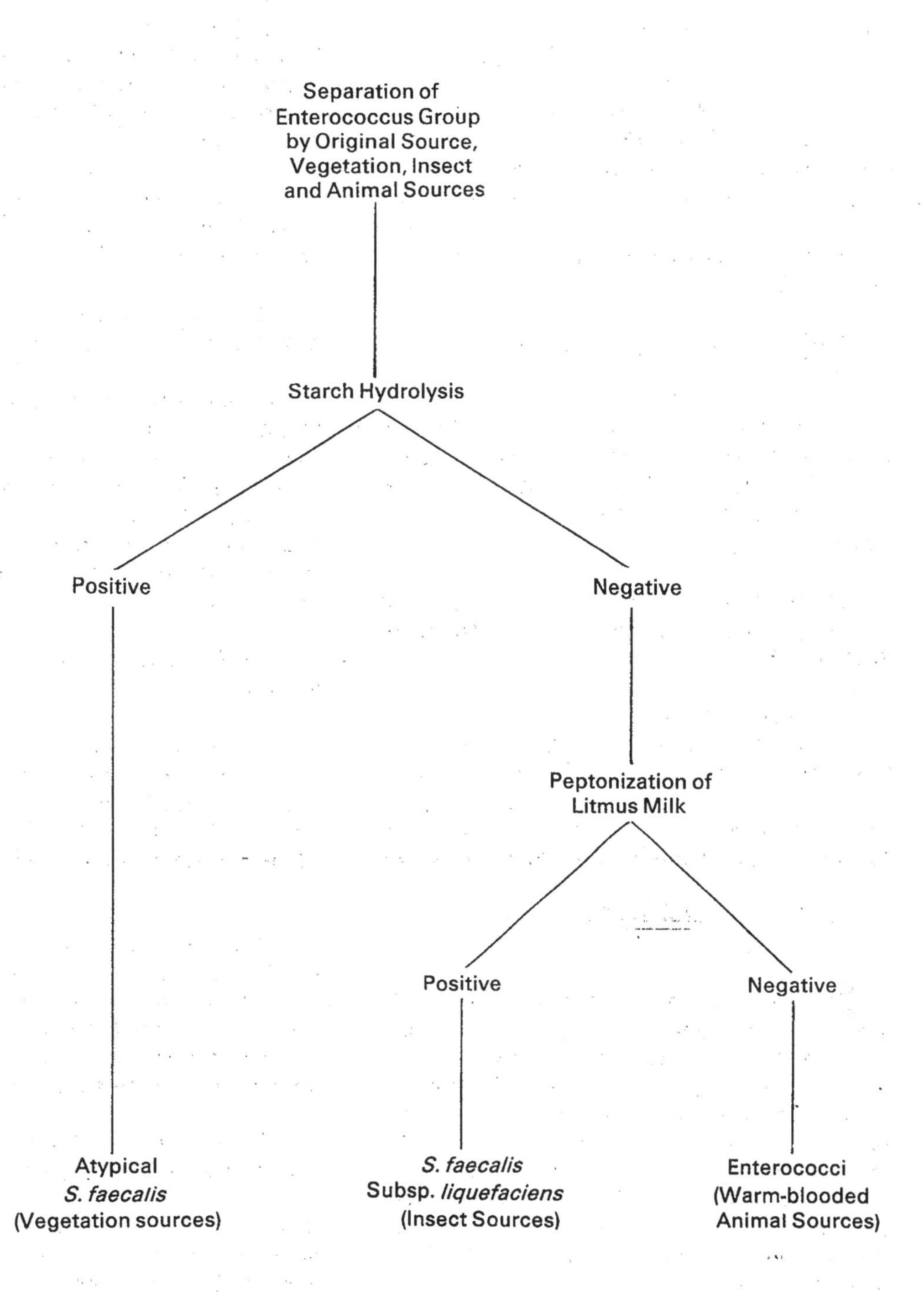

FIGURE III-D-4. Identification of Fecal Streptococci, Separation of Enterococci from Vegetation, Insect and Animal Sources.

dium (described in Part II-B, 5.1.8 above) which has been adjusted to pH 10 with sterile 38% sodium phosphate solution (see Part II-B, 5.4.8). Incubate the inoculated medium at 35 ± 0.5 C for at least 4 days. Growth indicates a positive test.

(c) Inoculate the growth from a 24 hour BHI slant into BHI broth containing 0.04% potassium tellurite and incubate at 35 ± 0.5 C for 7 days. Group Q organisms do not grow on this medium.

(d) Stab-inoculate gelatin with a small amount of growth from the 24 hour BHI slant. Incubate the culture at 35 ± 0.5 C for 2–14 days, according to the rate of growth. After incubation, place the tube in a cold water bath or refrigerator for 15–30 minutes to determine whether or not the gelatin will solidify. Uninoculated control must be done in parallel, especially when prolonged incubation periods are used.

(e) Group Q organisms do not hydrolyze starch. Perform starch hydrolysis test as in 7.4.5 (a) or (b) above.

(f) *Streptococcus avium* sp. constitute Lancefield's Group Q. Consequently, Group Q antiserum may be used in the precipitin test to provide further identification. However, the Q antigen is not demonstrable in all strains, therefore, identification in those cases will depend solely on physiological characteristics.

7.4.7 Separation and Speciation of *S. bovis* and *S. equinus*

(a) *S. bovis* and *S. equinus* were separated from the Enterococci and Group Q Streptococci by growth at 45 C but no growth at 10 C (see Figure III-D-2). *S. bovis* and *S. equinus* can be separated by the lactose fermentation test in which *S. bovis* produces acid and *S. equinus* produces no change, (c) below.

(b) Starch Hydrolysis Test: Perform starch hydrolysis test as in 7.4.5 (a) or (b).

Positive starch hydrolysis test confirms *S. bovis* and *S. equinus.*

(c) Lactose Fermentation Test: To differentiate between *S. bovis* and *S. equinus* by lactose fermentation, transfer a small amount of growth from the 24 hour BHI agar slant to a purple broth base tube containing 1% lactose and an inverted fermentation tube. Do not penetrate the agar slant surface and carry over agar to the carbohydrate medium. Incubate the inoculated carbohydrate medium at 35 ± 0.5 C and observe the reaction over a 4 day incubation period. *S. bovis* gives an acid reaction only; *S. equinus* shows no change.

REFERENCES

1. Nowlan, Sandra and R. H. Diebel, 1967. Group Q streptococci I. Ecology, serology, physiology and relationship to established enterococci. J. Bacteriol. 94, (2):291.

2. Kenner, B. A., H. F. Clark and P. W. Kabler, 1960. Fecal streptococci. II. Quantification of streptococci in feces. Am. J. Public Health. 50:1553.

3. Kenner, B. A., H. F. Clark and P. W. Kabler, 1961. Fecal streptococci. I. Cultivation and enumeration of streptococci in surface water. Appl. Microbiol. 9:15.

4. Pavlova, M. T., F. T. Brezenski and W. Litsky, 1972. Evaluation of various media for isolation, enumeration and identification of fecal streptococci from natural sources. Health Lab. Sci. 9:289.

5. Clausen, E. M., B. L. Green and W. Litsky, 1977. Fecal Streptococci: Indicators of Pollution, pp. 247–264. *In:* A. W. Hoadley and B. J. Dutka, Eds., Bacterial Indicators/Health Hazards Associated with Water, ASTM STP635, American Society for Testing and Materials, Philadelphia, PA.

6. Prescott, S. C., C-E. A. Winslow and M. H. McCrady, 1946. Water Bacteriology. (6th ed.) John Wiley and Sons, Inc., p. 46–50.

7. Geldreich, E. E., H. F. Clark and C. B. Huff, 1964. A study of pollution indicators in a waste stabilization pond. J. WPCF, 36 (11): 1372.

8. Geldreich, E. E., 1976. Fecal coliform and fecal streptococcus density relationships in waste discharges and receiving waters. *In:* CRC Critical Reviews in Environmental Control. p. 349.

9. Pavlova, M. T., W. Litsky and F. J. Francis, 1971. A comparative study of starch hydrolysis by fecal streptococci employing plate and tube techniques. Health Lab. Sci. 8:67.

PART III. ANALYTICAL METHODOLOGY

Section E *Salmonella*

Recommended methods are presented for recovery of *Salmonella* from water and wastewater and their subsequent identification. The methods are particularly useful for recreational and shellfish-harvesting waters. No single method of recovery and identification of these organisms from waters and wastewaters is appropriate for all sampling situations. The method selected depends on the characteristics of the sample and the microbiologist's experience with the techniques. Multiple option techniques are described for sample concentration, enrichment, isolation and identification. The Section is divided as follows:

1. The Genus, *Salmonella*
2. Methods for Concentration of *Salmonella*
3. Primary Enrichment for *Salmonella*
4. Isolation of *Salmonella*
5. Biochemical Identification of *Salmonella*
6. Serological Test for *Salmonella*
7. Quantitative Techniques
8. Optional Fluorescent Antibody Screening Technique

1. The Genus, *Salmonella*

1.1 Definition

The genus *Salmonella* is comprised of a large number of serologically related, gram-negative, nonspore-forming bacilli that are 0.4-0.6 μm in width × 1-3 μm in length, and which occasionally form short filaments. They are motile with peritrichous flagella or are non-motile. Ordinarily salmonellae do not ferment lactose, sucrose, malonate or salicin but do ferment numerous carbohydrates including glucose, inositol and dulcitol. These bacteria are positive for lysine and ornithine decarboxylase and negative for urease and phenylalanine deaminase. Usually they produce hydrogen sulfide and do not liquify gelatin. All of the known *Salmonella* species are pathogenic for warm-blooded animals, including man. They cause enteritis, (via contaminated water, food or food products) enteric fevers and are found in reptiles, amphibians and mammals. Edwards and Ewing have published an authoritative work on the isolation and characterization of *Salmonella* (1).

In Bergey's 8th edition (2), the salmonellae have been reclassified tentatively into 4 subgenera containing II subdivisions. However, the problem of Kaufman's listing of many serotypes, and the lack of agreement as to what constitutes the genus *Salmonella* or its species, leaves the taxonomy in a fluid state.

1.2 Identification Schemes

A comprehensive scheme of the recommended isolation, detection and identification methods is outlined in sequence in Figure III-E-1.

When space and equipment are limited, the number of options at each stage depicted in Figure III-E-1 can be reduced and salmonellae isolated successfully. One such simplified scheme is outlined in Figure III-E-2.

The procedures outlined in these schema are described in the following subsections 2–8.

2. Methods for Concentration of *Salmonella*

The initial steps for detection of salmonellae in water and wastewater require concentration of the organisms by one of several methods: the gauze swab, diatomaceous earth, the cartridge filter, or the membrane filter technique. The volume of sample tested is directly related to the level of pollution.

2.1 Swab Technique Modified After Moore's Method (3)

2.1.1 Summary: In this method sterile gauze swabs are immersed for about 5 days just below the surface of a water or wastewater. After the exposure period during which bacterial concentration occurs, the gauze swabs are retrieved, placed in sterile bags, iced and returned to the laboratory for examination. The swab, portions of the swab, or the expressed liquid from the swab are added to enrichment media for selective growth of salmonellae and suppression of coliforms and other non-salmonellae.

2.1.2 Scope and Application

(a) Advantages: The gauze swab technique is superior to grab sampling, because salmonellae concentration occurs in the swab permitting improved detection. Although this technique is not quantitative, it has proved effective in detection of low numbers of salmonellae in waters and wastewater. The technique is simple and inexpensive.

(b) Limitations: This is not a quantitative procedure, since some salmonellae may pass through the swab, others may desorb from the swab during the exposure period, and the water volume sampled is unknown. It is not possible to predict the salmonellae concentration in the water or wastewater from the concentration in the swab nor does the procedure reflect changes or cycling of salmonellae concentrations at the sample site.

2.1.3 Apparatus and Materials

(a) Cheesecloth roll, 23 cm wide.

(b) Paper cutter or large pair of shears.

(c) Length of 16 gauge wire.

(d) Sterile 250 ml flasks or jars, screw-cap, containing enrichment media, Part III-E, 3.3.10.

(e) Sterile plastic bags (e.g., Whirl-pak or heavy-duty food freezer bags).

(f) Insulated container with ice (optional).

2.1.4 Procedure

(a) Prepare a swab from a length of cheesecloth 180 cm long × 23 cm wide by folding the length 5 times to form a pad 36 cm × 23 cm. Cut the folds at one end. From this end, cut the pad into 5 parallel strips, 4.5 cm wide and 26 cm long, leaving an uncut top section of 10 cm (see Figure III-E-3).

(b) Bind the top of the gauze swab with 16 gauge wire to form a mop-head shape, with the strips hanging free (see Figure III-E-4).

(c) Wrap swab in kraft paper and autoclave.

(d) Place the swab just below the surface of the water or wastewater to be examined and secure with the wire to a solid support.

(e) Leave the swab in place for about 5 days.

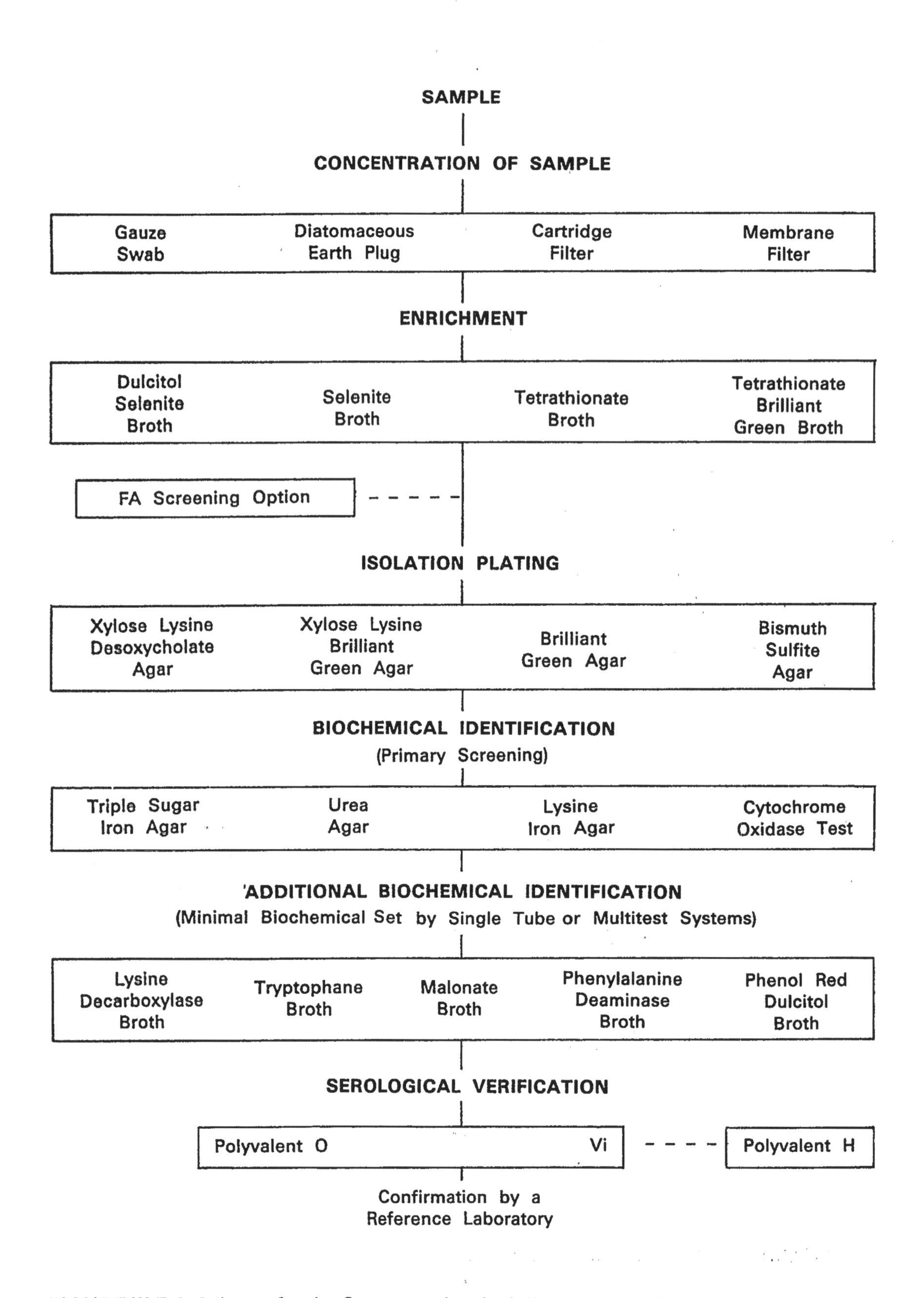

FIGURE III-E-1. Scheme for the Concentration, Isolation and Identification of *Salmonella.*

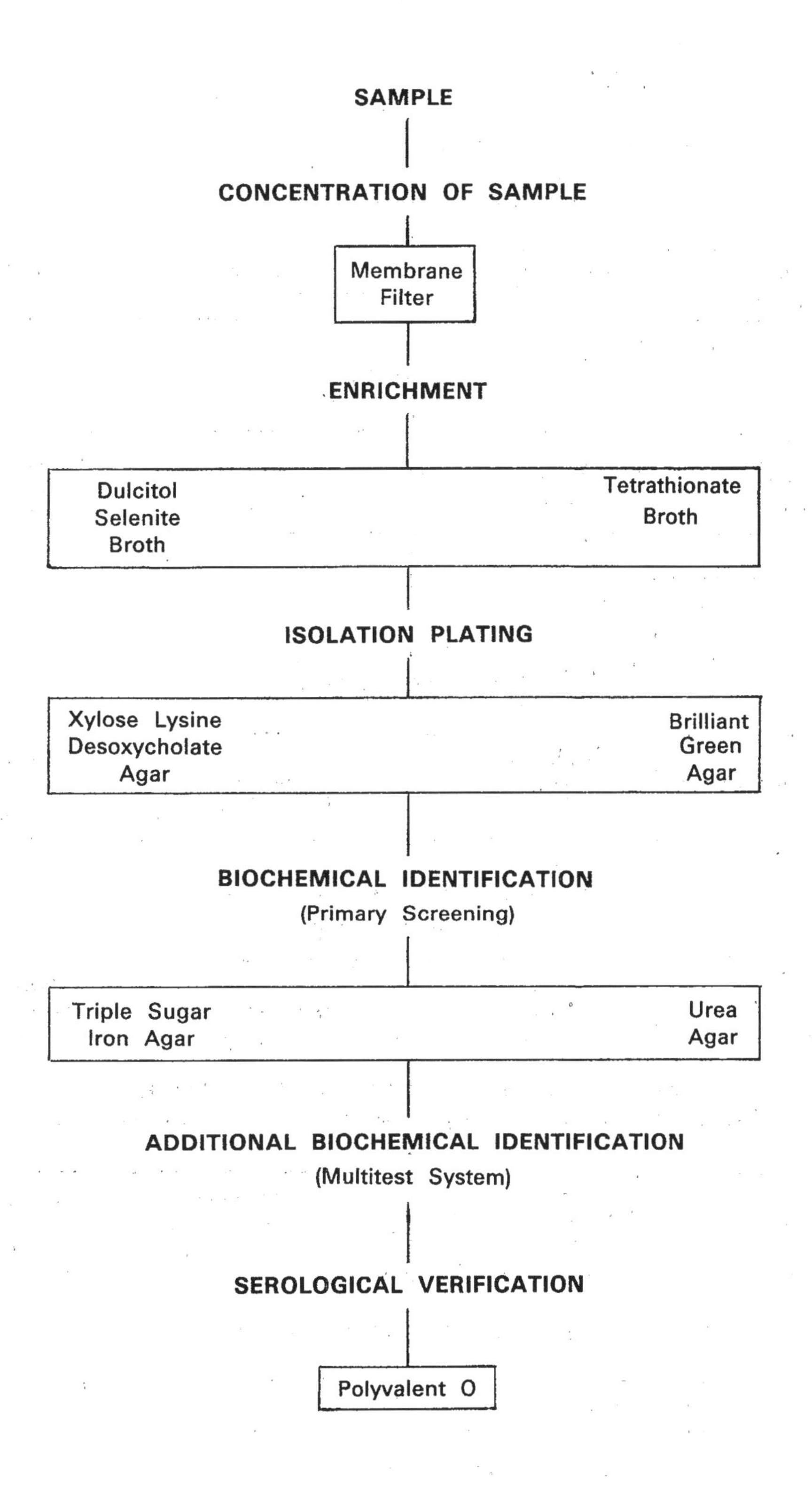

FIGURE III-E-2. Simplified Scheme for Concentration, Isolation and Identification of *Salmonella.*

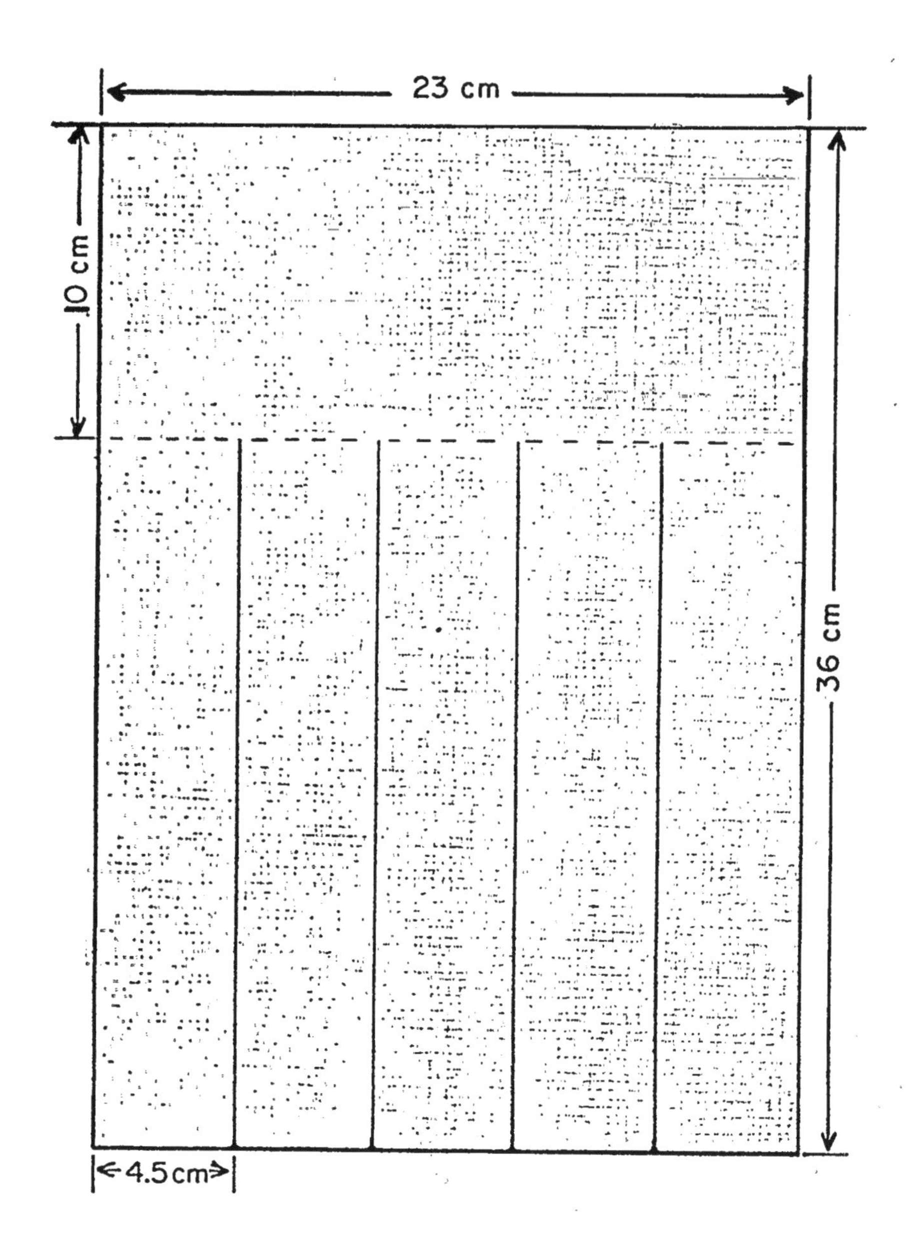

FIGURE III-E-3. Dimensions of the Gauze Swabs.

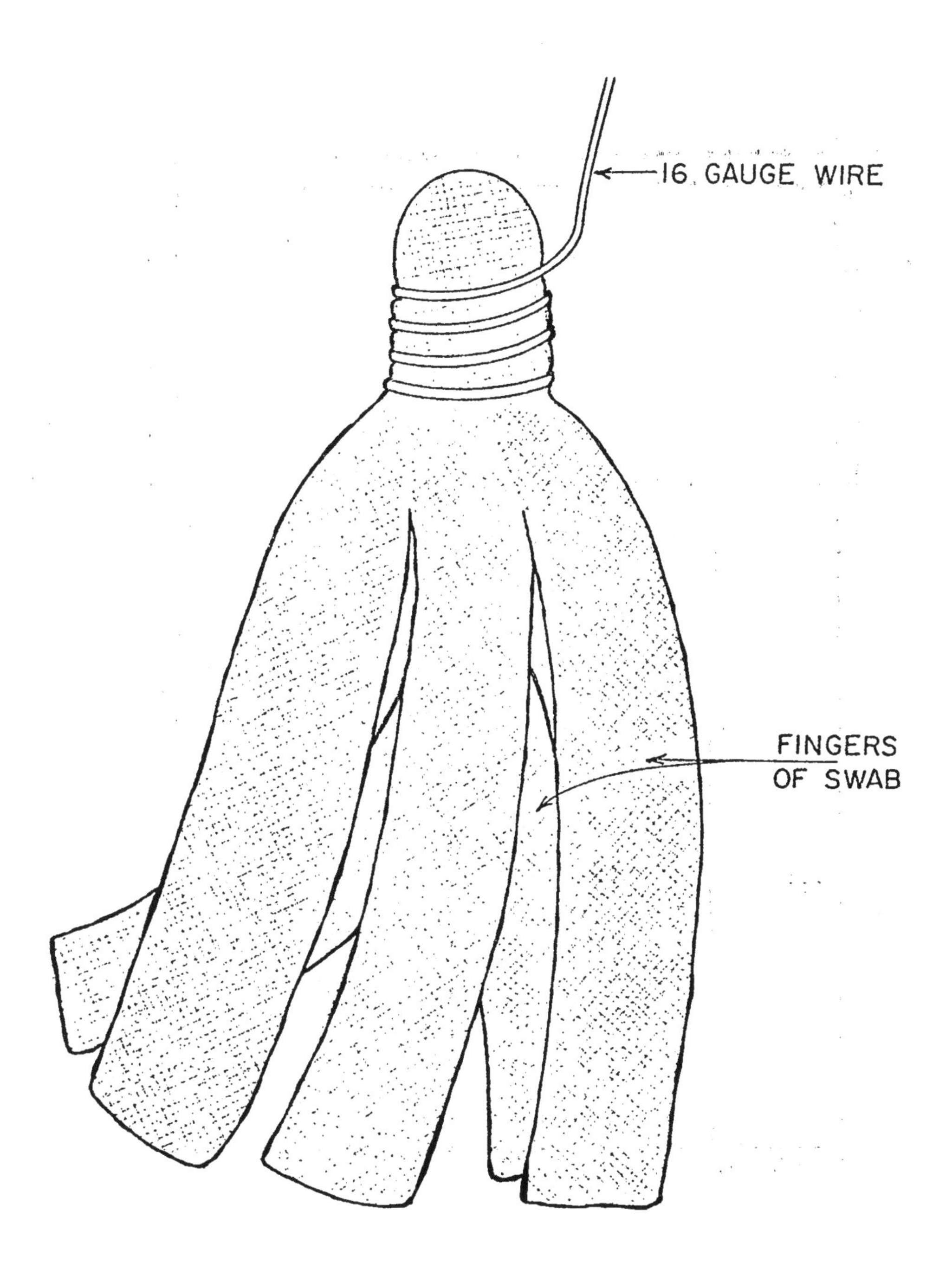

FIGURE III-E-4. The Gauze Swab in Position.

(f) After the exposure period, retrieve the swab and place it directly into a sterile plastic bag (Whirl-Pak), ice and return to the laboratory for examination. See enrichment step 3.4.1. Alternatively, the swabs may be placed in enrichment media and incubated on-site, then iced and returned to the central laboratory.

2.2 Diatomaceous Earth Technique

2.2.1 Summary: The filtering action of diatomaceous earth is used to concentrate the organisms (4). Diatomaceous earth is loosely packed on top of an absorbent pad in the funnel of an assembled membrane filtration unit. One to two liters of a water sample are passed through the diatomaceous earth using vacuum and portions of the diatomaceous earth plug are added aseptically to enrichment broth.

2.2.2 Scope and Application

(a) Advantages: Although it is not enumerative, this method is quantitative in the sense that known volumes of water or wastewater are filtered through the diatomaceous earth.

(b) Limitations: The diatomaceous earth filter is easily clogged with the suspended material found in turbid waters. This slows the filtering process but may not prevent its use. Also salmonellae may pass through due to improper formation of the diatomaceous filter.

2.2.3 Apparatus and Materials

(a) Diatomaceous earth (Johns-Manville's "Celite" or equivalent).

(b) Sterile membrane filter unit (filter base and funnel), plastic, glass or stainless steel, wrapped with aluminum foil or kraft paper.

(c) Vacuum source.

(d) Vacuum flask, 2-liter, to hold the filter base, with appropriate tubing. An alternative is a commercially-available manifold.

(e) Safety trap flask between the filtering flask and the vacuum source.

(f) Sterile spatula.

(g) Forceps.

(h) Sterile graduated cylinders, 1000 ml size, covered with aluminum foil or kraft paper.

(i) Containers for glass pipets.

(j) Sterile absorbent pads of cellulosic paper, 47 mm in diameter.

(k) Sterile T.D., 10 ml Mohr pipets, glass or plastic.

(1) Rinse water, sterile phosphate buffered water, prepared in large volumes, See Part II-B, 7 for preparation.

2.2.4 Procedure

(a) Assemble the membrane filter unit, substituting an absorbent pad for the membrane filter.

(b) With the funnel in place, loosely pack approximately 2.5 cm thickness of diatomaceous earth on top of the absorbent pad.

(c) Add enough sterile buffered water with a 10 ml pipet to saturate the diatomaceous earth. Draw water through filter under vacuum (15 lb.).

(d) Filter 1 liter or more of the sample under vacuum. The volume of sample filtered will depend upon the estimated amount of pollution in the water or wastewater.

(e) Rinse the funnel at least twice after sample filtration with 20–30 ml of the rinse water.

(f) Remove the funnel. Add the absorbent pad, the diatomaceous earth plug or halves of each to flasks of selected enrichment broths. See the enrichment step 3.4.2.

2.3 Membrane Filtration of Samples

2.3.1 Summary: The water or wastewater sample is passed under positive pressure through a 0.45 µm pore, 142 mm diameter filter in a pressure filtration unit. After filtration, the membrane filter is aseptically divided into portions that are added to enrichment broths.

2.3.2 Scope and Application

(a) Advantages: Membrane filtration is used when the sample volume desired is larger than the 1–2 liters employed with the diatomaceous earth technique, when the 5 day sample period for the swab technique is not feasible, or when the sample is not too turbid to permit passage of the desired volume. The method is useful for water of very low organic and particulate matter content. It is quantitative in that it retains all cells from the filtration of a known volume of water or wastewater.

(b) Limitations: Use of this technique is somewhat limited because the sample may clog pores and prevent filtration. Also, the limitations of membrane filtration cited in Part II-C, 3.2.2, are applicable.

2.3.3 Apparatus and Materials

(a) Membrane filter holder, stainless steel, 142 mm, autoclavable for use in pressure filtration. (Gelman Disc, Filter Holder, 11872, Millipore SS Filter Holder, YY22 14200, or equivalent).

(b) Dispensing pressure vessel, 10/12 liter size. (Gelman 15108, Millipore XX67 00003, or equivalent).

(c) Pressure pump, capable of maintaining high pressure necessary for pressure filtration (7 kg/cm^2 maximum pressure).

(d) Forceps, with smooth tips.

(e) Sterile shears to divide membrane.

(f) Dilution water, phosphate buffered, prepared in large volumes.

(g) Membrane filter, white, grid-marked, 142 mm in diameter with 0.45 $\pm$ 0.02 µm pore size.

(h) Prefilter (Millipore AP15–142–50 or equivalent).

2.3.4 Procedure

(a) Aseptically add the sterile membrane filter to the sterile filter holder.

(b) Add a 2–20 liter sample to the dispensing pressure vessel. The sample size is limited by the level of solids or turbidity in the sample.

(c) If sample is turbid use a prefilter ahead of the membrane filter.

(d) Using positive pressure, force the sample through the membrane filter.

(e) After filtration, the membrane and prefilters are added to enrichment media. See enrichment step 3.4.3.

2.4 Cartridge Filter Technique (5)

2.4.1 Summary: The water or wastewater sample is drawn under negative pressure through a filter of borosilicate glass microfibers bonded with epoxy resin. The volume filtered may be measured. After filtration, the filter is placed aseptically into enrichment broth.

2.4.2 Scope and Application

(a) Advantages: Because this technique can be used to filter 20 liters or more of sample, it is applicable to waters with low concentrations of organisms. It can be combined with an MPN procedure for a quantitative estimate of bacterial density.

(b) Limitations: The presence of high numbers of background organisms may make recovery of *Salmonella* very difficult by this technique. The autoclave must be sufficiently large to hold sample reservoir containers. As with other filtration procedures, high turbidity in the sample lengthens filtration time.

2.4.3 Apparatus and Materials

(a) Vacuum pump, capable of operation at a pressure differential up to 69 kN/m^2.

(b) Vacuum gauge.

(c) Water trap (heavy-walled flask or bottle, with at least a 5-liter capacity) closed by a 2-holed rubber stopper, with 2 short pieces of glass tubing inserted.

(d) Glass and rubber vacuum tubing, with shut-off valve.

(e) Optional for MPN technique: manifold, capable of 5 simultaneous filtrations.

(f) Sterile, 20 liter, heavy-walled pyrex glass carboy, calibrated from 1–20 liters (Corning CWG No. 434490).

(g) Balston (Lexington, MA) type AA cartridge filter, 2.5 × 6.4 cm or equivalent and a type 90 filter holder. Filters, filter holders and tubing are wrapped and sterilized by autoclave at 121 C for 15 minutes.

2.4.4 Procedure

(a) Connect the parts of the filtration apparatus as depicted in Figure III-E-5.

(b) Collect or transfer water sample to sterile, calibrated container.

(c) Insert filter and filter holder into sample container. Protect container from contamination by sealing glass tubing in neck of container with sterile cotton and aluminum foil.

(d) Start vacuum pump and filter desired volume of water. To avoid rupture of filter, the pressure differential should not exceed 69 kN/m^2.

(e) After filtering sample, aseptically remove filter and filter holder. Separate filter and place in appropriate enrichment medium. See 3.4.4.

(f) Place another sterile filter in the filter holder and repeat filtration procedure.

(g) When a new sample is to be filtered, replace glass tubing, filter, filter holder and sample container with new sterile units.

3. Primary Enrichment for *Salmonella*

Selenite and tetrathionate broths are used for primary enrichment. The selenite may be combined with dulcitol to improve the selectivity for salmonellae. The tetrathionate may be used as a broth base alone or combined with brilliant green dye which enhances its selectivity for salmonellae other than *S. typhi*.

3.1 Summary: After concentration of salmonellae, the gauze swabs, diatomaceous earth plugs, membrane filters or cartridge filters are placed in flasks of selenite, tetrathionate or other selected broth. The broth encourages salmonellae to grow while inhibiting other bacteria. Multiple flasks of the enrichment broth are incubated for 3–4 days at selected temperatures. After incubation, cultures are streaked onto differential plating media for salmonellae isolation. These differential plates should be streaked from the enrichment media every 24 hours over a 3–4 day period.

3.2 Scope and Application

3.2.1 Advantages: The enrichment methods described provide an optimal environment for salmonellae and other enteric pathogens, and to some extent suppress the other organisms that are present.

3.2.2 Limitations: Although enrichment broths provide an optimal environment for salmonellae development, recovery is not quantitative.

3.3 Apparatus and Materials

3.3.1 Incubators set at 35 ± 0.5 C, 41.5 ± 0.5 C, and optionally at 37 ± 0.5 and 43 ± 0.5 C.

3.3.2 Sterile shears and spatula.

3.3.3 Sterile forceps.

3.3.4 Sterile beakers, 500 ml size covered with aluminum foil or kraft paper.

3.3.5 Sterile erlenmeyer flasks, 125 ml size, to hold 50 ml of enrichment broth.

3.3.6 Bunsen, Fisher burner or electric incinerator.

3.3.7 Sterile petri dish.

3.3.8 Sterile aluminum foil.

3.3.9 Alcohol, 95% ethanol, in a vial.

3.3.10 Media

(a) Selenite broth (Difco 0275, BBL 11608, or equivalent). (See Part II-B, 5.5.1).

(b) Tetrathionate broth (Difco 0104, BBL 11706, or equivalent). (See Part II-B, 5.5.2).

(c) Dulcitol selenite broth (see Part II-B, 5.5.3).

(d) Tetrathionate brilliant green broth (same as (b) above with the addition of 0.01 gram of brilliant green per liter of medium). (See Part II-B, 5.5.4).

3.4 Procedures for Enrichment

Select at least two enrichment media, preferably one selenite and one tetrathionate type, for each sample, The actual choice of medium is based on the experience of the analyst.

Selenite and tetrathionate enrichment broths are useful for all *Salmonella* including *S. typhi.* Although the addition of brilliant green to tetrathionate broth base increases its selectivity for salmonellae, it inhibits the recovery of *S. typhi.* It is reported that tetrathionate broth is toxic to salmonellae at a temperature of 43 C (6). Dulcitol selenite medium may not completely recover *S. typhi, S. cholerae-suis, S. enteritidis* bioserotypes Paratyphi A and Pullorum, from some samples because these species ferment dulcitol slowly (7).

3.4.1 The Gauze Swab: Squeeze the gauze swab in the plastic bag to express the liquid from the swab into a sterile 500 ml beaker. After this, the entire swab or portions of it may be placed directly into the enrichment broth using sterile forceps and shears to divide the swab if necessary into the number of portions required for the enrichment media to be used. For example, 2 enrichment media incubated at 2 temperatures, 35 C and 43 C, would require 4 portions. The pieces of swab and measured volumes of the liquid can be apportioned into 125 ml flasks containing 50 ml of enrichment broth made double strength to compensate for the dilution effect of liquid.

3.4.2 The Diatomaceous Earth Plug: Remove the diatomaceous earth plug from the membrane filtration funnel by carefully tapping the funnel on a sheet of sterile metal foil. Tap the sides of the funnel and shake it gently to dislodge the plug. Transfer the entire plug to an enrichment broth or divide the plug into halves using a sterile spatula. Add one half of the plug to each 125 ml flask of medium used. Inoculate at least 2 media in flasks containing 50 ml of sterile single-strength broth. Incubate at 2 temperatures. Halve the absorbent pad with sterile shears, and add a portion to each of the previously inoculated flasks.

3.4.3 The Membrane Filter: Remove the membrane filter aseptically from the filter base with sterile forceps, and hold over the bottom of a sterile petri dish. Cut the membrane in quarters with the sterile shears, and let the quarters fall grid-side up into the petri dish bottom. Insert aseptically each quarter of a membrane filter into a 125 ml flask containing 50 ml of single-strength medium.

3.4.4 The Cartridge Filter: Remove the cartridge filter from the holder aseptically and place in the selected enrichment medium. More than I filter may be used in succession on a single sample and placed in the same culture medium. Repeat for the second enrichment medium.

3.4.5 After inoculation in 3.4.1, 3.4.2, 3.4.3 or 3.4.4, incubate enrichment flasks at 35 C, 41.5 C and other selected temperatures for at least 24 hours. However, some salmonellae are slow-growing and recovery may be increased by incubating for successive 24-hour periods up to 96 hours before streaking on isolation agars.

3.4.6 If a sample must be collected in an area some distance from the central laboratory responsible for *Salmonella* identifications, the scheme can be interrupted at different points. First, the samples can be concentrated, iced and transported back to the central laboratory. Second, the samples can be concentrated, placed in enrichment media, incubated at the selected temperature for 18-24 hours at a field laboratory then iced and transported to the central laboratory.

3.5 Incubation Temperature as a Selection Technique for Salmonellae

3.5.1 Historically 37 C has been used for the isolation of enteric microorganisms because it is representative of the gut temperature of humans and other warm-blooded animals. This temperature was considered optimal for the detection and isolation of these enterics.

3.5.2 Some workers have used 35 C in place of 37 C because the latter temperature is close to the upper tolerance limit for the group and might prevent the growth of some desired species. This inhibitory effect would be most significant if the 37 C temperature were not well-controlled and might be exceeded by 2-3 C. Consequently, double temperatures of 35 C and 37 C came into common usage (1, 8) for isolation and identification of Enterobacteriaceae.

3.5.3 Because the large populations of normal gut microorganisms interfere with the isolation of pathogens such as *Salmonella* from humans and other warm-blooded animals, temperatures above 37 C were proposed to reduce the background microorganisms.

3.5.4 The 43 C temperature was proposed as an aid to the isolation of *Salmonella*, particularly from various heavily contaminated materials (9, 10). The 41.5 C temperature has been recommended by others for the detection and isolation of *Salmonella* from water (11, 12, 13).

3.5.5 Although investigators have used different elevated temperatures, the results are inconclusive for a single incubation temperature. The general conclusion of most studies on comparison of recoveries of salmonellae at different incubation temperatures is that a greater number of isolates and more species are isolated at 41–43 C than at 35–37 C (9, 14–19). This manual recommends two temperatures in enrichment and isolation procedures for *Salmonella*, 35 C and 41.5 C.

Optional Fluorescent Antibody Screening Technique

The FA technique is a rapid screening method that can be used after primary enrichment but it is tentative and optional. The detailed procedure is described in Part III-E, 8 following the conventional methods for detection and identification of Salmonella.

4. Isolation of Salmonella

4.1 Summary: The organisms that develop in the primary enrichment broth media are isolated and differentiated on solid media. The enriched cultures are streaked every 24 hours for 3–4 days onto the surface of XLD, BG, XLBG and bismuth sulfite media. After incubation, the plates are examined for typical colonies of salmonellae which are picked and characterized biochemically and serologically.

4.2 Scope and Application

4.2.1 Advantages: Pure cultures of *Salmonella* can be isolated by the careful selection and use of plating media and incubation temperatures. Brilliant green agar is favored for the development and identification of salmonellae except for *S. typhi* and a few other species whereas bismuth sulfite agar allows the growth of most *Salmonella* including *S. typhi*.

4.2.2 Limitations: Bacteria other than salmonellae may compete with the salmonellae on the secondary differential media, thus interfering with their isolation and identification. The use of brilliant green agar at an elevated temperature of 41.5 C reduces the number of interfering organisms, but also inhibits development of some serotypes of *Salmonella*.

4.3 Apparatus and Materials

4.3.1 Incubators set at 35 ± 0.5 C, 41.5 ± 0.5 C, and optionally at 37 ± 0.5 C and 43 ± 0.5 C.

4.3.2 Water bath set at 44–46 C for tempering agar.

4.3.3 Petri dish canisters for glass petri dishes.

4.3.4 Thermometer certified by National Bureau of Standards or one of equivalent accuracy.

4.3.5 Inoculating needle and 3 mm loop.

4.3.6 Colony Counter, Quebec darkfield model or equivalent.

4.3.7 Bunsen/Fisher burner, or electric incinerator.

4.3.8 Sterile 100 mm × 15 mm petri dishes, glass or plastic.

4.3.9 Sterile phosphate buffered or peptone dilution water in bottles, 99 ± 2 ml volumes (see Part II-B, 7 for preparation).

4.4 Media: The following agar media are dispensed in bulk quantities in screw-capped bottles or flasks. (See Part II-B, 4 for preparation).

4.4.1 Xylose lysine desoxycholate (XLD) agar (see Part II-B, 5.5.7).

4.4.2 Brilliant green (BG) agar (see Part II-B, 5.5.5).

4.4.3 Xylose lysine brilliant green (XLBG) agar (see Part II-B, 5.5.6).

4.4.4 Bismuth sulfite agar (see Part II-B, 5.5.8).

4.5 Procedure

4.5.1 Prepare two selected media (XLD, BG, XLBG or bismuth sulfite agars) in petri dishes. As a minimum xylose lysine desoxycholate (XLD) and brilliant green (BG) or xylose lysine brilliant green (XLBG) agars are recommended. Bismuth sulfite agar permits the presumptive detection of *S. typhi* and/or *S. enteriditis*.

4.5.2 Streak the surface of a previously poured and solidified agar with a loopful of the enrichment culture.

4.5.3 Duplicate plates should be streaked from each enrichment culture every 24 hours for 3–4 days.

4.5.4 Inoculate duplicate plates from each enrichment culture and incubate, one each, at 35 C and 41.5 C (and optionally at 37 C and 43 C). Incubate the XLD and XLBG plates for 24 hours, and the BG agar and bismuth sulfite agar plates for 48 hours.

4.5.5 After incubation, examine plates for colony appearance. Table III-E-1 describes the appearance of colonies on XLD, XLBG, BG and bismuth sulfite agars. The salmonellae colonies on BG agar are pinkish white with a red background. Lactose fermenters will form greenish colonies or other colorations. Occasionally, slow lactose fermenters such as *Proteus, Citrobacter, Pseudomonas* and *Aeromonas* mimic *Salmonella*.

4.5.6 Pick growth from the centers of well-isolated colonies that have the characteristic appearance of salmonellae, and streak onto the screening media described in This Section, 5.4.1 and 5.5.1. Isolated, single colonies from a plate where all colonies appear alike may be assumed to be pure. At least 2 colonies of each type suspected to be *Salmonella* should be picked.

4.5.7 The suspected colonies of *Salmonella* should now be characterized by the sin-

TABLE III-E-1

Colonial Appearance of *Salmonella* and Other Enterics on Isolation Media

Colony Appearance	*Salmonella*	Other Enterics
	1. Bismuth Sulfite Agar	
Round jet black colonies with or without sheen	*S. typhi*	
" " "	*S. enteritidis* ser Enteritidis	
" " "	*S. enteritidis* ser Schottmuelleri	
Flat or slightly raised green colonies	*S. enteritidis* ser Typhimurium	*Proteus* spp.
" " "	*S. enteritidis* bioser Paratyphyi	
" " "	*S. cholerae-suis*	
	2. Brilliant Green Agar	
Slightly pink-white, opaque colonies surrounded by brilliant red medium	*Salmonella* spp	
Yellow-green colonies surrounded by yellow-green zone		*Escherichia, Klebsiella, Proteus* spp. (lactose or sucrose fermenters)
	3. XLD or XLBG Agar	
Red, black centered colonies	*Salmonella* spp.	
Red colonies		*Shigella* spp.
Yellow colonies		*Escherichia* spp. and biotypes
"		*Citrobacter* spp.
"		*Klebsiella* spp.
"		*Enterobacter* spp.
"		*Proteus* spp.

gle biochemical tests or multitests in Part III-E, 5.6. An O-1 bacteriophage screening test may also be used on the isolate for a rapid (4–5 hours) determination of the tentative identification of *Salmonella*. Results must be verified (20, 21).

5. Biochemical Identification of *Salmonella*

5.1 Summary: Salmonellae can be identified to genus by determining their reactions in a series of biochemical tests. These tests can be made in single tube media or in commercial multitest systems. This Section offers the minimal required set of tests, a series of additional, optional tests and a brief description of available multitest systems. These tests require specific training for valid results.

5.2 Scope and Application

5.2.1 Advantages: The biochemical reactions characterize the *Salmonella*-like isolate, permit a separation from closely related bacteria, and provide presumptive identification as *Salmonella*. Confirmed identification requires additional serological tests.

The multitest systems have a number of advantages over single tube media prepared in the laboratory:

(a) The different designs of these systems permit the user to select tests to fit his needs, facilities and budget.

(b) The multitest systems may be used with confidence in the laboratory or in field situations as rapid and convenient screening methods for cultures suspected to be *Salmonella*. Large numbers of cultures can be examined in a relatively short time; and those tentatively identified as *Salmonella* can then be held for later serological confirmation.

(c) These systems offer the advantages of minimal space requirements, immediate availability, and economy when compared with the preparation and use of tubed media. They are ideally suited to field work and to small laboratories that do not routinely perform these identification tests.

(d) The systems offer these same advantages for the identification and differentation of other enteric bacteria such as *E. coli, Klebsiella, Enterobacter, Citrobacter* and *Shigella.*

5.2.2 Limitations: Whether accomplished by individual tube or multitest methods, biochemical identification of large numbers of cultures is expensive and time-consuming. It should not be attempted independently without previous training and experience in reading reactions and interpreting results.

5.3 Apparatus and Materials

5.3.1 Incubator set at 35 ± 0.5 C.

5.3.2 Bunsen/Fisher burner, or electric incinerator.

5.3.3 Inoculating needle and loop.

5.3.4 Culture tubes, 100 × 13 or 150 × 20 mm.

5.3.5 Fermentation tubes, 75 × 10 mm.

5.3.6 Closures to fit culture tubes.

5.3.7 Thermometer certified by National Bureau of Standards, or one of equivalent accuracy.

5.4 Media

5.4.1 Screening Media

(a) Triple Sugar Iron Agar (TSI) (Difco 0265, BBL 11749). See Part II-B, 5.5.9.

(b) Lysine Iron Agar (LIA) (Difco 0849, BBL 11363). See Part II-B, 5.5.10.

(c) Motility Sulfide Medium (Difco 0450). See Part II-B, 5.5.16.

(d) Urea Agar Base Concentrate (Difco 0284). See Part II-B, 5.5.11.

(e) Cytochrome Oxidase Test Reagents

Reagent A. Weigh out 1 gram alphanapthol and dissolve in 100 ml of 95% ethanol.

Reagent B. Weigh out 1 gram of paraaminodimethylaniline HCl (or oxylate) and dissolve in 100 ml of laboratory pure water. Prepare frequently and store in refrigerator.

5.4.2 Minimal Biochemical Set

(a) Phenylalanine Agar (Difco 0745, BBL 11537). See Part II-B, 5.5.12.

(b) Ferric chloride reagent: Prepare a 10% solution (w/v) of ferric chloride in laboratory pure water. Store in a brown bottle in the refrigerator.

(c) Indole test (tryptophane broth). See Part II-B, 5.1.9.

(1) Tryptone (Difco 0123).

(2) Trypticase (BBL 11921).

(3) Indole Test Reagent: Dissolve 5 grams paradimethylamino benzaldehyde in 75 ml isoamyl or normal amyl alcohol, ACS grade. Slowly add 25 ml conc HCl. The reagent should be yellow and have a pH below 6.0; if the final reagent is dark in color it should be discarded. Examine the reagent carefully during preparation because some brands are not satisfactory after aging. Both amyl alcohol and benzaldehyde should be purchased in a small amount consistent with the volume of work anticipated. Refrigerate the reagent in a glass-stoppered bottle.

There has been some difficulty in obtaining amyl alcohol. If this problem occurs, Gillies describes an alternate paper strip test for indole production (22).

(d) Malonate Broth, Modified (Difco 0569, BBL 11436). See Part II-B, 5.5.13.

5.4.3 Optional Biochemical Tests

(a) Carbohydrate Utilization

Purple Broth Base (Difco 0092, BBL 11506). See Part II-B, 5.1.7 containing:

Dulcitol (Difco 0162),
Lactose (Difco 0156), or
Inositol (Difco 0164)

(b) Decarboxylase Activity

Decarboxylase Medium Base (Difco 0872). See Part II-B, 5.5.14 containing:

Lysine HCl (Difco 0705),
Arginine HCl (Difco 0583), or
Ornithine HCl (Difco 0293)

(c) ONPG Reagents

(1) Monosodium Phosphate Solution

Dissolve 6.9 grams $NaH_2Po_4 \cdot HOH$ in 45 ml of laboratory pure water. Add 3 ml of 30% NaOH and adjust pH to 7.0. Bring to 50 ml with laboratory pure water and store in refrigerator.

(2) ONPG Solution

Dissolve 80 grams of o-nitrophenyl-β-D-galactopyranoside (ONPG) in 15 ml pure water at 37 C. Add 5 ml of 1 M NaH_2PO_4 from (1) above. This 0.75 M solution of ONPG should be colorless. Store in a refrigerator. A portion of the buffered solution sufficient for the number of tests to be done should be warmed to 37 C before use.

5.4.4 Multitest Systems (optional to Single Test Series)

(a) API Enteric 20 (Analytab Products, Inc.).

(b) Enterotube (Roche Diagnostics).

(c) Inolex (Inolex Biomedical Division of Wilson Pharmaceutical and Chemical Corp.)

(d) Minitek (Baltimore Biological Laboratories, Bioquest).

(e) Pathotec Test Strips (General Diagnostics Division of Warner-Lambert Company).

(f) r/b Enteric Differential System (Diagnostic Research, Inc.)

5.5 Procedures: Since cultures are found which react atypically, an isolate should not be eliminated because of a single anomalous reaction, rather the biochemical reactions should be considered as a group. For example, LIA tubes not showing H_2S production, but having alkaline slants and alkaline butts may be atypical *S. typhi.* These tubes should be retained for further characterization.

Further, to insure that the media are yielding proper reactions, the analyst is urged to incorporate both positive and negative control cultures into Single Test and Multitest Procedures.

5.5.1 Screening Tests: Pick growth from the center of a single isolated colony on a selective plating medium and inoculate into the primary screening medium.

Fermentation tube reaction code for TSI and LIA Agars:

Report slant/butt where *K, A* and *N* indicate alkaline, acid and neutral reactions respectively; *G, g* indicate large and small amounts of gas production, respectively; and H_2S 1+ to 4+ indicate levels of blackening due to hydrogen sulfide production. For example, K/Ag is an alkaline slant and an acid butt with a small amount of gas.

- (a) Triple Sugar Iron Agar

(1) Inoculate by stabbing the butt and streaking the slant.

(2) Incubate at 35 C for 18–24 hours with cap loose.

(3) Read and record reactions. Color of slant or butt is yellow for an acid reaction or red for an alkaline reading. Gas production is evidenced by bubbles in the medium, and H_2S production by blackening of the medium.

(4) Typical reactions:

Salmonella: K/Ag with H_2S, 1+ to 4+.
S. typhi: K/A with H_2S, trace to 1+.
Citrobacter: K/Ag or A/Ag with H_2S, 1+ to 3+.

(5) Atypical reactions: TSI tubes showing alkaline slants and acid butts without H_2S production should be inoculated into Motility Sulfide Medium to verify the negative H_2S reaction. If still H_2S negative, perform serological testing to confirm an atypical *Salmonella.*

- (b) Lysine Iron Agar

(1) Inoculate by stabbing the butt twice and streaking the slant.

(2) Incubate for 18–24 hours, and if negative for an additional 24 hours, at 35 C.

(3) Read and record reactions. The slant or butt is yellow from an acid reaction and blue/purple for an alkaline reading. Gas production is evidenced by bubbles in the medium and H_2S production by blackening of the medium along the stab line. *Proteus* has a distinctive red slant caused by oxidative deamination and a yellow butt.

(4) Typical reactions:

Salmonella: K/K or K/N with H_2S +(–).
S. typhi: K/K with H_2S –(+).
Citrobacter: K/A with H_2S – or +.
Proteus: R(red)/A with H_2S –(+).

- (c) Urea Agar (Christensen)

(1) Inoculate slant only, using a heavy inoculum.

(2) Incubate for 18–24 hours at 35 C with cap loose. A positive reaction for *Proteus* may be recorded in 2–4 hours, but all negative tests at 2–4 hours must be held for 18–24 hours.

(3) Reactions are red for urease positive and yellow for urease negative. *Salmonella* give negative urease reactions. Cultures of

Citrobacter may yield weak delayed positive reactions at 18–24 hours.

• (d) Cytochrome Oxidase Test

The cytochrome (indophenol) oxidase test can be done with prepared paper strips or the following test on a nutrient agar slant:

(1) Inoculate nutrient agar slant and incubate at 35 C for 18–24 hours. Older cultures should not be used.

(2) Prepare reagents as in 5.4.1 (e).

(3) Add 2–3 drops of reagent A and reagent B to the slant, tilt to mix and read reaction within two minutes.

(4) A strong positive reaction (blue color slant or paper strip) occurs in 30 seconds. Ignore weak reactions that occur after two minutes. Pseudomonads, aeromonads and vibrios are positive. *Salmonella* is negative.

5.5.2 Minimal Biochemical Set: After reading the TSI reaction, use growth from the slant to inoculate the minimal biochemical set (in 5.4.3 above) with a straight wire needle. Sufficient culture to inoculate all of the minimal biochemical media is provided by one application of the tip of the needle to the TSI growth.

• (a) Phenylalanine Agar

(1) Inoculate surface of slant heavily.

(2) Incubate for 18–24 hours at 35 C with cap loose. A positive reaction for *Proteus* may be recorded in 4 hours but negative tests must be held for 18–24 hours.

(3) Test for phenylalanine deaminase by allowing 4–5 drops of a 10% solution of ferric chloride to run down over the growth on the slant.

(4) A dark green color on the agar slant and in the fluid is a positive reaction. A yellow or brown color is negative. *Salmonella* and *Citrobacter* give negative reactions.

• (b) Indole Test

(1) Inoculate the tryptophane broth lightly from the TSI agar slant culture.

(2) Incubate the broth at 35 $\pm$ 0.5 C for 24 $\pm$ 2 hours, with cap loose.

(3) Add 0.2-0.3 ml indole test reagent to the culture, shake and allow the mixture to stand for 10 minutes.

(4) Observe and record the results.

(5) A dark red color in the amyl alcohol layer on top of the culture is a positive indole test; the original yellow color of the reagent is a negative test.

(6) With rare exceptions, *Salmonella* and *Citrobacter* are indole-negative.

• (c) Malonate Broth Test

(1) Inoculate from the 18–24 hours TSI agar slant culture.

(2) Incubate for 48 hours at 35 C. Observe tubes after 24 and 48 hours. Positive reactions are indicated by a change in color of the medium from green to a deep blue. Lots of malonate medium should be checked with positive and negative cultures.

(3) *Salmonella arizonae* and some strains of *Citrobacter* utilize malonate. Other *Salmonella* do not.

• (d) Fermentation of Dulcitol in Phenol Red Broth Base

(1) Inoculate the Dulcitol broth lightly using a 24-hour culture.

(2) Incubate at 35 C and examine daily for 7 days.

(3) A positive reaction is production of acid with yellow color.

(4) Most salmonellae and some *Citrobacter* utilize dulcitol. Some that do not use it or

use it slowly include: *S. typhi, S. cholerae-suis, S. enteritidis* bioser Paratyphi A and Pullorum, and *S. entertidis* ser Typhimurum.

5.5.3 Optional Biochemical Tests: If the minimal set of biochemical tests has not satisfactorily identified cultures as *Salmonella*, or variable reactions have been observed, proceed with the following optional tests.

• (a) Fermentation of Lactose in Phenol Red Broth and Inositol in Phenol Red Broth

(1) Inoculate the broth lightly using a 24-hour culture.

(2) Incubate at 35 C and examine daily for 7 days.

(3) A positive reaction is production of acid and yellow color with or without gas production.

(4) *Salmonella* do not utilize lactose, but some strains of *Citrobacter* do.

Some *Salmonella* do utilize inositol.

• (b) Decarboxylase tests (lysine, arginine and ornithine)

(1) The complete decarboxylase test series requires tubes of each of the amino acids and a control tube containing no amino acids.

(2) Inoculate each tube lightly.

(3) Add sufficient sterile mineral oil to the broth to make 3–4 mm layer on the surface and tighten the screw cap.

(4) Incubate for 18–24 hours at 35 C. Negative reactions should be reincubated up to 4 days.

(5) Positive reactions are deep purple and negative reactions remain yellow. Read the control tube without amino acid first; it must be yellow for the reactions of the other tubes to be valid. Positive purple tubes must have growth as evidence by turbidity because uninoculated tubes are also purple, nonfermenters may remain alkaline throughout incubation.

(6) *Salmonella* are positive for lysine; positive or delayed positive for arginine; and positive for ornithine. *S. typhi* and bioser Gallinarum are negative for ornithine. *Citrobacter* and bioser Paratyphi A are negative for lysine. Strains are variable for arginine and ornithine.

• (c) Gelatin Liquefaction

(1) Inoculate nutrient gelatin tube by stabbing.

(2) Incubate at 35 C for 5 days.

(3) Cool tubes to 20 C and inspect. Failure to solidify is a positive reaction.

(4) All Group I Enterobacteriaceae are negative, except for *S. arizonae* which shows a delayed positive reaction (see Table III-E-5).

• (d) ONPG Test, o-nitrophenyl-β-D-galoctopyranoside (23)

(1) Emulsify a large loopful of growth from each culture in 0.25 ml of physiological saline in a 10 × 75 mm fermentation tube.

(2) Add one drop of toluene to each tube and shake well. Let tubes stand for 5 minutes in 35 C water bath.

(3) Add 0.25 ml of buffered 0.75 M ONPG solution to each tube (see 5.4.3, (c) (1)) and incubate again in 35 C water bath.

(4) Read tubes at ½, 1 and 24 hours. A positive result is development of yellow color. Lactose fermenters have β galactosidase so *Citrobacter* are positive and most *Salmonella* negative.

• (e) Motility Test: A test for motility is used in serology but is not recommended in the biochemical reactions because:

(1) Most members of the family Enterobacteriaceae are motile, so the test would not add much to a characterization of the isolate.

(2) A negative test cannot be considered conclusive until the culture is passed once or twice through a motility or broth medium and the isolate retested.

5.6 Multitest Systems: Multitest systems are available which use tubes containing prepared agar media, plastic units containing dehydrated media, media-impregnated discs and reagent-impregnated paper strips. Some of the systems use numerical codes to aid identification of bacteria. Others provide computerized identification of bacteria. A number of independent investigators have compared one or more multitest systems with conventional biochemical tests. Some of the earlier systems have been improved. Most of the recent studies report the correct identification of high percentages of isolates (24–29).

5.6.1 The following systems for the identification of Enterobacteriaceae are commercially available. These have been quality tested by the manufacturers and others and can be used with confidence.

(a) API Enteric 20 consists of 20 small chambers (called cupules) in a plastic strip, each containing dehydrated medium. An isolated colony is used to prepare a cell suspension to inoculate the media. The inoculated media are incubated for 18 hours at 35 C in a special plastic chamber. A numerical identification system based on thousands of reaction combinations is available. The identification system is updated periodically. Computer services may be obtained which are more comprehensive and accurate than the manual system (27,28).

(b) Improved Enterotube with 8 compartments of agar media in a single plastic tube provides tests for 11 biochemical reactions. The media are inoculated by touching one end of a wire to an isolated colony and drawing the wire containing the inoculum through the media. The Enterotube is incubated for 18–24 hours at 35 C. A manual numerical identification aid, ENCISE, is part of the system.

(c) The Inolex system (formerly Auxotab) is comprised of a test card unit containing 10 reagent-filled capillary chambers. A single isolated colony is picked into broth and cultured for 3½ hours at 35 C. After incubation, the broth tube is centrifuged, the cells resuspended in water and inoculated into each capillary chamber on the card. Each card is incubated for 3 hours at 35 C in its own plastic container. Isolates can be identified in 7 hours unless additional tests are required. A numerical binary code named Var-ident is part of the system.

(d) The r/b Enteric Differential System consists of 4 Beckford tubes, 2 basic and 2 expander tubes. The 4 tubes contain agar media and are constricted to form upper and lower compartments which provide 14 biochemical reactions. The tubes are stabbed, streaked and incubated at 35 C for 24 hours. A color chart of typical tube reactions is provided for identification.

(e) The Pathotec Test Strips are individual biochemical paper strips impregnated with reagents that test for enzymes or end products characteristic of certain bacteria. A cell suspension prepared from isolated colonies is used as the inoculum to demonstrate the biochemical reactions. An incubation period of 4 hours at 35 C may be required for the cell suspension. The earlier Rapid I-D System has been discontinued. A new identification system for enterics is planned for marketing in 1978–1979.

(f) The Minitek Microorganism Differentiation System consists of 10 impregnated discs that test for 12 biochemical reactions. A single colony suspended in special broth serves as the inoculum. The 10 basic discs are part of 34 discs and 37 reactions offered to identify aerobic and anaerobic bacteria. The inoculated discs are contained in a plastic tray within a humidor and incubated at 35 C for 18–24 hours. Accessory equipment required to process the discs includes plastic inoculum plates, a 10-disc dispenser, a special pipetter, pipet tips, a pipet tips organizer, an incubation humidor, a color comparator card set and inoculum broth. The system stresses the biochemical options and flexibility of the system. A flow diagram is provided for identifications.

5.6.2 Factors for Selection: The 6 multitest systems briefly described above were designed for the biochemical identification of members of the family, Enterobacteriaceae. Most of the isolates suspected as *Salmonella* could be identified by any of the multitest systems. Some of the factors that should be considered in selecting a multitest system are:

(a) Biochemical Reactions in the Multitest System: Since analysts working with *Salmonella* develop a series of tests that yield good results for them, they should consider a system which fits their preferred test pattern.

(b) Need to Identify Atypical Salmonella: Because it is important for the laboratory to identify typical and atypical *Salmonella* from a series of samples, systems that use numerical identification should be selected (API, Enterotube, and Inolex). Further, systems containing the most tests have a better chance for identification of typical and atypical *Salmonella*. Only one system, API, requires no additional tests.

(c) Multitests: The production rate and the time span required to identify typical *Salmonella* varies with the system as shown in Table III-E-2.

(d) Refrigerated Storage: Refrigerated space is required for some systems. This can be a problem in purchase of a large supply. Table III-E-3 shows the reported shelf-life of the multitest systems, with and without refrigeration.

(e) Purchase of Special Equipment: Some of the test systems contain all materials and equipment necessary to do the analyses. Others require purchase of special items of equipment for full use. For example, Inolex requires a small centrifuge; API can be sight-read or can utilize a profile register for easy, rapid identification of atypical and typical strains; Minitek uses a starter kit which includes special plates, broth, pipetter, pipet tips, color comparator card set, incubation humidor, pipet tips, organizer and dispenser.

(f) Safety Considerations: The probability of laboratory-acquired infections is directly proportional to the amount and number of exposures to pathogens. Some multitest systems are more dangerous to handle than others because they require more opening, closing and manipulating of the test container which may pose added hazards to the worker. Some systems such as Enterotube and r/b use direct colony picks for reactions. API, Inolex, Pathotec and Minitek require culturing and additional handling of cell suspensions which are greater hazards.

(g) Cost and Source of Multitest Systems: The per unit cost of the multitests varies with the system, as shown in Table III-E-4.

5.7 Biochemical Characteristics of Enterobacteriaceae: The analyst may wish to differentiate salmonellae from the other Enterobacteriaceae by applying additional biochemical tests. Table III-E-5 is a chart of these biochemical reactions.

5.8 Serological Verification: The analyst should understand that completion of the biochemical tests does not yield identification of *Salmonella*. The cultures that have been biochemically confirmed should be verified serologically.

6. Serological Testing for *Salmonella*

6.1 Summary: Serological typing of *Salmonella* strains is done by using slide-agglutination for somatic (O) antigens and tube testing for flagellar (H) antigens.

6.2 Scope and Application: Serological testing completes the identification of *Salmonella*. It is the only testing which identifies to serotype and bioserotype levels. However, serological testing is an expensive, complex procedure that should be carried out only by trained personnel.

6.3 Apparatus and Materials

6.3.1 Small inverted fluorescent lamp.

6.3.2 Incubator set at 35 $\pm$ 0.5 C, and a water bath set at 50 C.

TABLE III-E-2

Production Rate and Time Requirements of Multitest Systems*

Multitest System	Analyst's Time Per Culture in Min.	Time Span of Test in Hours	Cultures Per Day Per Analyst
API	10	18–22	80
Enterotube	6	18–24	100
Inolex	30	7	10–15
r/b Diff.	8	24	80
Pathotec	15	4	20–30
Minitek	10	18–24	80

*Based on experience in EMSL-Cincinnati.

TABLE III-E-3

Reported Shelf-Life of Multitest Systems With or Without Refrigeration

System	Refrigeration Required	No Refrigeration*
API	12 months	—
Inolex	—	12 months
Pathotec	3 years**	12 months
Enterotube	7 months	—
r/b	6–12 months	—
Minitek	2 years	—

*Store in cool, dark place, ambient temperature.

**Refrigeration not required, but will extend the shelf-life.

TABLE III-E-4

Cost and Source of Multitest Systems[1]

Multitest System	Cost per Unit	Cost per Box	Address of Manufacturer
API Enteric 20	$2.05[2]	$51.25 (25/box)	Analytab Products, Inc. 200 Express Street Plainview, NY 11803
Improved Enterotube	$2.16	$54.00 (GSA) (25/box)	Roche Diagnostics Div. of Hoffmann-La Roche, Inc. Nutley, NJ 07110
Inolex	$0.91	$22.80 (25/box)	Inolex Biomedical Division Inolex Corporation 3 Science Road Glenwood, IL 60425
r/b Enteric Differential System	$1.90 3 tubes/set	$38.60 (20 sets/box)	Diagnostic Research, Inc. 25 Lumber Road Roslyn, Long Island, NY 11576
Pathotec Test Strips	$0.20	$20.00 (100 tests/box)	Warner-Lambert Company General Diagnostics Division Morris Plains, NJ 07950
Minitek	$1.80[3]	$90.00 (50 tests/kit)	Baltimore Biological Laboratories Cockeysville, MD 21030

1. As of October, 1977.
2. Plus $99.00/year, if the numerical identification system, Analytical Profile Inolex Service, is used.
3. Requires one time purchase of accessories for $94.10.

TABLE III-E-5

Biochemical Characteristics of the Enterobacteriaceae

Reaction	Group I					Group II				Group III	Group IV
	Escherichia	*Edwardsiella*	*Citrobacter*	*Salmonella*	*Shigella*	*Klebsiella*	*Enterobacter*	*Hafnia*	*Serratia*	*Proteus*	*Yersinia*
Catalase	+	+	+	+	D[a]	+	+	+	+	+	+
Oxidase	−	−	−	−	−	−	−	−	−	−	−
B-Galactosidase	+	−	+	D	d	+	+	+	+	−	+
Gas from Glucose at 35 C	+	+	+	+	−	d	+	+	d	D	−
KCN (growth on)	−	−	+	D	−	+	+	+	+	+	−
Mucate (acid)	+	−	+	D	−	d	d	−	−		
Nitrate reduction	+	+	+	+	+	+	+	+	+	+	+
Carbohydrates: (acid production)											
Adonitol	−	−	−	−	−	d	+	−	d	D	D
Arabinose	+	−	+	+		+	+	+	−	−	+
Dulcitol	d	−	d	D	d	d	−	−	−	−	−
Esculin	d	−	d	−			D	−	−	d	D
Inositol	−	−	−	d	−	+	D	−	d	D	−
Lactose	+ or ×	−	+ or ×	D	D	D	+	−	−	−	−
Maltose	+	+	+	+		+	+	+	+	D	+
Mannitol	+	−	+	+	D	+	+	+	+	D	+
Salicin	d	−	d	−	−	+	+	−	+	d	D
Sorbitol	+	−	+	+		+	+	−		−	D
Sucrose	d	−	d	−	D	+	+	− or ×	+	D	D
Trehalose	+	−	+	+		+	+	+	+	d	+
Xylose	d	−	+	+	D	+	+	+	d	D	D
Related C sources:											
Citrate	−	−	+	+	−	d	+	+	+		−
Gluconate		−		−	−	+	+	+	+		
Malonate	−	−	d	D	−	D	+		−		−
d-Tartrate	d	−	+	D		d	−	−			D
M.R.	+	+	+	+	+	D	−	−	−	+	+
V.P.	−	−	−	−	−	D	+	+	D	d	−
Protein reactions:											
Arginine	d	−	d	+	−	−	D	−	−	−	−
Gelatin hydrolysis	−	−	−	D	−	(d)	(+)	−	+	D	−
H_2S from TSI	−	+	D	+	−	−	−	−	−	D	D
Indole	+	+	D	−	D	d	−	−	−	D	D
Lysine decarboxylase	+	+	−	+	−	d	D	+	+	d	−
Ornithine	d	+	d	+	d	−	+	+	+	D	D
Urea hydrolysis	−	−	(+)	−	−	d	(d)	−	−	D	D
Glutamic acid	−	−	−	−	−	−				+	
Phenylalanine	−	−	−	−	−	−	−	−	−	+	−

[a] D = different reactions given by different species of a genus; d = different reactions given by different strains of a species or serotype; X = late and irregularly positive (mutative).

From Bergey's Manual of Determinative Bacteriology, Eighth Ed. (2)

6.3.3 Bunsen/Fisher burner or electric incinerator.

6.3.4 Inoculating loop (3 mm) and needle.

6.3.5 McFarland Barium Sulfate Standard #10 (Difco 0691).

6.3.6 Test tubes, 150 × 25 or 150 × 20 mm and 100 × 13 mm.

6.3.7 Glass microscope slides, 5.0 × 7.6 cm (2" × 3") cleaned to remove grease and oil.

6.3.8 *Salmonella* O Antiserum Poly (Difco 2264, BBL 40707, Sylvana 27–108A or equivalent).

6.3.9 *Salmonella* Vi Antiserum (Sylvana 27–106A, BBL 40708, Difco 2827, or equivalent).

6.3.10 *Salmonella* O Antisera Set A-I (Difco 2892) includes 1 vial each of Groups, A, B, C_1, C_2, D, E_1, E_1, E_2, E_4, F, G, H and I; Poly A-I and Vi.

Salmonella Grouping Serum Kit (BBL 40709) includes one vial each of Groups A, B, C_1, C_2, D and E Polyvalent and Vi.

6.3.11 *Salmonella* H Antiserum (BBL 407–99, Difco 2406–47 or equivalent).

6.3.12 Capillary pipets with rubber bulb.

6.4 Media

6.4.1 Brain heart infusion (BHI) broth (BBL 11058, Difco 0027 or equivalent) (Part II-B, 5.4.5).

6.4.2 Brain heart infusion (BHI) agar slant (BBL 11064, Difco 0418 or equivalent) (Part II-B, 5.4.6).

6.4.3 Motility medium (Difco 0869, BBL 11436, or equivalent) (Part II-B, 5.5.16).

6.4.4 H-broth (Difco 0451, BBL 11289, or equivalent) (Part II-B, 5.5.17).

6.4.5 Blood agar base (without blood) (Difco 0045–02, BBL 11036, or equivalent) (Part II-B, 5.4.18).

6.4.6 Nutrient agar (Difco 0001–02, BBL 11471, or equivalent) (Part II-D, 5.1.1).

6.4.7 Phenolized saline (0.6 grams phenol in 100 ml of 0.5–0.85% NaCl solution).

6.4.8 Formalinized saline (0.6 ml formalin in 100 ml of 0.5–0.85% NaCl solution).

6.5 Procedure: as described in Figure III-E-1, there are serological procedures for the 3 antigen groups, O, Vi and H. However, it is usually necessary only to test for the O and Vi antigens for verification of the *Salmonella* identification.

6.5.1 Slide Agglutination Test for O Grouping (30)

(a) Prepare a dense suspension of organisms from a fresh 24 hour BHI slant in 0.5 ml of phenolized saline solution. The suspension should be homogeneous and at least as concentrated as that of McFarland Barium Sulfate Standard #10, which corresponds to 3×10^9 cells/ml.

(b) Mark rectangular areas on an alcohol-cleaned glass slide with a wax pencil. Mark heavy continuous lines to prevent flow of suspension from one section to another as shown in the example below. For safety, it is recommended that the outside perimeter be inscribed with a wax line to prevent flow off the edge of the slide. Note that slide sections 1, 2 and 3 are controls and section 4 is the one complete test:

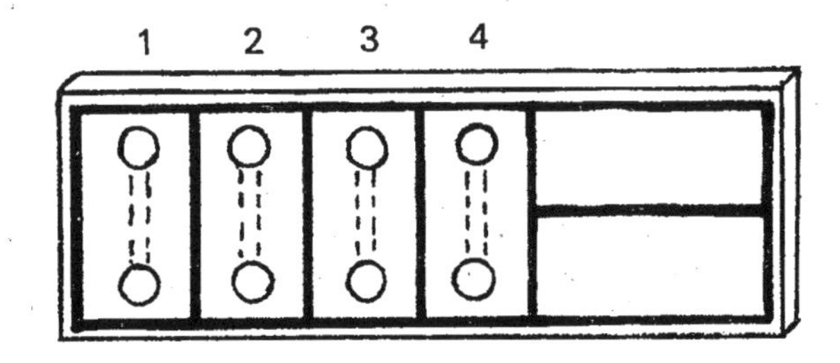

Section 1 – Add antiserum alone.

Section 2 – Combine antiserum and 0.85% NaCl.

Section 3 – Combine bacterial suspension and 0.85% NaCl.

Section 4 – Combine bacterial suspension, 0.85% NaCl and antiserum.

For large numbers of cultures, the 2 × 3 inch (50 × 75 cm) glass slide can be used to accommodate 12 or more sections per slide in 2 rows. It is recommended that only one culture be tested at one time against the series of sera to avoid premature drying. As in the following example:

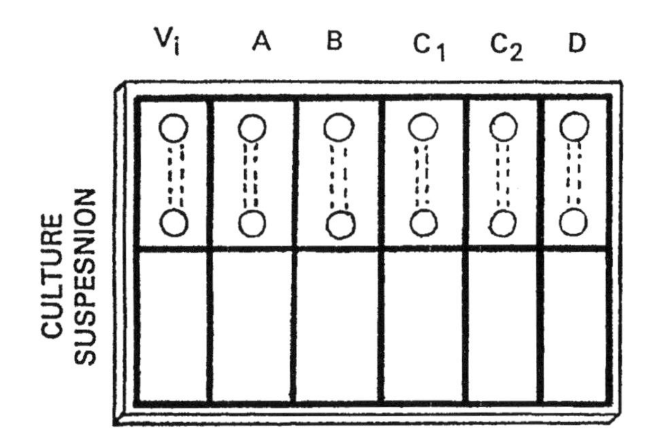

(c) Place a drop of *Salmonella* polyvalent O antiserum near the top of a rectangle and a drop of saline near the bottom.

(d) Using a wire loop, suspend a sufficient amount of growth from an isolated colony or a BHI slant into the drop of saline to produce a milky suspension.

(e) Mix the suspension with the antiserum, using the wire loop. Make a long, narrow track rather than a circular pool.

(f) Flame loop well between uses to prevent cross-reactions from contamination of sera.

(g) Mix the antigen and antiserum further by tilting the slide back and forth until agglutination (or clumping) is apparent. If agglutination is not evident or if it is weak at the end of 1 minute, consider the reaction negative. Compare reactions with controls. As a QC function, test antisera against cultures of known reactions, monthly or as indicated.

(h) If the polyvalent test is positive, test the culture with *Salmonella* O groups A, B, C, D and E antisera. Kits containing additional O group antisera are described in 6.3.

6.5.2 Slide Agglutination Test for Vi Antigen (*S. typhi*) (30): Occasionally *S. typhi* will be isolated without the capsule-like Vi antigen. The procedure described in 6.5.1 will identify it as *Salmonella*, O group D. If the biochemical reactions are characteristic of *S. typhi*, record the results presumptively as *S. typhi*, no Vi antigen detected and in this case, check other colonies from same plate for Vi antigen. If Vi antigen is present, it will block O agglutination and the procedure described in 6.5.1 will not identify the serogroup. If *S. typhi* is suspected, proceed as follows:

(a) To three rectangles on a glass slide, add respectively:

(1) Polyvalent *Salmonella* antiserum.

(2) Group D *Salmonella* O antiserum.

(3) Vi antiserum.

(b) Mix with antigen as in procedure 6.5.1.

(c) Reactions and interpretation:

(1) No agglutination in any rectangle – not a common *Salmonella*.

(2) Agglutination in polyvalent antiserum only – possibly a *Salmonella* other than Group D. Check antigen in other O groups.

(3) Agglutination in Vi antiserum only or in Vi and weakly in polyvalent – presumptive *S. typhi*.

(d) Follow-up on reactions (3) immediately above:

(1) Make a heavy suspension of antigen in 0.5 ml of phenolized saline.

(2) Heat the suspension for 15 minutes in a boiling water bath.

(3) Cool and retest suspension in the above three antisera. Include the three controls as in 6.5.1 (b).

(4) Compare results of slide agglutination obtained with live and heated antigens. *S. typhi* will give the following reactions:

Antiserum	Antigen	
	Live	Heated
Polyvalent *Salmonella*	+ or −	+
O group D	−	+
Vi	+	−

6.5.3 Alternative Procedure for Salmonella, Including *S. typhi*: Edwards & Ewing and Douglas & Washington describe this alternate procedure as a more rapid serological method (1, 30).

(a) Slide test the organism for agglutination in polyvalent O, Vi, and all of the common O group antisera at the same time. When indicated, heat antigen as described in 6.5.2 (d).

(b) Interpret results as outlined in 6.5.1 and 6.5.2.

6.5.4 Tube Test for H Antigen (1): Before tests are made for H (flagellar) antigen, test for motility by inoculating the pure cultures into Motility Test Medium (5.5.16), incubating and reading. If the tests are negative, transfer the cultures again through the motility medium before performing the flagellar antigen test. Motility medium in large diameter tubes or small petri dishes may be inoculated on one side and motile descendants picked on the other side after 24 hours. The test procedure follows:

(a) Inoculate H-broth from an 18–24 hour pure culture and incubate 18 hours at 30 C or with heavy inoculum in 35 C water bath for 4 hours.

(b) Dilute the 18 hour culture 1:1 with formalinized saline and mix. Allow to stand at room temperature for 1 hour.

(c) Pipet 0.5 ml of the formalinized culture to small test tubes (13 × 100 mm), 1 tube/antiserum.

(d) Prepare dilution of polyvalent H antisera according to directions of manufacturer (usually 1:1000). Add 0.5 ml to each tube using a fresh pipet for each tube. Mix by pipetting the solution up and down several times.

(e) Incubate tubes for 1 hour in 50 C water bath. If the H-antigen is present, flocculation/agglutination may occur in 5–10 minutes but will occur within 1 hour. The positive test is indicated by a diffused, fluffy sediment in the bottom of the tube. A negative reaction gives a tight "button-like" group of cells in the bottom of the tube.

6.5.5 Confirmation of *Salmonella* to serotype by an official typing center or state health laboratory is recommended when required for tracing sources of contamination, for enforcement or for other Agency actions. This service is usually available if cultures are of public health significance. See II-C, 6 for details on proper shipment procedures.

7. Quantitative Techniques

7.1 Summary: These quantitative methods are time-consuming, but necessary when it is desirable to determine salmonellae densities in recreational waters, shellfish-raising waters, stormwater run-off, wastewaters and sludges.

7.1.1 Samples are concentrated if necessary by the standard techniques described in this Section.

7.1.2 Samples or sample concentrates are inoculated into enrichment media in the standard 5-tube, 3-dilution sequence of the MPN. See Part II-C.

7.1.3 After incubation, each tube culture is streaked on selective plating media.

7.1.4 Colonies reacting as *Salmonella* are picked and confirmed biochemically or serologically.

7.1.5 Based on the confirmation above, the number of tubes confirmed as positive for *Salmonella* are tabulated and the MPN index calculated from the regular MPN table.

7.2 Concentration by MPN (15)

The need for concentration and sample volumes required vary with the anticipated salmonellae densities. High density samples such as sewage may be inoculated directly as 10 ml portions into double-strength, or as 1 ml or less portions into single strength enrichment media in MPN tubes (15). Low density samples may be concentrated by MF, diatomaceous earth, or cartridge filters before analysis by the MPN procedure.

7.3 Concentration by Membrane Filter or Diatomaceous Earth Filter (12, 31, 32, 33)

Another quantitative technique uses either a large MF (142 mm diameter, 0.45 μm pore) with filter aid (12) or diatomaceous earth with a support pad. One liter or larger sample is either mixed with 1% Celite and filtered through an MF or is filtered through the diatomaceous earth and pad. After filtering, the Celite/MF or diatomaceous plug and pad are placed in a sterile 1 pint (473 ml) blender jar containing 100 ml sterile 0.1% peptone water and blended at about 5000 RPM for 1 minute on a Waring blender (see Part II-C, 1.3). Beginning with 10 ml of the homogenate, serial tenfold dilutions (5 replicates per dilution) are inoculated into enrichment media. Verification tests are done on each positive enrichment tube and the MPN is calculated from the number of tubes which verify.

7.4 Concentration by Cartridge Filter (5)

The cartridge filter is particularly useful for concentrating large sample volumes (5). The sample is placed in a sterile, calibrated container and measured volumes (e.g. 10, 1 and 0.1 liters) are passed through separate filters as described in this Section, 2.4. To speed the analyses, 5 replicate portions are filtered simultaneously through 5 cartridge filters using a manifold. If turbidity is a problem, successive filters may be used. The filters are removed aseptically and placed in the selected enrichment media. The concentration procedure is repeated for each sample volume. Verification tests are done on each enrichment culture and the MPN is calculated from the positive verification tests.

8. Optional Fluorescent Antibody Screening Technique

The fluorescent antibody (FA) technique may be used as a rapid screening method for the detection of salmonellae particularly from large numbers of samples, after primary enrichment cultures (18-24 hours). Because this technique does not require pure cultures or serological typing, positive FA results should be confirmed by the isolation procedures described in this Section, 4.5. If the FA test is negative, the salmonellae detection procedure may be terminated.

8.1 Summary: An optional FA screening technique concentrates bacteria from a water sample in diatomaceous earth (Celite' or equivalent). See this Section, 2.2. The diatomaceous earth with the entrapped bacteria is added to enrichment broth. After incubation of the broth, loopfuls of enrichment culture are transferred to selective plating media to produce spot cultures on the agar surfaces. After the plates are incubated for 3 hours, slide impression smears are prepared from the micro-colonies and stained with a *Salmonella* polyvalent antiserum labeled with fluorescent dye. Fluorescent cells indicate a positive reaction and the possible presence of *Salmonella*.

8.2 Scope and Application

8.2.1 Advantages: Rapid screening of large numbers of samples for *Salmonella* eliminates negative samples from further testing.

8.2.2 Limitations: Careful interpretation of fluorescence is critical for this technique but difficult to attain. Positive FA results must be confirmed by the conventional cultural techniques described above (see Part III-E, 4, 5 and 6). The cost of equipment for fluorescence microscopy is approximately $5,000.

8.3 Apparatus and Materials

8.3.1 Equipment

(a) Light microscope and low autofluorescence optics suitable for UV microscopy.

(b) Condenser, Cardioid Dark-Field, oil immersion objective 95×, with iris diaphragm and with a numerical aperture (N.A.) at least 0.05 less than N.A. of objective lens.

(c) Filter Systems

Heat Filters – BG22, KG-1, KG-2 or equivalent.

Excitation Light Filters – UG1 and BG12 or equivalent.

Barrier Filters – GG-9 and OG1 or equivalent.

(d) Intense Light System: Fluorescence illuminator with power source, voltage regulator and mercury arc, quartz iodide or tungsten light source.

(e) Incubator set at 35 C ± 0.5 C.

8.3.2 Materials

(a) Non-drying low fluorescence immersion oil.

(b) Fluorescent antibody pre-cleaned micro slides, 2.5 × 7.6 cm.

(c) Cover glasses for FA slides, .16-.19 mm thick.

(d) FA Kirkpatrick fixative (Difco 3188, or equivalent).

(e) Petri dishes, 100 × 15 mm, pyrex glass.

(f) Nail polish.

(g) Phosphate buffered saline, pH 8.0. Add 10 grams of dry FA buffer (Difco 2314–33, or equivalent) to 1 liter of fresh laboratory pure water. Dissolve completely and adjust pH to 8.0 with NaOH.

(h) FA mounting fluid (Difco 2329–57, or equivalent), reagent grade glycerine adjusted to pH of 9.0–9.6.

(i) Laboratory pure water.

(j) Methanol or ethanol, 95%, for sterilizing forceps.

(k) Staining assembly consisting of jar, cover and slide rack with handle. At least 5 are needed: I each for fixative, ethanol and laboratory pure water and 2 for buffered saline rinses.

(l) FA *Salmonella* polyvalent conjugate is a fluorescein conjugated anti-salmonellae globulin (Difco 3187, 3185; Sylvana 27–100A, or equivalent).

When the conjugate is rehydrated, prepare 0.2–0.3 ml aliquots and freeze for future use. Rehydrated conjugates stored in the refrigerator are not stable. (See 8.4.5 for titration of conjugates).

(m) Moist chamber used to hold slides containing conjugated-stained smears. A simple chamber consists of a culture dish bottom (150 × 20 mm) placed over wet toweling. A larger dish for this purpose may be prepared by placing moist towels onto a flat tray, then placing the slides face-up on this surface and covering the tray with an inverted glass baking pan or similar metal pan.

8.3.3 Media

(a) Brain heart infusion (BHI) broth (Difco 0038, BBL 11059, or equivalent). See Part II-B, 5.4.5..

(b) Brilliant green agar (BG) (BBL 11073, Difco 0285–01). See Part II-B, 5.5.5.

(c) Xylose lysine desoxycholate agar (XLD) (BBL 11838, Difco 0788–01). See Part II-B, 5.5.7.

(d) Xylose lysine brilliant green agar (XLBG) XL agar base (BBL 11836, Difco 0555). See Part II-B, 5.5.6.

8.4 FA Staining Techniques (13, 34, 35)

8.4.1 Preparation

Collect samples and concentrate them by cartridge filter, membrane filter or diatomaceous earth technique as described in this Section, 2.2, 2.3 and 2.4.

After concentration, place whole or half plugs and pads, cartridge filters or membrane filters into flasks of selenite broth, tetrathionate broth, dulcitol selenite broth or tetrathionate brilliant green broth. Incubate flasks for 18-24 hours at 35 and 41.5 C. See this Section, 3.

8.4.2 Spot Culture Plates: After the primary enrichment has been incubated 18-24 hours, prepare spot culture plates on the differential media: brilliant green (BG) agar, xylose lysine desoxycholate (XLD) agar, and xylose lysine brilliant green (XLBG) agar as follows:

(a) Mark the bottom of the plate to locate the drops of inoculum then place 1 drop of enrichment culture at 4 separate points on the surface of each solid medium.

(b) Space the drops 6–7 cm apart on the plate so that when an FA microscope slide is later placed over 2 of the drops, both inoculation points will be included on the slide. This is essential, since glass slide impression smears of the inoculated points will be made after incubation of the plates.

(c) After the spots have been placed on the agar surface of the differential plates, incubate the plates at 35 C for 3 hours.

(d) Remove the plates after incubation and make impression smears.

8.4.3 Impression Slides: Place a clean FA 1 × 3 inch glass slide (nonfluorescent) over 2 of the inoculated points on the agar. Press down lightly, without moving the glass slide to either side. Too much pressure will cause movement of the slide and accumulation of agar on the slide. Repeat the process for the remaining 2 inoculation points with another clean slide. Make impression slides for each differential plate.

8.4.4 Fixing Slides

(a) Air dry the smears.

(b) Fix for 2 minutes in Kirkpartick's Fixative (mix 60 ml absolute ethanol, 30 ml chloroform, and 10 ml formalin).

(c) Rinse the slides briefly in 95% ethanol.

(d) Air dry. Do not blot.

8.4.5 Titration of Conjugates

After the slides have been fixed, stain with *Salmonella* FA conjugate. However, each lot number of conjugate must be titered before use.

Based on the previous experience of the analyst or other workers, the conjugate and test components must have been proven with water samples. Sources of polyvalent conjugates known to be acceptable (CDC Approved List) are: Burroughs Wellcome & Company, Clinical Sciences Inc., Difco Laboratories and Sylvana Company.

The Difco Bacto-FA Poly Conjugate contains antibodies against the major O, Vi and H antigens. The Sylvana *Salmonella* Polyvalent OH Globulin represents the somatic antigens of *Salmonella* O groups A through S. The Difco Panvalent conjugate is not recommended because of excessive cross-reactions.

Titer the conjugate as follows:

(a) Inoculate a known *Salmonella* culture into BHI broth. Incubate at 35 C for 18–24 hours.

(b) From this broth culture, make smears on clean FA glass slides. Prepare enough smears for each conjugate dilution and controls. Include a known 4+ *Salmonella* control to be used as a reading standard. Air dry the smears and follow the instructions for fixing slides, Part III-E, 8.4.4.

(c) Prepare the following dilutions of the conjugate in buffered saline: 1:2, 1:4, 1:6, 1:8 and 1:10. Most batches are effective at the 1:4 dilution.

(d) Cover a smear with one of the conjugate dilutions or the undiluted conjugate. Place slides in moist chamber and proceed as in Part III-E, 8.4.6 (b)-(f) and 8.4.7 (a)-(b). Use the known 4+ *Salmonella* control as the standard.

(e) The titer of conjugate to be used is the second highest dilution which gives 4+ fluorescence. For example, if the conjugate dilutions outlined above in 8.4.5 (c) above gave fluorescence intensity ratings (in order) 4+, 4+, 4+, 2+, 1+, the conjugate dilution to be employed would be 1:4. This value should be marked on the conjugate bottle. Prepare an amount of chosen conjugate dilution sufficient for each day's run.

8.4.6 Staining Impression Slides: After the slides have been fixed, they are stained with *Salmonella* FA conjugate.

(a) Cover each smear with one drop of the predetermined dilutions of conjugate (Difco FA Salmonella Polyvalent, or Sylvana Polyvalent OH Globulin). Include a known 4+ *Salmonella* culture as a procedure control. Place the slides in a moist chamber to prevent evaporation of the staining reagent. Tap off excess reagent onto paper towel.

(b) After 30 minutes, wash away the excess reagent by dipping the slides into a bath of saline buffered at pH 7.2.

(c) Transfer the slides to a second bath of buffered saline for 10 minutes.

(d) Replace the rinse solutions in (b) and (c) above, after each use.

(e) Remove slides, rinse in a laboratory pure water bath and allow to drain dry. Do not blot.

(f) Place a small drop of FA mounting fluid (Difco 2329–57) onto the smear and cover with a coverslip. Seal the edges of the coverslip with nail polish. Such slides may be stored for up to 1 year in a freezer at –70 C with minimal loss of fluorescence.

(g) Examine the slides under a fluorescence microscope (900–1000×) fitted with the proper filters. Use a known 4+ slide for comparison. Read slides within 2½ hours if slides have not been sealed and stored in a deep-freeze.

8.4.7 Interpretation of Fluorescence, Reporting Results

(a) Fluoresence results are recorded as shown in the following table. The number of fluorescing cells and the degree of fluorescence/cell are the criteria on which positive results are based. Smears with large numbers of strongly fluorescent cells (3+ or 4+) are positve; weakly fluorescent cells are negative. Smears with few fluorescent cells should be examined carefully.

(b) Cultures giving positive FA reactions (4+ or 3+) must be isolated on differential plating media, confirmed by biochemical identification, and verified serologically. Cultures displaying negative FA results may be discarded.

Rating	Intensity	Description
Positive	4+	Brilliant yellow-green fluorescence, cell sharply-outlined
	3+	Bright yellow-green fluorescence, cell sharply outlined with dark centers
Negative	2+	Dull yellow-green fluorescence, cells not sharply outlined
	1+	Faint green discernible in dense areas, cells not outlined
	0	No fluorescence

REFERENCES

1. Edwards, P. R. and W. H. Ewing, 1972. Identification of Enterobacteriaceae (3rd ed). Burgess Publishing Co., Minneapolis, MN.

2. Buchanan, R. E. and N. E. Gibbons, 1974. Bergey's Manual of Determination Bacteriology (8th ed). The Williams and Wilkins Company, Baltimore, MD.

3. Moore, B., 1948. The detection of paratyphoid carriers in towns by means of sewage examination. Monthly Bulletin of the Ministry of Health and the Public Health Laboratory Service (G. Brit.) 1:241.

4. Brezenski, F. T. and R. Russomanno, 1969. The detection and use of salmonellae in studying polluted tidal estuaries. J. WPCP 41:725.

5. Levin, M. A., J. R. Fischer and V. J. Cabelli, 1974. Quantitative large-volume sampling technique. Appl. Microbiol. 28:515.

6. McCoy, J. H., 1962. The isolation of *Salmonella*. J. Appl. Bacteriol. 25:213.

7. Raj, H., 1966. Enrichment medium for selection of *Salmonella* from fish homogenate. Appl. Microbiol. 14:12.

8. Dunn, C. and W. J. Martin, 1971. Comparison of media for the isolation of salmonellae and shigellae from fecal specimens. Appl. Microbiol. 22:17.

9. Harvey, R. W. S. and T. H. Price, 1968. Elevated temperature incubation of enrichment media for the isolation of *Salmonella* from heavily contaminated materials. J. Hyg. Camb. 66:377.

10. Kampelmacher, E. H. and L. M. van Norle Jansen, 1971. Reduction of *Salmonella* in compost in a hog-fattening farm oxidation vat. J. Wat. Poll. Cont. Fed. 43:1541.

11. Spino, D. F., 1966. Elevated temperature technique for the isolation of *Salmonella* from streams. Appl. Microbiol. 14:591.

12. Cheng, C. M., W. C. Boyle and J. M. Goepfert, 1971. Rapid, quantitative method for *Salmonella* detection in polluted water. Appl. Microbiol. 21:662.

13. Cherry, W. et al., 1972. Salmonellae as an index of pollution of surface waters. Appl. Microbiol. 24:334.

14. Buffer, J. R., 1971. Comparison of the isolation of salmonellae from human feces by enrichment at 37 C and at 43 C. Zbl. Bakt. I. Abt. 217:35.

15. Phirke, P. M., 1974. Elevated temperature technique for enumeration of salmonellae in sewage. Indian J. Med. Res. 62:6:938.

16. Harvey, R. W. and T. H. Price, 1976. Isolation of salmonellae from sewage-polluted river water using selenite F and Muller-Kauffmann tetrathionate. J. Hyg. Camb. 77:333.

17. Edel, W. and E. H. Kampelmacher, 1939. Comparative studies on *Salmonella* isolation in eight European laboratories. Bull. World Health Org. 39:487.

18. Kampelmacher, E. H. and L. M. Van Noorle Jansen, 1973. Comparative studies on isolation of *Salmonella* from effluents. Zbl. Bakt. Hyg., I. Abt. Org. 157:71.

19. Vanderpost, J. M. and J. B. Bell, 1975. A bacteriological investigation of meat-packing plant effluents in the Province of Alberta with particular emphasis on *Salmonella*. Environmental Protection Service, Environment Canada, Report No. EPS 3-WP-76-9.

20. Cherry, W. et al., 1954. A simple procedure for the identification of the genus *Salmonella* by means of a specific bacteriophage. J. Lab. Clin. Med. 44:51.

21. Wolkos, S., M. Schreiber and H. Baer, 1974. Identification of *Salmonella* with O-1 bacteriophage. Appl. Microbiol. 28:618.

22. Gillies, R. R., 1956. An evaluation of two composite media for preliminary identification of *Shigella* and *Salmonella*. J. Clin. Pathol. 9:368.

23. Media and Tests for Differentiation of Enterobacteriaceae, 1970. Center for Disease Control, USDHEW, PHS, Atlanta, GA.

24. Nord, C. E., A. A. Lindberg and A. Dahlback, 1974. Evaluation of five test—kits—API, Auxotab, Enterotube, Pathotec and r/b—for identification of Enterobacteriaceae. Med. Microbiol. 159:211.

25. Shayegani, M., M. E. Hubbard, T. Hiscott and D. McGlynn, 1975. Evaluation of the r/b and minitek systems for identification of Enterobacteriaceae. J. of Clin. Microbiol. 1:6:504.

26. Tomfohrde, K. M., D. L. Rhoden, P. B. Smith and A. Balows, 1973. Evaluation of the redesigned Enterotube—a system for the identification of Enterobacteriaceae. Appl. Microbiol. 25:301.

27. Kiehn, T. E., K. Brennan and P. D. Ellner, 1974. Evaluation of the minitek system for identification of Enterobacteriaceae. Appl. Microbiol. 28:668.

28. Robertson, E. A. and J. D. MacLowry, 1974. Mathematical analysis of the API enteric 20 profile register using a computer diagnostic model. Appl. Microbiol. 28:691.

29. Robertson, E. A. and J. D. MacLowry, 1975. Construction of an interpretive pattern directory for the API 10 S kit and analysis of its diagnotstic accuracy. J. of Clin. Microbiol. 1:6:515.

30. Douglas, G. W. and J. A Washington, 1973. Identification of Enterobacteriaceae in the clinical laboratory, Center for Disease Control, Public Health Service, USDHEW, Atlanta, GA.

31. Presnell, M. N. and W. H. Andrews, 1976. Use of the membrane filter and a filter aid for concentrating and enumerating indicator bacteria and *Salmonella* from estuarine waters. Water Research 10:549.

32. Kaper, J. B., G. S. Sayler, M. M. Baldini and R. R. Colwell, 1977. Ambient-Temperature primary nonselective enrichment for isolation of *Salmonella* spp. from an estuarine environment. Appl. & Environ. Microbiol. 33:4:829.

33. Olivieri, V. P. and S. C. Riggio, 1976. Experience on the assay of microorganisms in urban runoff, in proceeding of workshop on microorganisms in urban stormwater, EPA-60012-76-244, Municipal Environmental Research Laboratory, EPA, Cincinnati, OH.

34. Thomason, B. M., 1971. Rapid detection of *Salmonella* microcolonies by fluorescent antibody. Appl. Microbiol. 22:1064.

35. Thomason, B. M. and J. G. Wells, 1971. Preparation and testing of polyvalent conjugates for fluorescent-antibody detection of salmonellae. Appl. Microbiol. 22:876.

PART III. ANALYTICAL METHODOLOGY

Section F Actinomycetes

1. Summary of Method

The pour-plate technique is the prinicipal method for measuring actinomycete populations. Selective media are employed that favor actinomycete development over fungi and other bacteria. The mass of branching filaments characteristic of this bacterial group offers distinctive features for identification. To facilitate identification and counting of the actinomycete colonies, plates are prepared by the two-layer agar technique. Since only the upper layer is inoculated, the method assures a predominance of surface colonies.

2. Definition

2.1 The actinomycetes are a group of microorganisms with cells ranging from 0.5-2.0 µm diameter but normally less than 1.0 µm diameter, which usually develop as non-septate hyphae in branching mycelial masses. The actinomycetes are generally saprophytic but some are parasitic or pathogenic to plants, animals and man.

2.2 The actinomycetes are fungal in morphology and in spore formation, but lack a membrane around nuclear materials. They have a sensitivity to bacterial antibiotics, are susceptible to specific phage and have other biochemical characteristics which class them as filamentuous, branching bacteria (1). Although actinomycetes are found in water and sediments, the greatest natural reservoir for these organisms is the soil.

3. Scope and Application

3.1 Actinomycetes are of interest in water treatment and waste treatment facilities because of the taste and odor problems they cause in potable waters and fish, and the foaming problems they can cause in waste treatment plants.

3.2 The taste and odor problems result from volatile products characterized by an intense earthy-musty odor (2, 3). Evidence points to 2 highly odoriferous metabolites, geosmin and 2-methylisoborneol, as the sources of the problem (4, 5, 6). It appears that the relative abundance of these 2 metabolites in natural waters is linked to an ecological factor not yet resolved. Traces of these products can impart a disagreeable persistent odor to a municipal water supply, which is extremely difficult to treat. These natural odorants, prevalent in many parts of the world, can also affect commercial fishing.

3.3 Actinomycete distribution in waters with an earthy-musty odor shows a correlation between actinomycete counts and odor levels, indicating such enumeration to be a useful parameter in measuring quality of water. The genus of most interest is *Streptomyces*.

3.4 Actinomycetes have also been recognized as the cause of disturbances in the operation of activated sludge wastewater treatment plants where massive growths of these organisms can produce thick foams (7, 8). The genus involved is *Nocardia*.

3.5 Because of diverse nutrient requirements, no single medium has been devised that will support the growth of all actinomycetes. Moreover, the culture media that have proven useful in their isolation are not necessarily the preferred media for encouraging abundant growth. The isolation media are restrictive and nutritionally poor. They act by depressing growth of other microorganisms and by favoring a higher proportion of actinomycete colony development.

3.6 The pour-plate method does not indicate whether the isolated colony originated from individual spores, spore aggregates, small mycelial fragments or a mycelial mat.

4. Apparatus and Materials

4.1 Incubator set at 28 $\pm$ 0.5 C.

4.2 Water bath set at 44–46 C for tempering agar.

4.3 Electric oven set at 105–110 C.

4.4 Hand tally or electronic counting device (optional).

4.5 Thermometer which has been checked against a National Bureau of Standards-Certified Thermometer.

4.6 Pipet containers of stainless steel, aluminum or pyrex glass for glass pipets.

4.7 Petri dish canister of stainless steel or aluminum for glass dishes.

4.8 Erlenmeyer flasks, pyrex, screw-cap, 250 and 500 ml volume.

4.9 Sterile T.D., bacteriological or Mohr pipets, glass or plastic, of appropriate size.

4.10 Sterile 100 mm $\times$ 15 mm petri dishes, glass or plastic.

4.11 Screw-cap culture tubes, borosilicate glass, 25 $\times$ 150 mm.

4.12 Dilution bottles (milk dilution), pyrex, marked at 99 ml, screw-cap with neoprene rubber liner.

5. Media

5.1 Sterile starch-casein agar or equivalent agar prepared in 17 ml volumes in screw-cap tubes and in 250–300 ml volumes in 500 ml, screw-cap bottles or flasks. See Part II-B, 5.6.

5.2 Sterile buffered dilution water in bottles containing 99 $\pm$ 2 ml volumes. See Part II-B, 7.

5.3 Cycloheximide Stock Solution. Weigh out 100 mg cycloheximide (antifungal antibiotic), bring to 100 ml with distilled water, pour into a screw-cap flask and mix until dissolved. Sterilize for 15 minutes at 121 C (15 lbs. pressure). Cycloheximide is available as *Actidione* from the Upjohn Company.

6. Sample Preparation

6.1 Water Samples

6.1.1 Fill sample bottle only 3/4 full so that ample airspace is left in the bottle for thorough mixing of the water sample.

6.1.2 Mix water sample by shaking vigorously about 25 times. Using a pipet, transfer 11 ml immediately after mixing to a 99 ml water blank.

6.1.3 Repeat for desired dilutions. A dilution of 10^{-3} is usually sufficient for plating water samples.

6.2 Soil Samples

6.2.1 Mix soil sample thoroughly and weigh out a 50 gram sample in a tared weighing pan. Dry at 105–110 C to constant weight.

The final weight is used in calculating numbers of organisms/gram dry soil.

6.2.2 Prepare the initial dilution by weighing out 11 grams of soil and adding to a 99 ml volume of buffered dilution water for a 1:10 dilution. Shake dilution bottle vigorously for 1 minute.

6.2.3 Transfer a 11 ml sample of the 1:10 dilution to a second dilution bottle containing 99 ml buffered dilution water and shake vigorously about 25 times. Repeat this process until the desired dilution is reached. Dilution of 10^{-3} to 10^{-6} are usually necessary for enumeration of soil samples.

7. Plating Procedures

Prepare triplicate plates for each test dilution using the two-layer agar technique as follows:

7.1 Cool flask of starch casein agar to 44–46 C and pour 15 ml layer in each petri dish. Allow to harden.

7.2 Melt starch-casein agar in tubes and cool to 44–46 C. Add 2 ml of sample or sample dilution and 1 ml of cycloheximide solution. Mix tube contents well.

7.3 Pipet immediately 5 ml of the inoculated agar over the solid agar base layer in each plate with gentle swirling to evenly distribute the inoculated agar. Each 5 ml contains 0.5 ml of the particular dilution used. This 0.5 factor must be taken into consideration in calculating the final dilution.

7.4 After plates are solidified, invert and incubate at 28 C for approximately 7 days and count.

8. Counting of Colonies

8.1 Select plates for counting with 30–300 colonies.

8.2 Rules for making plate counts are given in Part III-A, 5.6.2.

8.3 Examine plates macroscopically holding toward a light source. Actinomycete colonies appear dull or chalky when covered with aerial mycelium. The edges of the colonies are less dense producing a halo effect. Colonies adhere strongly to the agar and have a tough, leathery texture. In contrast, bacterial colonies appear shiny or opalescent with a soft texture, adhere weakly to the agar, and show no general distinction between the edge and the colony as a whole (see Figure III-F-1).

8.4 Actinomycete colonies can be verified at a magnification of 100×. Because of their filamentous growth, they typically have fuzzy borders which contrast sharply to the smooth borders characteristic of bacteria. (See Figures III-F-2 and 3).

8.5 Addition of cycloheximide to the isolation medium suppresses development of fungal colonies. If fungi do develop, they can be differentiated from actinomycetes by their woolly appearance and much larger cell diameter. With a little experience in examining fungal and actinomycete colonies, it is fairly easy to distinguish them macroscopically.

9. Reporting Results

9.1 Calculate the actinomycete density in water samples in counts/ml according to the following equation:

$$\frac{\text{Sum of Replicate Plate Counts}}{\text{Total Volume of Original Sample Tested, in ml}} =$$

$$\frac{\text{Sum of Replicate Plate Counts}}{\text{No of Replicates} \times \text{Sample Dilution Tested} \times \text{Agar Plate Dilution Factor (See 7.3)}}$$

$$= \text{Actinomycete Count/ml}$$

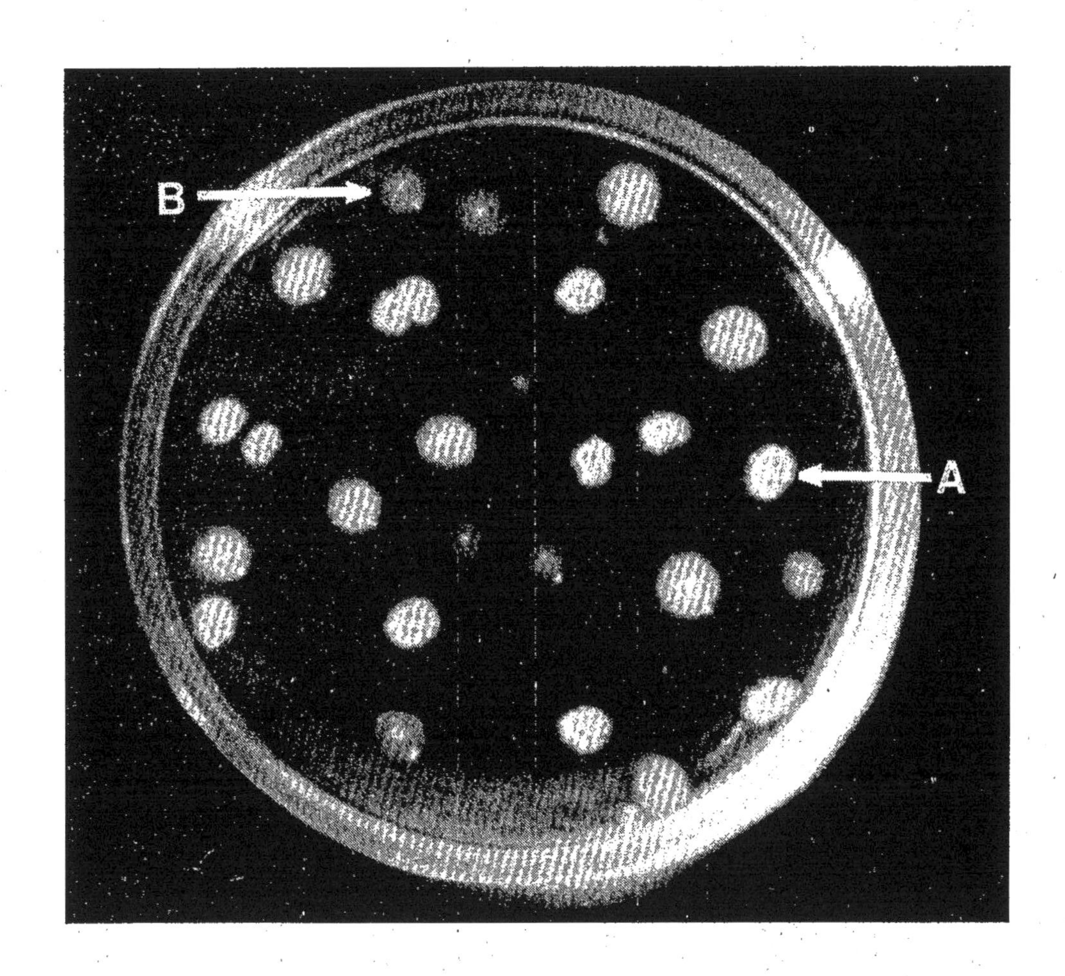

FIGURE III-F-1. A plate containing bacterial and actinomycete colonies.

A— The dull, powdery appearance of a sporulating actinomycete colony.

B— The smooth, mucoid appearance of a bacterial colony.

FIGURE III-F-2. An actinomycete colony showing the branching filaments that cause the fuzzy appearance of its border. × 225

FIGURE III-F-3. A bacterial colony with its relatively-distinct, smooth border. × 225

For example, if the triplicate sample volumes tested at a 1:10 dilution yielded 45, 39 and 42 actinomycete colonies, the calculation would be:

$$\frac{126}{3(\text{rep}) \times 0.1\ (\text{dil}) \times 0.5\ (\text{A.P.D. factor})} =$$

$$\frac{126}{0.15} = \begin{array}{c}840\\ \text{Actinomycetes/ml}\end{array}$$

9.2 Correct the actinomycete counts from soil and mud samples for water contents. See 6.2, this Section. Calculate the counts as follows:

$$\frac{\text{wt. of collected soil}}{\text{wt. of dried soil}} = \text{conversion factor}$$

conversion factor $\times$ count/gm collected soil = count/gm dry soil

Determine count/gm collected soil by the equation given in 9.1.

REFERENCES

1. Lechevalier, H. A. and M. P. Lechevalier, 1967. Biology of Actinomycetes. Am. Rev. Microbiol. 21:71.

2. Adams, B. A., 1929. *Cladothrix dichotoma* and allied organisms as a cause of an "indeterminate" taste in chlorinated water. Water & Water Eng. 31:327.

3. Thaysen, A. C., 1936. The origin of an earthy or muddy taint in fish. Ann. Appl. Biol. 23:99.

4. Rosen, A. S., C. I. Mashni and R. S. Safferman, 1970. Recent developments in the chemistry of odour in water: the cause of earthy/musty odour. Water Treatment Exam. 19:106.

5. Piet, G. J., B. C. J. Zoeteman and A. J. A. Kraayeveld, 1972. Earthy-smelling substance in surface waters of the Netherlands. Water Treatment Exam. 21:281.

6. Yurkowski, M. and J. L. Tabachek, 1974. Identification, analysis and removal of geosmin from muddy-flavored trout. J. Fish. Res. Board Can. 31:1851.

7. Lechevalier, H. A., 1975. Actinomycetes of sewage-treatment plants. U.S. Environmental Protection Agency, Environmental Protection Technology Series, EPA-600/2–75–031, Cincinnati, OH.

8. Lechevalier, M. P. and H. A. Lechevalier, 1974. *Nocardia amarae*, sp. nov., an actinomycete common in foaming activated sludge. Int. J. Syst. Bacteriol. 24:278.

PART IV. QUALITY ASSURANCE

Regulatory agencies making decisions on water quality standards and wastewater discharge limits require formal analytical quality control programs for their laboratories and program participants to assure validity of their data. For example, water quality regulations include provisions for quality control and testing procedures (the Safe Drinking Water Act, National Interim Primary Drinking Water Regulations published in Federal Register, Title 40 Part 141, 59566, December 24, 1975). Further, the Act states that analyses for maximum contaminant levels must be conducted by laboratories approved in a formal certification program. The quality control procedures specified in the Manual for the Interim Certification of Laboratories Involved in Analyzing Public Drinking Water Supplies (Appendix B) are a recommended minimal program and are not equivalent to the comprehensive system described in Part IV of this Manual.

A laboratory quality control program is the orderly application of the practices necessary to remove or reduce the errors that occur in any laboratory operation due to personnel, equipment, supplies, sampling procedures and the analytical methodology in use.

A quality control program must be practical, integrated and require only a reasonable amount of time or it will be by-passed. When properly administered, a balanced, conscientiously applied quality control program will assure the production of uniformly high quality data without interfering with the primary analytical functions of the laboratory. This within-laboratory program should be supplemented by participation of the laboratory in an interlaboratory quality control program such as that conducted by EPA and detailed in Part V. The major considerations for quality control are discussed under three separate Sections:

Section A Laboratory Operations

1. Sample Collection and Handling
2. Laboratory Facilities
3. Laboratory Personnel
4. Laboratory Equipment and Instrumentation
5. General Laboratory Supplies
6. Membrane Filters
7. Culture Media

Section B Statistics for Microbiology

1. Measures of Control Tendency
2. Measures of Dispersion
3. Normal Distribution
4. Poisson Distribution
5. Measures of Performance

Section C Analytical Quality Control Procedures

1. Quality Control on Routine Analyses
2. Quality Control in Compliance Monitoring
3. Comparative Testing of Methodologies

PART IV. QUALITY ASSURANCE

Section A Laboratory Operations

Section A describes the checks and monitoring procedures that should be performed on materials, supplies, instrumentation and the physical facility. These checks should be documented completely and recorded as performed. See Part V-A, 1.2 for details on this documentation.

1. **Sample Collection and Handling**
2. **Laboratory Facilities**
3. **Laboratory Personnel**
4. **Laboratory Equipment and Instrumentation**
5. **General Laboratory Supplies**
6. **Membrane Filters**
7. **Culture Media**

1. Sample Collection and Handling

The acquisition of valid data begins with collection of a representative water sample or other environmental material being tested. Samples must be maintained as closely as possible to original condition by careful handling and storage.

Sample sites and a sampling frequency are selected to provide data representative of the characteristics and the variability of the water quality at that station. The most important quality control factor in sampling is the immediate analysis of the sample. If the sample cannot be analyzed at once, it should be refrigerated and analyzed within six hours. Recommended procedures for collecting, transporting and handling water and wastewater samples are described separately in Part II-A of this Manual.

2. Laboratory Facilities

2.1 Ventilation: Laboratories should be well-ventilated and free of dust, drafts and extreme temperature changes. Central air conditioning has advantages: 1) The incoming air is filtered, reducing contamination of the laboratory and culture work. 2) The uniform temperature control of air conditioning permits stable operation of incubators. 3) With closed windows, drafts and air currents which can cross-contaminate surroundings and work areas during warm weather are minimized. 4) Low humidity reduces moisture problems with media, chemicals, analytical balances and other instrumentation.

2.2 Space Utilization: Ideally the areas provided for the preparation and sterilization of media, glassware and equipment should be separated from the laboratory working area but located close enough for convenience. In public health laboratories that analyze many different types of samples a separate work area is desirable for water analysis. Special work areas such as an absolute barrier or vented laminar flow hood (see Part V-C) are often used for dispensing and preparing sterile media and tissue cultures, for transferring microbial cultures or for working with pathogenic materials. In smaller laboratories it may be necessary to carry out these separate activities in different sections of the same room. However, limited facilities and restricted work space may seriously hamper the quality of the work and influence the validity of results. Visitors and through traffic should be discouraged in work areas. Through traffic can be prevented by laboratory design.

2.3 Laboratory Bench Areas: Sufficient clean bench space should be available for the analyses to be performed efficiently. For routine work, 6 linear feet is the recommended minimum work area for each analyst. Research work or other analyses using specialized equipment may require significantly more space per worker. These estimates of bench space are exclusive of work areas used for preparatory and supporting activities. Laboratory lighting should be even, screened to reduce glare, and provide about 100 footcandle light intensity at working surfaces.

Bench tops should be set at heights of 36–38 inches with a depth of 28–30 inches. This height is comfortable for work in a standing or sitting position. Desk tops or sit-down benches are set at 30–31 inches height to accommodate microscopy, plate counting, calculations and writing activities. Bench tops should be stainless steel, epoxy plastic or other smooth impervious material which is inert, corrosion-resistant and has minimum seams.

2.4 Walls and Floors: Walls should be covered with waterproof paint, enamel or other surface material that provides a smooth finish which is easily cleaned and disinfected. Floors should be covered with good quality tiles or other heavy duty material which can be maintained with skid-proof wax. Bacteriostatic agents contained in some wall or floor finishes increase the effectiveness of disinfection.

2.5 Monitoring for Cleanliness in Work Areas: High standards of cleanliness should be maintained in work areas. The laboratory can be monitored for cleanliness by one or more of the procedures described below. Since these monitoring procedures cannot recover all of the microorganism populations present, absolute limits are difficult to develop. Rather, the tests should be used regularly on a weekly or other basis to monitor counts in the same work areas over time or to make comparisons between different work areas.

2.5.1 RODAC Agar Plates (1)

Work areas and other surfaces can be checked by RODAC plates which contain general growth media for total counts or selective media for coliforms, enteric pathogens, streptococci, staphylococci, or other microorganisms cultured in the laboratory. RODAC is an acronym for Reproducible Organism Detection and Counting. The RODAC dish has a test area of about 25 cm with Quebec style grids embossed in the plate. It is specifically designed to enumerate the microbial population of flat solid surfaces by contact techniques.

(a) Purchase RODAC plates prefilled with desired test medium or prepare plates by filling the center well with about 16 ml of appropriate agar. When preparing agar in the laboratory, add 0.07% soy lecithin or 0.5% polysorbate (Tween 80) to the agar to neutralize the effect of the disinfectant on test surfaces. Cooling leaves a raised bed of agar about 1 mm higher than the rim of the dish, allowing contact of the sterile agar to the test surface for direct counts.

(b) To sample an area, remove the plastic cover and carefully press the agar to the solid surface being sampled. It is important that the entire agar layer contacts the test surface. Use a rolling motion with uniform pressure on the back of the plate to insure complete contact. Replace the cover and incubate in an inverted position for the appropriate time and temperature.

(c) Count the colonies with a Quebec colony counter and report as the number of colonies per RODAC plate or number of colonies per 25 sq. cm.

2.5.2 Swab Method (1)

(a) The swab contact method can be used to monitor the contamination of work areas and especially those with cracks, corners, crevices and rough surfaces. Dacron swabs rinse out more easily than cotton ones. These may be purchased from Econ Microbiological Laboratory, 2716 Humboldt Avenue, South, Minneapolis, MN. Rayon swabs are available from Fuller Pharmaceutical Company, Minneapolis, MN, and Consolidated Laboratories Inc., Chicago Heights, IL. Swabs made of calcium alginate which is soluble in water, can also be used. These are available from Consolidated Laboratories, Inc., Chicago Heights, IL.

Rinse Solution Vials – Add 1.25 ml stock phosphate buffer solution, 5 ml of 10% aqueous sodium thiosulfate, 4 grams Asolectin (Associated Concentrates, 32–30 61st Street, Woodside, Long Island, NY 11377) and 10 grams Tween 20 or Tween 80 (Hilltop Research, Inc., Miamiville, OH 45147) to 500 ml distilled water; heat to solution in boiling water bath. Cool and make up to 1 liter. Dispense in screw-capped vials in 10 ml or other volumes and sterilize for 15 minutes at 15 lb. pressure (121 C). Because Asolectin is hygroscopic, store it in a desiccator and weigh quickly.

(b) To sample an area, open a sterile swab container, grasp the end of the stick and remove aseptically. Open a 10 ml vial of neutralizing buffer, moisten swab head, and press out excess solution against the inside of the vial. Hold the swab handle at a 30° angle against the sampling area. Rub the swab head slowly and throughly over approximately 8 sq. in. of surface area. Repeat procedure 3 times over the same area, turning the swab and reversing directions. Place the swab head in the neutralizing buffer vial, rinse briefly in solution and press out the excess liquid.

Test four more 8 sq. in. areas of surface, rinsing the swab in solution after each swabbing. After the fifth area has been swabbed, remove the excess liquid, place swab head in vial and break or cut with a sterile scissors leaving the swab head in the vial. Replace the cap and store vial at 4 C or on ice until analysis.

When using alginate swabs, rinse solution vials should contain 4.5 ml after sterilization. Prepare a 10% solution of sodium hexametaphosphate in appropriate vials and steam sterilize. Follow the above swabbing procedures but after swabbing the fifth 8 sq. in. area, deposit the swab head in the rinse vial and add 0.5 ml of sterile hexametaphosphate solution.

To begin plating, shake the vial vigorously to dislodge the bacteria from the swab. Pipet 1 ml and 0.1 ml aliquots of the rinse solution. The 1 ml portions represent a 1:10 dilution and the 0.1 ml portions represent a 1:100 dilution. Pour the plates with Standard Methods Agar. Mix and cool to solidify and incubate at 35 C for 48 hours.

(c) Count colonies and convert the value to the number/ml which equals the count per 8 sq. in. of area.

2.5.3 Air Density Plates (1)

The number of microorganisms in the laboratory air is directly proportional to the amount and kind of activity. These organisms affect results when the suspended cells contact materials and equipment or when they settle on exposed test materials and surfaces.

(a) The numbers and types of airborne microorganisms can be determined by exposing petri plates for a specified time at points where inoculating, filtering, plating and transfer work is done. This exposure method can be used to monitor total bacteria or the specific organisms being tested such as coliforms, enteric pathogens, streptococci, staphylococci, yeasts or molds.

(b) Pour the petri dishes with the appropriate agar and allow to harden. Store poured plates in the refrigerator if they are not used on the same day.

(c) Remove petri dish covers and place dishes top side up on sterile towels. Expose the plates in selected work sites for 15 minutes and mark them with sample site identification. Replace covers. Incubate Standard Methods Agar at 35 C for 48 hours and other media for the specified time periods.

(d) Count the colonies and report the number per square foot. The number of organisms which settle in 15 minutes of exposure on a petri dish is equivalent to that for 1 sq. ft./minute because the area of a standard-size petri dish is approximately 1/15 sq. ft. The microbial density should not normally exceed 15 colonies per sq. ft. Air density plates should be taken weekly during peak work periods for routine monitoring.

2.6 Laboratory Maintenance: Laboratory benches, shelves, floors and windows should be cleaned on a scheduled basis. Proper maintenance is evidenced by lack of dust and soil build-up on shelves, in corners, etc. Floors should be wet-mopped and treated with a disinfectant solution to reduce contamination of air in the laboratory. Sweeping or dry-mopping should not be permitted in a microbiology laboratory. Work benches should be wiped down with disinfectant before and after each use and at the end of the day.

Every attempt should be made to maintain the laboratory areas free of clutter. Laboratory benches, space under benches, shelves and drawers accumulate equipment, materials and supplies at a steady rate. The disorder can interfere with laboratory operations. It can be controlled by making a directed effort to clean up work areas immediately after each use and by conducting a weekly clean-up of the laboratory. Discard unneeded materials and store equipment and supplies not in current use. The most important step in maintaining an orderly laboratory is to have ample storage space near-by.

3. Laboratory Personnel

Microbiologists, technicians and support personnel in the microbiology laboratory must have training and experience appropriate to the laboratory's analytical program. The variety and complexity of the tasks and tests performed determine the professional and on-the-job-training required.

3.1 Professional Microbiologist: The microbiologist directs and participates in the collection and storage of samples, the preparation of glassware, equipment and media, the analysis of a variety of waters and wastewaters and the evaluation of procedures as required. He takes part in the quality control program. He counts, tabulates and summarizes data and prepares or helps to prepare reports from the results. He should have at least a BS degree in microbiology or a BS/BA degree in biology with a minor in microbiology.

The professional microbiologist initially carries out routine tasks and progresses to more difficult work as he gains experience. When he reaches the senior level he performs the most complex duties.

3.2 Senior Grade and Supervisory Microbiologists: A senior grade microbiologist performs the duties described in 3.1 but carries out a wider range of assignments and more difficult tests. He participates in program planning and laboratory management and solves significant microbiological problems that require a higher than bench-level skill and knowledge. The experienced microbiologist consults and advises on procedural problems and trains personnel. He provides expert testimony, interprets results, prepares reports and recommends microbiological standards or actions for regulatory programs.

The supervisory microbiologist performs administrative functions of: planning and directing water quality monitoring programs, designing field surveys, advising administrative officers on policy matters related to microbiology and inspecting laboratory facilities. He is recognized for his authoritative scientific competence.

3.3 Technicians: The technician performs semi-professional and professional duties of limited scope and complexity. Typically, he assists professionals by doing the routine tests.

Under supervision, he performs tasks involving a series of steps. Technicians learn through on-the-job instruction and by performing standard tasks repetitively.

Technicians' tasks begin at the simplest level but can progress to the more detailed procedures performed by the professional microbiologist. However, these higher levels of skill and knowledge are generally limited to the specific areas in which they have received on-the-job-training.

3.4 Support Personnel: Laboratory aides and clerks provide the necessary support services to the laboratory. Aides prepare glassware, make media and sterilize materials. One or more aides are recommended for every bench microbiologist. Clerical duties include recording data, filing records, typing reports, ordering supplies, maintaining inventories, distributing mail and answering the telephone.

3.5 On-the-Job-Training and Experience: In addition to the formal academic training, the professional microbiologist should have experience in aquatic or environmental microbiology. Technical or on-the-job-training for at least two weeks is required for each parameter tested in water microbiology. The formal training of the technician ranges from high school to technical or academic training short of a degree in microbiology. Technical and scientific experience is often substituted for advanced formal training.

The microbiologist and technician should be encouraged to attend courses at centers of expertise such as universities, commercial manufacturers, US EPA facilities, the Center for Disease Control (CDC) of USPHS, FDA and other federal and state governmental agencies. Courses might include fluorescent antibody techniques, isolation and identification of the Enterobacteriaceae and laboratory safety.

3.6 Laboratory Supervison: Ideally, the laboratory should be directed by a professional microbiologist. However, in a small laboratory where the staff consists of a single non-professional technician, an approved consultant microbiologist should be available for guidance and assistance.

Work assignments in the laboratory should be clearly defined. The analyst should be trained in basic laboratory procedures and should perform well at a given level of responsibility before new assignments are made. The supervisor should periodically review procedures such as sample collecting and handling, media and glassware preparation, sterilization, routine testing procedures, counting, data handling, and quality control techniques. Problem areas should be identified and solved by the staff. Improved laboratory results will be the measure of effective personnel practice and training.

4. Laboratory Equipment and Instrumentation

Quality control of laboratory apparatus includes servicing and monitoring the operation of incubators, waterbaths, hot-air sterilizing ovens, autoclaves, water stills, refrigerators, freezers, etc. Each item of equipment should be tested to verify that it meets the manufacturer's claims and the user's needs for accuracy and precision. Maintenance should be performed on a regular basis by a technician who is familiar with the equipment.

4.1 A summary of recommended monitoring procedures for laboratory equipment is given in Table IV-A-1.

4.2 Monitoring Procedures for UV Lamps

4.2.1 Spread Plate Irradiation

This test should be run when a new UV lamp is installed and rerun quarterly to measure continuing effectiveness of the UV irradiation.

(a) Prepare 100 ml or more of plate count agar. Pour 10–15 ml of the melted agar into each 100 mm petri dish as needed. Keep covers opened slightly until agar has hardened and moisture and condensation have evaporated. Close dishes and store in refrigerator until use.

(b) Prepare a series of dilutions of a coliform culture so that 0.5 ml of inoculum will give a 200–250 colony count (see Geldreich and Clark).

(c) Pipet 0.5 ml of the selected dilution to each agar plate.

(d) Remove a glass spreader-rod from the alcohol container and ignite by passing through flame. Let burn completely and cool for 15 seconds. Test glass rod on edge of agar to verify safe temperature before use. Glass spreaders may also be autoclaved.

(e) Place sterile, cool glass spreader on agar surface next to inoculum. Position spreader so that the tip forms a radius from the center to the plate edge. Holding stick motionless, rotate plate several revolutions, or hold the plate and move the stick in a series of sweeping arcs to spread the inoculum uniformly over the entire surface of the agar.

(f) Lift the glass spreader from the agar and place in disinfectant solution.

(g) Repeat the spread plate procedure with other petri plates.

(h) With cover removed, place agar spread plates under UV lamp at points where sterility is desired.

(i) Place one inoculated plate under ordinary laboratory lighting as a control.

(j) Expose plates for two minutes.

(k) Close plates and incubate at 35 C for 24 hours. Remove and examine for growth.

(1) Count all plates and record results. Control plate should contain 200–250 colonies. UV-irradiated plates should show 99% reduction in the count of the control plate. If count reduction is less than 80%, replace lamp.

4.2.2 Measurements of UV Light Intensity

To monitor performance of UV lamps, it is necessary to measure UV light intensity using the UV light meter manufactured for this purpose. The shortwave UV meter Model J-225 is available from UV Products, Inc., San Gabriel, CA 91778 for about $125. This instrument or equivalent is recommended.

(a) Measure and average light intensities of a new UV lamp at proper distances of use. Record results in QC log, noting readings and dates of installation.

(b) Monitor lamp intensity quarterly thereafter. Replace bulb when light is down to 80% of original intensity.

(c) Measure and average light intensities of replacement lamp at proper distances. Record readings and date of installation.

5. General Laboratory Supplies

5.1 Laboratory Glassware

5.1.1 A summary of procedures for maintenance of glassware is given in Table IV-A-2).

5.1.2 Glassware pH Check: Traces of some cleaning solutions are difficult to remove completely. Before using a batch of clean glassware, test several pieces for an alkaline or acid residue by adding and dispersing a few drops of bromthymol blue or other pH indicator and observing the color reactions. Bromthymol blue is particularly advantageous for this check because it shows color changes of yellow to blue green to blue, in the pH range 6.5 to 7.3.

To prepare bromthymol blue indicator, add 16 ml of 0.01 N NaOH to 0.1 gram of bromthymol blue. Dilute to 250 ml with distilled water (0.04% solution).

5.1.3 Test Procedure for Suitability of Detergents Used in Washing (2)

(a) Wash and rinse six, 100 mm diameter petri dishes in the usual manner. These are Group A.

(b) After normal washing, rinse a second group of 6 petri dishes 12 times with successive portions of non-toxic distilled water. These are Group B.

(c) Wash 6 petri dishes with the detergent wash water using detergent concentrations normally employed, and dry without rinsing. These are Group C.

(d) Sterilize dishes in the usual manner.

(e) Add the proper dilution (usually two different dilutions are used) of a water sample yielding 30–300 colonies to triplicate petri dishes from each Group (A, B and C). Proceed according to the Standard Plate Count Procedure in Part III-A.

(f) The results are interpreted as follows:

(1) Differences in bacterial counts of less than 15% among all Groups indicate the detergent has no toxicity or inhibitory effect.

(2) Differences in bacterial counts of 15% or more between Group A and B demonstrate that inhibitory residues are left on glassware after the normal washing procedure used.

(3) Disagreement in averages of less than 15% between Groups A and B, and greater than 15% between Groups A and C indicates that detergent used has inhibitory properties which are eliminated during routine washing.

5.1.4 Sterility Checks on Glassware

After sterilization of a load of bottles, flasks or tubes, test items of each type for sterility by adding to one of each aerobic or anaerobic broth medium (e.g., lauryl tryptose or fluid thioglycollate broth). Incubate and check for growth.

5.2 Pure Water Quality: Pure water systems are meant to produce the best possible water; however, the potential quality of the water will vary with the type of system used.

The quality of water obtainable from a pure water system differs with the system used and its maintenance. The acceptable limits are given in Table IV-A-3.

5.3 Water Suitability Test

5.3.1 Summary: The water suitability procedure of Geldreich and Clark (3) is a sensitive test for determination of toxic or stimulatory effects of distilled or deionized water on bacteria. It is based on the growth of *Enterobacter aerogenes* in a chemically defined medium. Reduction of 20% or more in the bacterial population compared to a control, is judged toxic. Increased growth greater than 300% is called stimulatory.

5.3.2 Scope and Application: The test is recommended for periodic use and as a special measure of water suitability. This test called the Test for Bacteriological Properties of Distilled water, in the 14th edition of *Standard Methods* is required annually of laboratories in the Interim Certification Program for Drinking Water Supplies.

However, the test is not easily done on an infrequent basis because it requires: work over a four day period to complete, an ultrapure control water and very pure reagents, and absolute cleanliness of culture flasks, petri dishes, test tubes and pipets, etc. It is a complex method that requires skill and experience, is very sensitive to toxicants and cannot be related directly to routine analytical results.

5.3.3 Apparatus and Materials

(a) Incubator set at 35 ± 0.5 C.

(b)Water bath for tempering agar set at 44-46 C.

(c) Colony counter, Quebec dark field model or equivalent.

(d) Hand tally or electronic counting device (optional).

(e) Pipet containers of stainless steel, aluminum or pyrex glass for glass pipets.

(f) Petri dish containers of stainless steel or aluminum for glass petri dishes.

(g) Thermometer certified by National Bureau of Standards or one of equivalent accuracy, with calibration chart.

(h) Sterile TD (To Deliver) bacteriological or Mohr pipets, glass or plastic of appropriate volumes, see Part II-B, 1.8.1.

(i) Sterile 100 mm × 15 mm petri dishes, glass or plastic.

(j) Sterile pyrex glass flasks, 100 ml, 500 ml, and 1000 ml volume.

(k) Dilution bottles (milk dilution), pyrex glass, marked at 99 ml volume, screw cap with neoprene rubber liner.

(l) Bunsen/Fisher gas burner or electric incinerator.

(m) Inoculation loops, at least 3 mm diameter, or needles, nichrome or platinum wire, 26 B&S gauge, in suitable holder.

5.3.4 Media

(a) Sterile Plate Count Agar (Tryptone Glucose Yeast Agar) dispensed in tubes (15 to 20 ml per tube) or in bulk quantities in screw cap flasks or dilution bottles. See Part II-B, 5.1.5.

(b) Sterile buffered dilution water, 99 ± 2 ml volumes, in screw capped dilution bottles. See Part II-B, 7.

(c) Nutrient agar slants with slopes approximately 6.3 cm (2½ in) length in 125 × 16 mm screw cap tubes.

5.3.5 Reagents: Sensitivity of the test depends on the purity of the reagents used. Use reagents of the highest purity and prepare them in water freshly redistilled from a glass still.

(a) Sodium citrate solution: Dissolve 0.29 g sodium citrate, $Na_3C_6H_5O_7 \cdot 2H_2O$, in 500 ml redistilled water.

(b) Ammonium sulfate solution: Dissolve 0.60 g ammonium sulfate, $(NH_4)_2SO_4$, in 500 ml redistilled water.

(c) Salt mixture solution: Dissolve 0.26 g magnesium sulfate, $MgSO_4 \cdot 7H_2O$; 0.17 g calcium chloride, $CaCl_2 \cdot 2H_2O$; 0.23 g ferrous sulfate, $FeSO_4 \cdot 7H_2O$; and 2.50 g sodium chloride, NaCl, in 500 ml redistilled water.

(d) Phosphate buffer solution: Stock phosphate buffer solution, diluted 1:25 in redistilled water, see Part II-B, 7.1 in this Manual.

Boil reagent solution 1-2 min.to kill vegetative cells. Store in sterilized glass-stoppered bottles in the dark at 5 C for several months but test for sterility before each period of use. Prepare the salt-mixture solution without the ferrous sulfate for long-term storage. To use the mixture, add an appropriate amount of the freshly prepared and freshly boiled iron salt. Solutions with a heavy turbidity should be discarded and a new solution prepared. Bacterial contamination may cause turbidity in the phosphate buffer solution. Discard if this occurs.

5.3.6 Culture: Isolate a pure culture of *Enterobacter aerogenes* (IMViC type --++) from a polluted stream or sewage sample.

5.3.7 Procedure

(a) Collect 200 ml each, of the unknown Test Water and the Control Water (laboratory pure reference water) in sterile 500 ml screw cap flasks. Boil for 2 minutes and cool to room temperature.

(b) Label five, sterile 200 ml screw cap flasks as A, B, C, D, and E. Add the Test Water, Control Water, and reagents to each flask as described in the following table:

Reagents	Control Test (ml)		Optional Tests (ml)		
	Control Water A	Unknown Test Water B	Nutrient Available C	Nitrogen Source D	Carbon Source E
Sodium citrate solution	2.5	2.5	-	2.5	-
Ammonium sulfate solution	2.5	2.5	-	-	2.5
Salt-mixture solution	2.5	2.5	2.5	2.5	2.5
Phosphate buffer (7.3 ± 0.1)	1.5	1.5	1.5	1.5	1.5
Test Water	-	21.0	21.0	21.0	21.0
Control Water	21.0	-	5.0	2.5	2.5
Total volume	30.0	30.0	30.0	30.0	30.0

(c) Perform a Standard Plate Count on prepared reagents, Control Water, and Test Water as a check on contamination.

(d) On the day before performing the distilled-water suitability test, inoculate a strain of *E. aerogenes* onto a nutrient agar slant. Streak the entire agar surface to develop a continuous-growth film and incubate 18-24 hours at 35 C. To harvest viable cells, pipet 1-2 ml sterile dilution water from a 99 ml water blank onto the 18-24 hour culture. Emulsify the growth on the slant by gently rubbing the bacterial film with the pipet, being careful not to tear the agar; then pipet the suspension back into the original 99 ml water blank.

(e) Make a 1:100 dilution of the original bottle into a second water blank, a further 1:100 dilution of the second bottle into a third water blank, then 10 ml of the third bottle into a fourth water blank, shaking vigorously after each transfer. Pipet 1.0 ml of the fourth dilution (1:10^{-6}) into each of Flasks A, B, C, D, and E. This procedure should result in a final dilution of the organisms to a range of 30-80 viable cells from each ml of test solution.

(f) Variations among strains of the same organism, different organisms, media, and surface area of agar slopes will possibly necessitate adjustment of the dilution procedure in order to arrive at a specific density range between 30 and 80 viable cells. To establish the growth range numerically for a specific organism and medium, make a series of plate counts from the third dilution to determine the bacterial density. Then choose the proper volume from this third dilution, which, when diluted by the 30 ml in Flasks A, B, C, D, and E, will contain 30-80 viable cells/ml. If the procedures are standardized as to surface area of the slant and laboratory technic, it is possible to reproduce results on repeated experiments with the same strain of microorganism.

(g) Add a suspension of *Enterobacter aerogenes* (IMViC type −−++) of such density that each flask will contain 30-80 cells/ml, prepared as directed above. Cell densities below this range result in ratios that are not consistent, while densities above 100 cells/ml decrease sensitivity to nutrients in the Test Water. Make an initial bacterial count by plating triplicate 1 ml portions from each culture flask in plate count agar. Mix well and incubate Test A through E at 35 C for 24 $\pm$ 2 hours. Prepare final plate counts from each flask, using dilutions of 1, 0.1, 0.01, 0.001, and 0.0001 ml.

5.3.8 Calculations

(a) For growth-inhibiting substances:

$$\text{Ratio} = \frac{\text{colony count/ml Flask B}}{\text{colony count/ml Flask A}}$$

A ratio of 0.8 to 1.2 (inclusive) shows no toxic substances; a ratio of less than 0.8 shows growth-inhibiting substances in the water sample.

(b) For nitrogen and carbon sources that promote growth:

$$\text{Ratio} = \frac{\text{colony count/ml Flask C}}{\text{colony count/ml Flask A}}$$

(c) For nitrogen sources that promote growth:

$$\text{Ratio} = \frac{\text{colony count/ml Flask D}}{\text{colony count/ml Flask A}}$$

(d) For carbon sources that promote bacterial growth:

$$\text{Ratio} = \frac{\text{colony count/ml Flask E}}{\text{colony count/ml Flask A}}$$

(e) Do not calculate ratios b, c, or d when ratio a indicates a toxic reaction. Ratios b, c, or d, in excess of 1.2 indicate an available source for bacterial growth.

5.3.9 Interpretation of Results

(a) The colony count from Flask A after 20-24 hours at 35 C, will depend on the number of organisms initially planted in Flask A and on the strain of *E. aerogenes* used in the test procedures. This is the reason the control, Flask A, must be run for each individual series of tests. However, for a given strain of *E. aerogenes* under identical environmental conditions, the terminal count should be reasonably constant when the initial plant is the same. The difference between 30 and 80 cells in the initial plant in Flask A will produce a three-fold difference in the final counts, providing the growth rate remains constant. Thus, it is essential that the initial colony counts on Flask A and Flask B be approximately equal to secure accurate data.

(b) When the ratio exceeds 1.2, it may be assumed that growth-stimulating substances are present. However, this procedure is extremely sensitive and ratios up to 3.0 would have little significance in actual practice. Therefore, when the ratio is between 1.2 and 3.0, Tests C, D, and E do not appear to be necessary except in special circumstances.

(c) Usually Flask C will be very low and Flasks D and E will have a ratio of less than 1.2 when the ratio of Flask B to Flask A is between 0.8 and 1.2. The limiting factors of growth in Flask A are the nitrogen and organic carbon present. An extremely large amount of ammonia nitrogen with no organic carbon could increase the ratio in Flask D above 1.2, or the absence of nitrogen with high carbon concentration could give ratios above 1.2 in Flask E, with a B:A ratio between 0.8 and 1.2.

(d) A ratio below 0.8 indicates that the water contains toxic substances, and this ratio includes all allowable tolerances. As indicated in the proceding paragraph, the ratio could go as high as 3.0 from 1.2 without any undesirable consequences.

5.4 Use Tests for Media, Membranes and Laboratory Pure Water

The Use Test, a pragmatic approach to evaluation of materials and supplies, uses the routine methods of analysis to compare current and new batches or lots. Such tests operate on the theory that if a stimulatory or toxic effect cannot be demonstrated in actual test use, there is no effect.

When a new shipment or lot of culture medium, membrane filters or a new source of laboratory pure water is to be used, or at annual testing period for water, conduct comparison tests of the current lot in use (reference lot) against the new lot (test lot) as follows:

5.4.1 Use a single batch of pure water, glassware, membrane filters, or other needed materials as specified to control all other variables except the one under study.

5.4.2 Use the reference lot and the test lot, to conduct parallel pour plate or MF plate count tests on five natural or treated water samples according to standard procedures in this Manual.

When comparing sources of pure water, conduct the tests in parallel using the reference water and test water separately for all water purposes in the tests (dilution, rinse, media preparation).

5.4.3 After incubation, compare the bacterial colonies from the two lots for size and appearance. If the colonies on the test lot are atypical or noticeably smaller than colonies on the reference lot plates, record the evidence of inhibition or other problem, regardless of count differences.

5.4.4 Count plates and calculate the individual counts per ml or per 100 ml, as required for final reporting values.

5.4.5 Transform the final reporting values to logarithms.

5.4.6 Compile the log-transformed results for the two lots in parallel columns and calculate the + or − difference, d, between the two transformed results for each sample.

5.4.7 Calculate the mean, $\bar{d}$, and the standard deviation s_d of these differences.

5.4.8 Calculate the Student's t statistic, using the number of samples as n:

$$t = \frac{\bar{d}}{s_d/\sqrt{n}}$$

5.4.9 Use the critical value, 2.78, selected from a Student's t table at the .05 significance level for five samples (4 degrees of freedom), for comparison against the calculated value.

5.4.10 If the calculated t value does not exceed 2.78, the lots do not produce significantly different results and the test lot is acceptable.

5.4.11 If the calculated t value exceeds 2.78, the lots produce significantly different results. If the test lot results exceed reference lot results, the test lot is more stimulatory. If the test lot results are less than the reference lot, the test lot is less stimulatory.

5.4.12 If condition 5.4.3 or 5.4.11 occurs, review test conditions, rerun the test and/or obtain different lots for testing and use.

5.5 Reagents in General: The quality of test reagents must be assured. They must be correctly prepared and properly stored. The following general rules should be followed:

5.5.1 Use ACS or AR grade chemicals that meet ACS specifications for preparing reagents. Impurities in uncertified or lesser grades of chemicals may inhibit bacterial growth, provide nutrients or fail to produce the desired reaction.

5.5.2 Date chemicals and reagents when received and when opened for use.

5.5.3 make reagents up to volume in volumetric flasks. For storage, transfer to good quality plastic (polyethylene, polypropylene or tetrafluoroethylene) or borosilicate glass bottles with polyethylene or other inert plastic stoppers or caps.

5.5.4 Identify prepared reagents with the generic name, the concentration, the date prepared and the initials of preparer.

5.5.5 Store reagents under the conditions recommended by the manufacturer.

5.5.6 Run positive and negative controls with each series of cultural or biochemical tests.

5.6 Serological Reagents

5.6.1 Evaluate serological antisera against known antigens and compare with antisera that have demonstrated acceptable reactivity. The quality of commercial serological reagents is subject to methodology changes and new knowledge in manufacturing processes. Continuity of reagent quality depends on compliance of each new reagent with minimum specifications.

5.6.2 Repeat quality control procedures each time that reagent batches are prepared, regardless of the expiration date given by the manufacturer.

5.6.3 Discard sera or antigens if contamination is discovered.

5.6.4 Select another working dilution if the level of activity has dropped.

5.7 Fluorescent Antibody Reagents

Highly specific reagents, antigens of high purity and very specific potent antibodies are required for fluorescent antibody techniques. The antisera must exhibit high staining intensities. Some sera may have high titers in one type of serological test, but demonstrate poor staining titer and vice versa.

5.7.1 Store desiccated fluorescent antibody (FA) reagents at 4 C. Prepare aliquots of the rehydrated conjugate in screw-cap glass vials, freeze and store at −20 C until working dilutions are prepared.

5.7.2 Test FA reagents for correct reactions with positive and negative controls before each use. Results of positive controls should be within one dilution of the average titer.

5.8 Dyes/Stains: Organic chemicals are used as selective agents (e.g., brilliant green in brilliant green lactose bile broth), as indicators in bacteriological media (phenol red lactose), and as bacteriological stains (gram stain). Dyes from any commercial supplier vary from lot to lot in percent dye, dye complex, insolubles and inert materials present. Because dyes for microbiological uses must be of proper strength and stability to produce the correct reactions, purchase only dyes which have been certified by the Biological Stain Commission for biological use.

5.8.1 Fluorescent Dyes: An important factor in the preparation of antisera-dye conjugates is the purity of the fluorescein dye. Infrared studies reveal that the purity of fluorescein isothiocyanate (FITC) commonly used in FA tests, ranges from 30 to 100% in commercial products. Ideally 100% pure FITC preparations are used. Dyes of less purity may be satisfactory if the weight used to label the protein component of the serum allows for the impurities, and if the impurities do not increase non-specific staining.

5.8.2 Check bacteriological stains before use with at least one positive and one negative control culture.

6. Membrane Filters

6.1 Government Specifications

6.1.1 The quality and performance of membrane filters vary with the manufacturer, type, brand, lot number and storage conditions. Membranes considered for purchase by federal agencies must conform to the government specifications published in 1974 Military Specifications (4). The detailed specifications required of the manufacturer by the federal government are an important advantage to other users of membrane filters because by meeting federal requirements, manufacturers realize it is most efficient to make all membranes to these specifications. Hence, membrane filters purchased from companies selling to federal agencies meet the federal specifications.

6.1.2 The specifications list requirements for pore size, porosity, flow rate, diffusibility, autoclavability, sterility, diameter, thickness, bacterial retention and bacterial cultivation. The gridlines must be easily visible, permanent and not imprinted into the surface so as to cause channelling of growth. The grid ink must not stimulate or inhibit growth.

6.1.3 The pads must have a specified diameter and thickness, limited acidity, no waxes, sulfites, and other stimulating or inhibitive materials, and must absorb uniformly a specified volume of medium.

6.2 Manufacturer's Quality Assurance

6.2.1 Some manufacturers certify that their membranes meet stated specifications on sterility, retention, recovery, pore size, flow rate, pH, total acidity, phosphate and other extractables.

6.2.2 Most manufacturers sell sterile membranes and those packaged for sterilization by the user. Membranes may be sterilized by autoclaving, ethylene oxide or irradiation.

6.3 User's Quality Assurance

6.3.1 In addition to the quality assurance provided by the manufacturer, the user should determine that the membranes perform satisfactorily by inspecting and testing each lot he orders. The membranes should not be misshapen or the gridlines distorted after autoclaving.

6.3.2 A membrane should allow good diffusibility of medium and be completely wetted within 15 seconds after it is placed on the medium substrate with no dry areas indicated by colorless or lighter-colored areas.

6.3.3 After incubation, the colonies should be of the expected size, with defined shape and clearly delineated edges. The color and morphology of the colonies must be typical of those defined by the test procedure.

6.3.4 The analyst should observe that the gridline ink does not "bleed" across the surface, or restrict colony development on or adjacent to the cross-hatching. Colonies should be evenly distributed across the membrane surface. Membranes containing sizable areas with no colony development are questionable.

6.4 ASTM Test Procedures

Because of variable recoveries of microorganisms on different membrane filters, there is a critical need for standard procedures for evaluating membrane filters. Procedures are being developed by Committee D-19.08.04 of ASTM for physical, chemical and microbiological characteristics of membrane filters. The procedures are intended for the larger laboratory and the manufacturer to test new batches or lots of MF's. Brief descriptions of these tests for bacterial retention, inhibitory effects, recovery, extractables and flow rate follow.

6.4.1 Bacterial Retention Test

The test is based on filtration of a standard culture through a 0.45 μm MF into a broth. Sterile equipment and aseptic techniques must be used. Five randomly-selected membranes from five randomly-selected packages should be tested. Control membranes should be taken from the same package as the test membranes.

(a) Add 140 ml of double strength Trypticase soy or Tryptic soy broth to a liter vacuum flask and attach vacuum tubing. Wrap flask in kraft paper and sterilize.

(b) Using an 18 hour broth culture of *Serratia marcescens,* prepare a final dilution in 0.1% peptone, containing 1000 cells per 100 ml.

(c) Assemble a membrane filtration apparatus using a sterile vacuum flask containing the broth. Insert the test filter, turn on vacuum and pour in 20 ml of sterile rinse water to set the membrane. Filter 100 ml of the culture suspension. Rinse funnel twice with 20 ml of peptone water.

(d) After the filtration remove funnel and transfer the membrane filter to a Trypticase soy or Tryptic soy agar plate. Then aseptically remove membrane filteration base from the flask and insert a sterile stopper. Repeat filtrations four more times with sterile MF assemblies, culture flasks and membranes.

(e) Incubate agar plates and flasks for 48 hours at 35 C. Examine membranes for growth outside of filtering areas which indicates a leaky filtration assembly that could cause a false positive test. Turbidity in the culture flask indicates bacterial growth and failure of the membrane to retain the 0.2–0.3 μm sized bacteria, but does not indicate the numbers of organisms passing the filter.

(f) Turbidity indicates filter failure. Repeat the test and control procedures.

6.4.2 Inhibitory Effects: The test for inhibitory effects compares membrane filter counts on one or more test lots of membrane filters, with spread plate counts on Plate Count Agar using a specific pure culture of *Escherichia coli* (IMViC + + − −).

(a) Prepare a dilution of the stabilized 24-hour culture in 0.1% peptone dilution water so that 0.2 ml contains 30–150 viable cells. Add about 20 ml of sterile dilution water to the funnel before filtration.

(b) Filter five replicate sample volumes of 0.2 ml through five replicate membrane filters of each MF test lot and place each on a plate of Tryptic soy agar.

(c) Prepare the corresponding spread plates in random fashion to avoid a time effect, following a randomization table.

(d) With the same pipet, deliver five additional 0.2 ml volumes to the surface of five plates of Tryptic soy agar. The samples are distributed evenly over the surfaces of the plates with sterile glass rods.

(e) Incubate the membrane filters at 44.5 C and agar plates at 35 C for 24 hours. Count the colonies. Compare recoveries on spread plates and membrane filters. Acceptable filters should recover some set percentage of the spread plate count. This percentage should be established by each laboratory based on previous performance by a known acceptable lot.

6.4.3 Recovery (5): The procedure for recovery compares the fecal coliform counts on test membranes to the counts on spread plates using M-FC agar substrate. Four polluted waters and one raw sewage sample are analyzed.

(a) To determine sample test volumes, prepare serial dilutions of each sample in 0.1% peptone water to produce a suspension containing approximately 20–60 fecal coliforms per 0.1 ml. Hold original samples at 4 C. Determine the fecal coliform density of each of the samples or dilutions by membrane filter or spread plate test. Read the results after 22–24 hours incubation at 44.5 ± 0.2 C.

(b) If the fecal coliform density in the raw water samples is less than 10 per 0.1 ml, seed the sample with raw sewage, allow to stabilize at 4 C for 24 hours and test as in (a). Then proceed with the standard test procedure.

(c) Use the optimum dilutions from (a) or (b) for the test membranes and the spread plate controls.

(d) Perform the membrane filter tests according to the procedure described in Part II-C, 3 of This Manual.

(e) Prior to beginning the spread plate tests, air dry the surface of the M-FC agar contained in 100 ml petri dishes. Aseptically deliver 0.1 ml of the selected sample dilution to the agar surface and spread with a sterile bent glass rod. Allow the sample aliquot to be completely absorbed before inverting the dish.

(f) Alternate MF tests with spread plate controls to randomize systematic errors.

(g) Insert petri dishes into waterproof bags or seal with waterproof tape, and submerge in the waterbath incubator. Incubate for 22–24 hours at 44.5 $\pm$ 0.2 C. Record the temperature continuously during the incubation period. After incubation, remove plates and examine.

(h) Count the blue colonies. If more than one dilution was prepared, select the plates with between 10 and 100 colonies, but preferably with 20–60 colonies. Calculate the arithmetic mean of the five replicate fecal coliform counts and the five replicate spread plates. Determine percent recovery:

$$\frac{\text{MF } \overline{X} \text{ Count}}{\text{Spread Plate } \overline{X} \text{ Count}} \times 100 = \% \text{ Recovery}$$

(i) Verification: Pick 20 blue colonies from each of 2 randomly-selected filters and 2 spread plates. If plates contain less than 20 colonies, pick all blue colonies. Verify the colonies in EC media as described in Part III-C, 4 of This Manual. To be acceptable, 80% of the colonies must verify. If the samples are known or suspected to have a high concentration of pseudomonads or aeromonads, perform the cytochrome oxidase test. As in the verification procedure, a minimum negative oxidase test confirmation of 80% should be achieved for the membranes to be considered acceptable.

6.4.4 Extractables

(a) Total Extractables

(1) Dry filters for 15 minutes at 70 C then bring to room temperature in a desiccator. Weigh to constant weight on a four-place analytical balance.

(2) Boil filters in 100 ml of distilled water for 20 minutes. Remove the filters and dry at 70 C for 60 minutes. Bring filters to room temperature in a desiccator and reweigh to constant weight. Weighings shall be to the nearest 0.1 milligram.

$$\frac{\text{Original Weight} - \text{Weight After Extraction}}{\text{Original Weight}} \times 100 = \text{Percent Extractables}$$

(b) Specific Extractables

(1) Immerse filter in ASTM Type 1 reagent grade water for 24 hours (6).

(2) Remove filter and assay the extract for metals, total organics, phosphorus and ammonia.

6.4.5 Flow Rate

Water flow rate shall be determined by timing the passage of 500 ml of particle-free distilled water through a filter at 25 C and at a differential pressure of 70 centimeters of mercury. Particle-free water for this test is produced by passage of a high purity water through a 0.2 μm membrane filter, three times in succession, at 25C and at a differential pressure of 177 cm of mercury.

$$\text{The flow rate} = \frac{\frac{60}{t} \times 500 \text{ ml}}{\text{EFA cm}^2}$$

$= \text{ml/min/cm}^2$

Where: t = experimental time in minutes for filtering 500 ml of particle-free distilled water

and

EFA = effective filtering area of a 47 mm diameter filter

An average flow rate reported by manufacturers for 0.45 μm pore size, 47 mm diameter filters is 65 ml/min/cm^2.

7. Culture Media

Since even the best cultural procedure is ineffectual if the medium is not prepared correctly, it is important to train personnel to use the best materials and techniques in media preparation, storage and application. Some factors that must be considered follow:

7.1 Ordering Media

7.1.1 Order media in quantities to last up to one year. Always use oldest stock first.

7.1.2 Whenever possible and practical, order media in 1/4 pound multiples rather than one pound bottles to insure sealed protection of the supply as long as possible. Most deterioration of media occurs after bottles are opened.

7.1.3 Maintain an inventory record of media: the dates received, sizes, number of units, etc. Review the inventory quarterly for necessary reordering. Date each bottle when received and when opened. Bottles should be inspected for color changes, caking or other indication of deterioration. Discard such bottles and reorder.

7.2 Holding Time Limits for Media

Because of the myriad of environmental conditions affecting media, and the unique composition and sensitivities of different media, it is impossible to establish universal time limits for holding unopened bottles of media. Therefore, a conservative and protective recommendation is to limit the storage of unopened bottles of cultural media to two years. This limit should insure good performance of media with proper storage conditions.

7.3 Preparation of Media

7.3.1 In high humidity areas, store opened bottles of media in a large hinged-door desiccator. Open bottles as briefly as possible during the weighing process and return to the desiccator immediately after use.

7.3.2 Discard opened bottles of media 6 months after initial use.

7.3.3 Weigh media to the nearest 0.1 gram on a single pan, top loader balance as quickly as possible.

7.3.4 Keep balance out of drafts and away from high humidity. Use a plastic shield around the balance to protect from drafts.

7.3.5 Clean the balance and surrounding area immediately after weighing media.

7.4 Solution of Media

7.4.1 Check cleanliness of glassware. Use bromthymol blue indicator to spot-check pH of glassware (5.1.2 in this Section).

7.4.2 Prepare media in a deionized or distilled water of proven quality. Ultra-pure water from a recirculating deionization system is recommended. Measure volumes with the greatest accuracy possible using the proper pipet or graduate (Part II-B, 1.8).

Check the pH of media after solution and after sterilization using a laboratory model pH meter. Enter results in the QC record book. The reading should be within 0.2 units of the stated value. If not, discard batch and remake. If the pH is still incorrect, use another bottle or batch. pH differences indicate a problem of distilled water quality, deterioration of medium or improperly-prepared medium. The problem must be identified and corrected. The problem should be carefully documented in the quality control record book and reported to the manufacturer if the data indicate the medium as the source of error.

7.4.3 Note in the QC record any unusual color development, darkening, or precipitation of media. Check the sterilization time and temperature. If there is a drastic change in appearance, discard medium and remake it. If the problem still exists, remake the batch using a different lot of medium.

7.4.4 Containers for preparation of batches of broth or agar should be twice the volume of the medium being prepared.

7.4.5 Media should be stirred continuously while being heated to avoid burning. Agar media are particularly susceptible to scorching and boilover. The only insurance against scorching or boilover is use of a boiling water bath for small batches of media or constant attention while heating larger batches on a hot plate or burner. A combination hot plate and magnetic stirrer is recommended for solution of media.

7.4.6 Bottles, tubes or plates of prepared media are identified and dated.

7.5 Sterilization by Autoclave

7.5.1 Media should be sterilized for the minimal time specified by the manufacturer. The amount of time required to sterilize a medium in an autoclave will vary with the type and volume of medium and the size and shape of containers. See the table in Part II-B, 4.3 on sterilization.

7.5.2 Since the potential for damage to media increases with increased exposure to heat, the amount of lag time before the autoclave is at full pressure and temperature can be a critical factor in whether media are damaged. The danger from an extended heating period is reduced with use of a double-walled autoclave which allows the operator to maintain full pressure and temperature in the jacket between loads. As soon as the autoclave is loaded and closed, steam can be admitted to the chamber and in a relatively short time, full pressure and temperature are developed in the chamber. The total exposure time for media sterilized 15 minutes at 121 C should not exceed 45 minutes.

7.5.3 Avoid overcrowding in an autoclave which reduces its efficiency. Large volumes of media should be preheated to reduce the lag period before placing them in the autoclave.

7.5.4 Remove sterile media from the autoclave as soon as pressure is at zero.

7.5.5 Media must be discarded if contamination is suspected. Reautoclaving is not permitted.

7.5.6 A preventive maintenance contract is recommended for autoclaves. It should include checks on the accuracy of pressure and temperature gauges and recorders and operability of safety valve.

7.5.7 Check the effectiveness of sterilization weekly, using strips or ampuls of *Bacillus stearothermophilus* spores. Commercial packages of these spores are available in ampuls of growth-indicator media. Sterilization at 121 C for 12–15 minutes will kill the spores. Incubate the autoclaved cultures at 55–60 C and if growth occurs, sterilization is inadequate.

7.6 Sterilization by Filtration

7.6.1 Non-autoclavable solutions can be sterilized by membrane filtration. With careful preparation of the sterile filtration and receiving apparatus, passage of a solution through a 0.2 µm membrane filter will produce a sterile solution.

7.6.2 The filtration and subsequent sterile dispensations should be performed in a safety cabinet or bio-hazard hood.

7.7 Gaseous Sterilization

Equipment, supplies or other solid or dry materials which are sensitive to heat can be sterilized by long exposure to ethylene oxide gas (ETO) using available commercial equipment. Check sterility with culture of *B. subtilus* var. *niger* spores.

7.8 Use of Agars, Broths, and Enrichment Media

7.8.1 Agars

(a) Agar plates to be used for streaking or spread plates are kept open slightly for 15 minutes after pouring or after taking out of refrigerated storage to evaporate free moisture which would cause confluent growth on streak plates.

(b) Agar plates used for MF and spread plate work must be free of lumps, uneven surfaces, pock marks, bubbles or foam which prevent good contact between the agar and the membrane or uniform growth on spread plates.

(c) Melted agars should be held in a tempering water bath at 44-46 C but no longer than three hours. As a safety precaution against the use of agars which are too hot and might kill cells, place a bottle of agar in the same boiling water or under the same autoclave conditions as the agars to be used. After the agar is melted, transfer agar to a tempering bath. Insert thermometer in the agar bottle and use it to determine when the temperature of the agars is at approximately 44-46 C and safe for use in pour plates.

7.8.2 Broths

(a) Handle sterile MPN fermentation tubes of lauryl tryptose broth or brilliant green bile broth carefully prior to use. Shaking can entrap air in the inner tubes and produce a false positive. Examine fresh tubes before use and discard any with a bubble.

(b) Reduced media such as thioglycollate broth oxidize in storage. Before use, these broths must be heated in boiling water for 20–30 minutes to reduce the medium.

7.8.3 Enrichments

(a) Bring the base medium to 44–46 C before addition of a labile constituent.

(b) Warm enrichments such as blood or serum to room temperature before adding to a base medium.

(c) Once labile material is added to a medium, prepare plates or tubes as soon as possible. Do not hold the batch medium in a water bath for more than 10 minutes.

7.9 Storage of Media

7.9.1 The recommended time limits for holding prepared media in the laboratory are:

MF broths in screw-cap flasks at 4 C	96 hours (Work Week)
MF agars in plates with tight-fitting covers at 4 C	Two Weeks
Agar or broths in loose-cap tubes, at 4 C	One Week
Agar or broths in screw-cap tubes, tightly closed, at 4 C	Three Months
Agar plates (non-MF) with loose-fitting covers, in sealed plastic bags, at 4 C	Two Weeks
Large volumes of agar in screw-cap flasks or bottles tightly-closed, at 4 C	Three Months

7.9.2 Store fermentation tube media in the dark at room temperature or 4 C. If refrigerated, incubate overnight at room temperature to detect false positive gas bubbles.

7.9.3 Since loss of moisture is a major problem of storage, screw-capped tubes and flasks are recommended. Prepoured agar plates can be sealed in plastic bags to retain moisture and refrigerated.

7.9.4 A simple check for loss of moisture in broth tubes can be made by marking the original level in several tubes of each batch and then observing the loss of moisture over time. If the estimated moisture loss exceeds ten percent, discard the tubes of broth.

7.9.5 Protect media containing dyes from light. If color changes are observed, discard the medium.

7.9.6 Prepared sterile broths and agars are available from commercial sources. Their use may be advantageous when analyses are done intermittently, when staff is not available for such preparation work, or when cost of their use can be balanced against other factors of laboratory operation. However, purchase of prepared media does not reduce the responsibility of the laboratory for checking the performance of the media, regardless of the stated quality control practices of the manufacturer.

7.10 Quality Control of Prepared Media

7.10.1 Maintain a book with a complete record of each batch of medium prepared. Include the date, the name of the medium and lot number, amount of medium weighed out and volume of medium prepared, the record of sterilization, pH measurements, pH adjustments made, special handling or preparation techniques, e.g., use of heat-sensitive compounds or components, and name of preparer.

7.10.2 Incubate five percent of each batch of medium for 2 days at 35 C and inspect for growth.

7.10.3 Check each batch of medium when used by inoculating 2 tubes or plates with pure cultures of species producing positive and negative reactions for that medium.

7.10.4 Test new batches of differential media by inoculating with organisms of known fermentative or other biochemical ability. Similarly, enrichment and selective media are tested for productivity of the desired microorganisms and inhibition of other microorganisms. Tables IV-A-4 and IV-A-5 list a group of organisms with the broths, agars, biochemical tests and reactions to which they can be applied.

7.10.5 Record sterility and positive/ negative performance checks in the media preparation portion of the quality control log.

TABLE IV-A-1

Monitoring Laboratory Equipment

Item	Monitoring Procedure
1. Balance	a. An analytical balance with a sensitivity of 1 mg or less at a 10 g load should be used for weighing 2 g or less. For larger quantities, a balance with accuracy of 50 mg at a 150 g load should be used.
	b. Check balance monthly with a set of certified class S weights.
	c. Wipe balance and weights clean after each use.
	d. Protect weights from laboratory atmosphere and corrosion.
	e. Contract with a qualified expert for balance maintenance on an annual basis.
2. pH Meter	a. Compensate for temperature with each use.
	b. Date standard buffer solutions when first opened and check monthly with another pH meter. Discard the buffer solution if the pH is more than $\pm$ 0.1 pH unit from the manufacturer's stated value or if it is contaminated with microorganisms.
	c. Standardize with at least one standard buffer (pH 4.0, pH 7.0, or pH 10.0) before each use.
	d. Do not re-use buffer solution.
	e. Have meters inspected at least yearly as part of a maintenance contract.
3. Water Deionizer	a. Monitor water for conductance daily. Monitor trace metals and other toxic or nutritive compounds monthly. See Table IV-A-3.
	b. Replace cartridges as indicated by manufacturer or as indicated by analytical results.
	c. Monitor bacterial counts at exit point of unit. Replace the cartridges when standard plate count exceeds 10,000/ml.
4. Water Still	a. Drain and clean monthly according to instructions from the manufacturer.
	b. Drain and clean distilled water reservoir quarterly.
	c. Check distilled water continuously or daily using a conductance meter. See Table IV-A-3.
	d. Conduct chemical tests on water to detect toxicity or stimulation effect. See Table IV-A-3.
	e. Conduct standard plate counts monthly on stored water and clean out reservoir if count $>$ 10,000/ml.

TABLE IV-A-1
(continued)

Monitoring Laboratory Equipment

Item	Monitoring Procedure
5. Dispensing Apparatus	a. Check accuracy of dispensation with an NBS class A, graduated cylinder at the start of each volume change and periodically throughout extended runs.
	b. Lubricate moving parts according to manufacturer's instructions or at least once per month.
	c. Correct immediately any leaks, loose connections or malfunctions.
	d. After dispensing each type of medium, pass a large volume of hot distilled water through dispenser to remove traces of agar or medium.
	e. At the end of the work day, break down unit into parts, wash well, rinse with distilled water and dry.
6. Ultraviolet Sterilizer	a. Remove plug from outlet and clean ultraviolet lamps monthly by wiping with a soft cloth moistened with ethanol.
	b. Test ultraviolet lamps with light meter quarterly; if they emit less than 80% of their rated initial output, replace them.
	c. Perform spread plate irradiation test quarterly, see This Section, 4.2.1.
7. Membrane Filter Equipment	a. Check funnel support for leaks.
	b. Check funnel and funnel support to make certain they are smooth. Discard funnel if inside surfaces are scratched.
	c. Clean thoroughly after each work day.
8. Spectrophotometer	a. Maintain quality control and calibration check as recommended by the manufacturer.
	b. Have inspected yearly by a factory maintenance man.
9. Centrifuge	a. Check brushes and bearings for wear every six months.
	b. Check rheostat control against a tachometer at various loadings every six months to ensure proper gravitational fields.

TABLE IV-A-1
(continued)

Monitoring Laboratory Equipment

Item	Monitoring Procedure
10. Microscope	a. Allow only trained technicians to use.
	b. Appoint one laboratory worker to be responsible for the care of the microscope.
	c. Clean optics and stage after every use. Use only lens paper for cleaning.
	d. Keep covered when not in use.
	e. Establish annual maintenance on contract.
	f. Maintain in one location if possible.
11. Microscope, Fluorescence	a. Allow only trained technicians to use microscope and light source.
	b. Keep a log of lamp operation time.
	c. Monitor lamp with meter. See Section 4.2.2. Replace the lamp when $<$ 80% of original fluorescence is observed.
	d. Check lamp alignment, particularly if bulb has been changed. Realign the fluorescent light source if necessary.
	e. Use known 4+ fluorescence slides as controls.
12. Safety Cabinet (Hood)	a. Check filters monthly for plugging or obvious dirt accumulation. Clean or replace filter as needed.
	b. Check cabinet for leaks and for rate of air flow every three months.
	c. Expose blood agar plates to air flow for one hour once per month to measure contamination.
	d. Remove plug from the outlet and clean ultraviolet lamps every two weeks by wiping with a soft cloth moistened with ethanol.
	e. Test ultraviolet lamps quarterly with a light meter. If lamp emits less then 80% of the rated output, replace lamp.
	f. Perform maintenance as directed by the manufacturer.
	g. Purchase and use a pressure monitor control device to measure efficiency of air flow.

TABLE IV-A-1
(continued)

Monitoring Laboratory Equipment

Item	Monitoring Procedure
13. Thermometers and Recording Devices	a. Check the accuracy of thermometers and temperature recording instruments, in the monitoring range, at least annually against a certified thermometer or one of equivalent accuracy. Thermometer graduations should not exceed the 0.2 or 0.5 C deviation permitted in the analytical method. Check mercury columns for breaks.
	b. Record calibration checks in quality control (QC) record. Mark NBS calibration correction on each thermometer or on the outside of the incubator, refrigerator or freezer containing the thermometer.
	c. Record daily temperature checks on charts and keep for at least three years. A simple, one year chart is shown in Figure IV-A-1.
14. Water Bath	a. Check and record temperature daily. Bath must maintain the uniform temperature needed for the test in use.
	b. Maintain accurate thermometer completely immersed in water bath.
	c. A recording thermometer and alarm system are recommended.
	d. Clean monthly.
	e. Use only stainless steel, rubber, plastic-coated, or other corrosion-proof racks.
15. Refrigerator at 4 C	a. Check and record temperature daily.
	b. Clean monthly.
	c. Require identification and dating of all material.
	d. Defrost unit and discard outdated materials in refrigerator and freezer compartments every three months.
16. Hot Air Oven	a. Test performance with spore strips or suspensions quarterly.
	b. Equip and monitor sterilization with a thermometer accurate in 160-180 C range.
17. Freezers	a. Check temperature and record daily.
	b. Use of recording thermometer and alarm system recommended.
	c. Require identification and dating of all materials.
	d. Clean and defrost freezer every six months. Discard outdated materials.

TABLE IV-A-1
(continued)

Monitoring Laboratory Equipment

Item	Monitoring Procedure
18. Autoclave	a. Record temperature and pressure for each run. Recording thermometer recommended. b. Verify that autoclave maintains uniform operating temperature. c. Check operating temperature with a min/max thermometer on a weekly basis. d. Test performance with spore strips or suspensions weekly. If evidence of contamination occurs, check until the cause is identified and eliminated. e. Procure semi-annual preventive maintenance inspections.
19. Incubator (Air/Water-Jacket)	a. Check and record temperature daily. b. If partially-submersible glass thermometer is used, bulb and stem must be immersed in water to the mark on stem. c. Measure temperatures daily on top and bottom shelves. Periodically measure temperature on all shelves in use. d. Expand test points proportionately for walk-in incubators. e. Recording thermometer and alarm system are recommended. f. Locate incubator where room temperature does not go outside of the 16–27 C range.

Instrument ______________________________ Temperature __________ Room _____

Read daily.
Record temperature in space provided.

Date	Jan.	Feb.	Mar.	Apr.	May	June	Jul.	Aug.	Sep.	Oct.	Nov.	Dec.	Date
1													1
2													2
3													3
4													4
5													5
6													6
7													7
8													8
9													9
10													10
11													11
12													12
13													13
14													14
15													15
16													16
17													17
18													18
19													19
20													20
21													21
22													22
23													23
24													24
25													25
26													26
27													27
28													28
29													29
30													30
31													31

FIGURE IV-A-1. Equipment Operation Temperature Record.

TABLE IV-A-2

Glassware Maintenance

Item	Monitoring Procedure
1. Utensils and Containers for Media Preparation	Use utensils and containers of non-corrosive and non-contaminating materials such as pyrex glass, stainless, steel or aluminum.
2. Glassware (Reusable)	a. With each use, examine glassware especially screw-capped dilution bottles and flasks, for chipped or broken edges and etched surfaces. Discard chipped or badly-etched glassware. b. Inspect glassware after washing. If water beads excessively on the cleaned surfaces, run the glassware through again. c. Test for acid or alkaline residues by adding bromthymol blue indicator to representative glassware items (see 5.1.2 in This Section). d. Test for residual detergent by the test in 5.1.3, This Section.

TABLE IV-A-3

Laboratory Pure Water for Bacteriological Testing

Parameter	Ideal Monitoring Frequency	Limit
<u>Chemical Tests</u>		
Conductivity	Continuously or with each use	1–2 μmhos/cm at 25 C
pH	With each use	5.5–7.5
Total Organic Carbon	Monthly	< 1.0 mg/liter
Trace Metal, Single	Monthly	< 0.05 mg/liter
Trace Metals, Total (Cd, Cr, Cu Ni, Pb, Zn)	Monthly	< 1.0 mg/liter
Ammonia/Amines	Monthly	< 0.1 mg/liter
Free chlorine	With each use	< 0.1 mg/liter
<u>Biological Tests</u>		
Standard Plate Count		
Fresh Water	Monthly	< 1000 bacteria/ml
Stored Water	Monthly	< 10,000 bacteria/ml
Water Suitability Test	Yearly and when conditions change	Ratio: 0.8–3.0
Water Use-Test	Yearly and when conditions change	Calculated t value < 2.78

TABLE IV-A-4

Quality Control of Media

Medium	Control Cultures	Expected Results
M-Endo MF Broth or Agar	*Escherichia coli*	Golden green metallic sheen
	Enterobacter aerogenes	Golden green metallic sheen
	Achromobacter sp	Red colonies
	Pseudomonas sp	Red colonies
	Salmonella sp	Red colonies if medium overheated
M-FC Broth or Agar	*E. coli*	Blue colonies
	K. pneumoniae	Blue colonies
	E. aerogenes	No growth
Brilliant Green Bile Lactose Broth	*E. coli*	Growth with gas
	E. aerogenes	Growth with gas
	C. freundii	Growth with gas
	Staph. aureus	No growth
Lauryl Tryptose Broth	*E. coli*	Growth with gas
	E. aerogenes	Growth with gas
	S. typhimurium	Marked to complete inhibition
	S. aureus	Marked to complete inhibition
Levine's Eosin Methylene Blue Agar	*E. coli*	Nucleated black colonies with golden green metallic sheen
	E. aerogenes	Pink colonies with dark centers
	C. freundii	Colorless colonies
	Salmonella sp	Colorless colonies
	Klebsiella sp	Large brown mucoid colonies
Xylose Lysine Desoxycholate Agar (XLD)	*Salmonella* sp	Red colonies, to red with black centers
	Klebsiella sp	Yellow colonies
	E. coli	Yellow colonies
	E. aerogenes	Yellow colonies
Bismuth Sulfite Agar	*Salm. typhosa*	Black colony with black or brownish-black zone, with or without sheen
	Other *Salmonella* sp	Raised green colonies
	Coliforms	Green colonies
Brilliant Green Agar	*Salmonella* sp	Pink-white opaque colonies surrounded by brilliant red zone
	E. coli	Inhibition or yellow green colonies
	P. vulgaris	Marked to complete inhibition or red colonies

TABLE IV-A-4
(continued)

Quality Control of Media

Medium	Control Cultures	Expected Results
KF Streptococcus Agar	*Strep. faecalis*	Pink to red colonies
	Strep. pyogenes	No growth
	S. aureus	No growth
	E. coli	No growth
PSE Agar	*S. faecalis*	Black colonies
	E. coli	No growth
	S. aureus	No growth

TABLE IV-A-5

Quality Control of Biochemical Tests

Test	Control Culture	Expected Results
BHI Broth at pH 9.6	*Strep. faecalis* *Strep. mitis-salivarius*	Positive: growth Negative: no growth
BHI Broth with 6.5% NaC1	*S. faecalis* *S. mitis-salivarius*	Positive: growth Negative: no growth
Arginine Dehydrolase (Moeller's medium)	*Salm. typhimurium* *Salm. flexneri*	Positive: alkaline reaction reddish violet color Negative: yellow color
Lysine Decarboxylase (Moeller's medium)	*S. typhimurium* *S. flexneri*	Positive: alkaline reaction reddish violet color Negative: acid, yellow reaction
Ornithine Decarboxylase (Moeller's medium)	*S. typhimurium* *S. flexneri*	Positive: alkaline reaction reddish violet color Negative: yellow color
Indole Production (Tryptophane Broth)	*Escherichia coli* *Salmonella* sp *Enterobacter aerogenes*	Positive: red color Negative: orange/yellow color Negative: orange/yellow color
Methyl Red (Buffered Peptone Glucose Broth)	*E. coli* *E. aerogenes*	Positive: red color Negative: no change
Voges-Proskauer (Buffered Peptone Glucose Broth)	*E. aerogenes* *E. coli*	Positive: pink color Negative: no color change
Citrate Utilization (Simmons Citrate Broth)	*E. aerogenes* *E. coli*	Positive: growth, change to blue color Negative: no color change, no growth
Urease Production (Christensen's Urea Agar)	*Proteus mirabilis* *Salmonella* sp	Positive: color change, pink to red Negative: no color change
Catalase (BHI agar slant)	*S. aureus* *S. faecalis*	Positive: bubbles Negative: no bubbles

TABLE IV-A-5
(continued)

Quality Control of Biochemical Tests

Test	Control Culture	Expected Results		
Cytochrome Oxidase (Alpha-napthol and para-amino-dimethylaniline oxalate)	*Pseudomonas aeruginosa* *E. coli* *S. aureus*	Positive: blue color Negative: no change Negative: no change		
Phenylalanine Deaminase (Phenylalanine Agar)	*P. mirabilis* *Salmonella* sp *E. coli*	Positive: green color Negative: no color change Negative: no color change		
Malonate Utilization (Malonate Broth)	*K. pneumoniae* *E. coli*	Positive: blue color change Negative: no growth or color change		
Milk, Methylene Blue, 0.1%	*S. faecalis* Group Q Streptococci *Strep. salivarius* *P. aeruginosa* *C. perfringens*	Positive: reduction of methylene blue Negative: no growth Negative: no growth Peptonization and digestion Acid, coagulation and gas		
Nitrate Reduction (Potassium Nitrate Broth)	*E. coli* *P. aeruginosa*	Positive: red color change Negative: no color change		
2, 3, 5-Triphenyl Tetrazolium Chloride in TG Agar	*S. faecalis* *Strep. faecium*	Positive: reduction of TTC (red color) Negative: no color change		
Tellurite Agar	*S. faecalis* *S. faecium*	Growth No growth		
Beta-Hemolysis in Blood Agar	*S. faecalis* var. *zymogenes* *S. faecalis*	Positive: lysis of red blood cells Negative: no lysis of red blood cells		
		Slant	Butt	H_2S Production
Hydrogen Sulfide (Triple Sugar Iron Agar)	*E. coli* *P. vulgaris* *S. typhimurium*	A A K	A G A G A G	– + +

TABLE IV-A-5
(continued)

Quality Control of Biochemical Tests

Test	Control Culture	Expected Results		
		Slant	Butt	H_2S Production
Lysine (Lysine Iron Agar)	*S. typhimurium*	K	K	+
	S. flexneri	K	A	–
Gelatin Liquefaction at 20 C (Nutrient Gelatin)	*S. marcescens*	Positive:	liquefaction	
	S. faecalis var. *liquefaciens*	Positive:	liquefaction	
	S. faecalis	Negative:	no liquefaction	
	C. perfringens	Positive:	liquefaction	
	E. coli	Negative:	no liquefaction	

REFERENCES

1. American Public Health Association, 1972. Standard Methods for the Examination of Dairy Products, (13 ed.) American Public Health Association, Inc., Washington, DC. RODAC: p. 192; Swab: pp. 43-44; and Air: p. 44.

2. American Public Health Association, 1975. Standard Methods for the Examination of Water and Wastewater (14 ed.) American Public Health Association, Inc., Washington, DC. p. 885.

3. Geldreich, E. E. and H. F. Clark, 1965. Distilled water suitability for microbiological applications. J. Milk and Food Tech. 28:351.

4. Military specifications for Disk, Filtering, Microporous, Hydrosol Type, 47 mm diameter 100s. MIL-D-37005 (DSA-DM) 5 September, 1973. Directorate of Medical Materiel, Defense Personnel Supply Center, Dept. of Defense, Philadelphia, PA.

5. Tentative method for evaluating water testing membrane filters for fecal coliform recovery. D3508-76T. 1978 Annual Book of ASTM Standards, Part 31, p. 1127.

6. Standard specification for reagent water. D1193-77. 1978 Annual Book of ASTM Standards, Part 31, p. 20.

PART IV. QUALITY ASSURANCE

Section B Statistics for Microbiology

The Section is divided into five major areas of statistical measure:

1. **Measures of Central Tendency**
2. **Measures of Dispersion**
3. **Normal Distribution**
4. **Poisson Distribution**

In this Section the computational formats for the more commonly used measures in statistics are described. In the following discussion let x_i denote a typical observed result so that $(x_1, x_2, ..., x_n)$ represents a sample of n observations. A good reference book for further details on these parameters is Dixon and Massey (1).

1. Measures of Central Tendency

1.1 The Arithmetic Mean: The most commonly used measure of central tendency is the arithmetic mean which is often simply called "the mean". Denote the sample mean by $\overline{X}$ and the population mean, of which $\overline{X}$ is an estimate, by μ. The computational formula is:

$$\overline{X} = \frac{\sum_{i=1}^{n} x_i}{n}$$

where $\sum_{i=1}^{n} x_i = x_1 + x_2 + \ldots x_n$

An example of microbiological data is shown in Table IV-B-1.

TABLE IV-B-1

Microbiological Results, Count/100 ml

79	220	330
110	230	490
130	280	490
130	330	790
170	330	950
220	330	1100

Arithmetic Mean, $\overline{X}$	=	372.7
Geometric Mean, $\overline{X}_g$	=	287.8
Median	=	305
Mode	=	330
Standard Deviation (S)	=	293.2
Range	=	1021

The sample mean is calculated as follows:

$$\overline{X} = \frac{\sum_{i=1}^{n} x_i}{n} = \frac{6709}{18} = 372.7$$

1.2 The Geometric Mean: A second measure of central tendency that is preferred for certain distributions such as the Poisson. It will be discussed later, in 4.3 but is defined as:

$$\overline{X}_g = \sqrt[n]{(x_1)(x_2)....(x_n)}$$

The geometric mean of the data in Table IV-B-1 is 287.8.

1.3 The Median: Another measure of central tendency is the median. The median is the value such that half of the other values are greater and half are less. To find the sample median, the data are arranged in ascending or descending order and the middle value is picked. When there is an even number of observations, the average of the two middle values is taken. For the data in Table IV-B-1, the median is 305.

1.4 The Mode: The mode, one other measure of central tendency, is the most frequently occurring value. In Table IV-B-1 the mode is 330, since this value occurs four times. The population mode is the value corresponding to the peak of the frequency distribution curve.

Frequency distributions with more than one peak are called multimodal. In a symmetrical frequency distribution, the mean, median, and mode are all equal.

2. Measures of Dispersion

2.1 The Standard Deviation: Of the several measures of dispersion, the most common is the standard deviation. Denote the sample standard deviation by S and the population value by σ(of which S is an estimate), when the computational formula is:

$$S = \sqrt{\frac{\sum_{i=1}^{n}(X_i - \overline{X})^2}{n - 1}}$$

However, the computation using this formula is tedious. It is relatively simple to show the following relationship:

$$S = \sqrt{\frac{\sum_{i=1}^{n}(X_i - \overline{X})^2}{n - 1}} = \sqrt{\frac{n\sum_{i=1}^{n} X^2 - (\sum_{i=1}^{n} X_i)^2}{n(n - 1)}}$$

The derived formula is preferable because of its adaptability to the desk calculator. The sample standard deviation of the microbiological data in Table IV-B-1 is calculated as follows:

$$S = \sqrt{\frac{(18)\ (3961541) - (6709)^2}{18\ (17)}}$$

$$= \sqrt{\frac{71307738 - 45010681}{306}}$$

$$= \sqrt{\frac{26297057}{306}}$$

$$= \sqrt{85939} = 293.2$$

Confidence Limit (95% and 99%): The range of values within which a single analysis will be included, 95% or 99% of the time.

$$95\%\ C.\ L. = \overline{X} \pm 1.96S$$

$$99\%\ C.\ L. = \overline{X} \pm 2.58S$$

2.2 The Variance: The sample value S^2 is referred to as the sample variance and is merely the square of the sample standard deviation. Often it is more convenient in conversation as well as computation to refer to the variance. This should not cause confusion if the above relationship is kept in mind.

The population variance is represented by σ^2. Its formula is:

$$\sigma^2 = \frac{\sum_{i=1}^{n}(X_i - \mu)^2}{n}$$

This is the same as the formula for S^2 except that the true population mean μ is used rather than its estimate $\overline{X}$ and the numerator is divided by n instead of n – 1.

In calculating the sample variance the true mean is not known and the estimate of the mean from the data is used instead. Because the sample mean is being used to calculate the variance of the same data, only n – 1 of the squared difference terms are independent. It can be shown that the estimate of the variance must be based upon the sum of independent squared terms, thus indicating the division by n – 1, which is called the number of degrees of freedom (d.f.) in the sample. As a rule, in any calculation, for every parameter that must be estimated, one degree of freedom is lost.

2.3 The Range: The range is also used as a measure of dispersion. It is the difference between the highest and lowest values in a set of data.

$$R = \max(x_i) - \min(x_i)$$

For the data in Table IV-B-1 the range is then:

$$R = 1100 - 79 = 1021$$

A rough estimate of S can be made by dividing the range of the sample by the square root of n, the number of observations, when $n \leq 10$:

$$S \simeq \frac{R}{\sqrt{n}}$$

Use of the range as a measure of dispersion is generally limited to instances where the labor of computing the standard deviation is impractical.

3. The Normal Distribution

3.1 The most important theoretical distribution in statistics is the familiar bell-shaped normal distribution which is symmetric about its peak (see Figure IV-B-1). The following assumptions give rise to this distributional form:

3.1.1 Values above or below the mean are equally likely to occur.

3.1.2 Small deviations from the mean are extremely likely.

3.1.3 Large deviations from the mean are extremely unlikely.

3.2 The normal distribution is completely defined by its mean, μ, and its standard deviation σ in the following manner:

3.2.1 The area under the normal curve between μ minus σ and μ plus σ is 68 percent of the total area, to the nearest 1 percent.

3.2.2 The area under the normal curve between μ minus 2σ and μ plus 2σ is 95 percent of the area, to the nearest 1 percent.

3.2.3 The area under the normal curve between μ minus 3σ and μ plus 3σ is 99.7 percent of the total area, to the nearest 0.1 percent.

3.3 If the frequency distribution of a sample is a good approximation to the normal curve, these characteristics of the normal curve can be used to develop a great deal of information about the underlying population.

4. Non-symmetric Distribution

4.1 Asymmetry: In some investigations one encounters distributions which are not symmetric. For example, distributions of bacterial counts are often characterized as having a skewed distribution because of the many low and a few extremely high counts. This characteristic leads to an arithmetic mean which is considerably larger than the median or the geometric mean. The frequency curves of these distributions have a long right tail, as shown in Figure IV-B-2, and are said to display positive skewness.

4.2 Logarithmic Transformation: For practical and theoretical reasons, statisticians prefer to work with symmetric distributions like the normal curve. Therefore, it is usually necessary to convert skewed data so that a symmetric distribution resembling the normal distribution results. An approximately normal

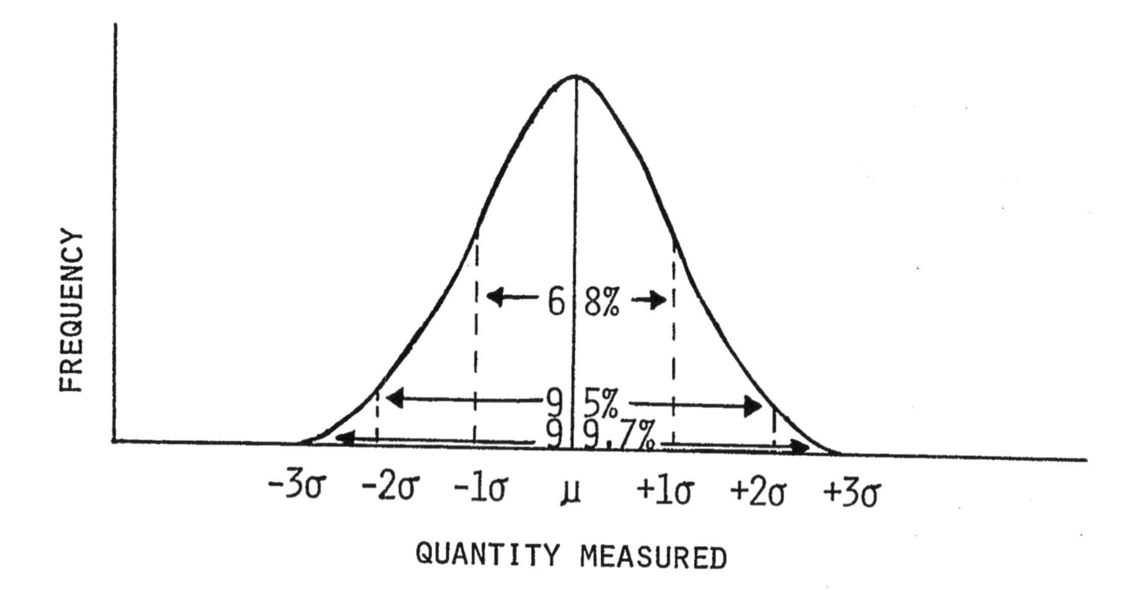

FIGURE IV-B-1. Normal Distribution Curve.

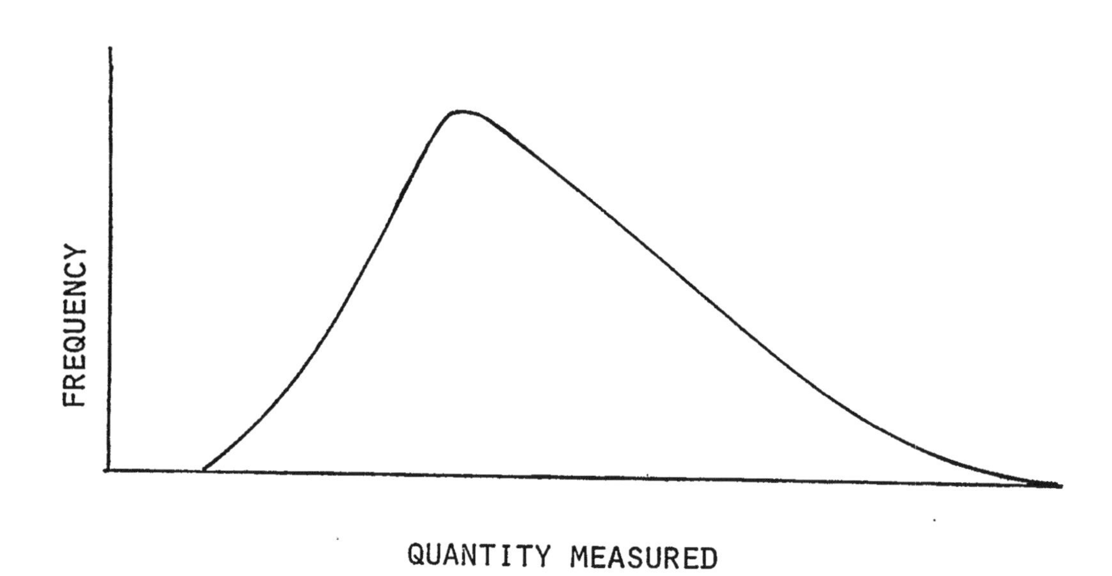

FIGURE IV-B-2. Positively-Skewed Distribution Curve.

distribution can be derived from positively-skewed distributions by expressing the original data as logarithms. An example of coliform counts and their logarithms are shown in Table IV-B-2. A comparison of the frequency tables for the original data and their logs in Tables IV-B-3 and 4, shows that the logarithms more closely approximate a symmetric distribution.

4.3 The Best Measure of Central Tendency for Microbiological Data: Assuming that the microbiological data has been normalized through a logarithmic transformation, the arithmetic mean is the best estimate of central tendency. However, there is a direct relationship between this mean and the geometric mean of the original data:

$$\overline{(\log X)} = \frac{\sum_{i=1}^{n} \log X_i}{n}$$

$$= \log \sqrt[n]{(X_1)(X_2)\ldots(X_n)} = \log \overline{X}_g$$

Therefore, the best measure of central tendency for microbiological data is the log-transform.

The mean of the log MPN data in Table IV-B-2 is:

$$\overline{\log X} = \frac{\sum_{i=1}^{n} \log X_i}{n} = \frac{32.737}{15} = 2.1825$$

and the true mean of the MPN data is:

$$= \text{antilog } (2.1825) = 152$$

REFERENCES

1. Dixon, W. J. and F. J. Massey, Jr., 1969. Introduction to Statistical Analysis, 3rd Edition, McGraw-Hill, Inc., New York, NY.

TABLE IV-B-2

Coliform Counts and Their Logarithms

Coliform Count/100 ml MPN	log MPN
11	1.041
27	1.431
36	1.556
48	1.681
80	1.903
85	1.929
120	2.080
130	2.114
136	2.134
161	2.207
317	2.501
601	2.779
760	2.881
1020	3.009
3100	3.491

TABLE IV-B-3

Comparison of Frequency of MPN Data

Class Interval	Frequency (MPN)
0 to 400	11
400 to 800	2
800 to 1200	1
1200 to 1600	0
1600 to 2000	0
2000 to 2400	0
2400 to 2800	0
2800 to 3200	1

TABLE IV-B-4

Comparison of Frequency of Log MPN Data

Class Interval	Frequency (log MPN)
1.000 to 1.300	1
1.300 to 1.600	2
1.600 to 1.900	1
1.900 to 2.200	5
2.200 to 2.500	1
2.500 to 2.800	2
2.800 to 3.100	2
3.100 to 3.400	0
3.400 to 3.700	1

PART IV. QUALITY ASSURANCE

Section C Analytical Quality Control Procedures

The Section on Analytical Quality Control is divided into three major areas of statistical usage:

1. **Quality Control on Routine Analyses**
2. **Quality Control in Compliance Monitoring**
3. **Comparative Testing of Methodology**
4. **Method Characterization**

1. Quality Control on Routine Analyses

Each laboratory must establish quality control over the microbiological analyses in use. Fifteen percent of total analyst time should be spent on quality control practices discussed in this Manual.

1.1 Duplicate Analyses: Run duplicate analyses on 10% of the known positive samples analyzed and a minimum of one per month. The duplicates may be run as split samples by more than one analyst.

1.2 Positive Control Samples: Test a minimum of one pure culture of known positive reaction per month for each parameter tested.

1.3 Negative (Sterile) Control: Include one negative control with each series of samples using buffered water and the medium batch at the start of the test series and following every tenth sample. When sterile controls indicate contamination, data on samples affected should be rejected and a request made for immediate resampling of those waters involved.

1.4 MF Colony Counting by More than One Analyst: At least once per month, two or more analysts should count the colonies on the same membrane from a polluted water source. Colonies on the membrane should be verified and the analysts' counts compared to the verified count.

1.5 Check Analyses on Water Supply Program by State Laboratories: In a local laboratory, a minimum number of the water supply samples should be analyzed by the State laboratory. For example, "laboratories that are required to test less than 100 samples per month should submit an additional 10% of the number to the State laboratory for analysis. Water systems with sample requirements above 100 per month should submit an additional 2% to the State Laboratory for analysis.

1.6 Reference Sample: Laboratories should analyze reference samples quarterly when available for the parameters measured.

1.7 Performance Sample: Laboratories should analyze at least one unknown performance sample per year when available, for parameters measured.

1.8 MF Verification: Five percent of the analyses performed should be verified.

1.8.1 Total Coliforms: Pick at least 10 isolated sheen colonies from each sample. Transfer into lauryl tryptose broth. Incubate and read. Transfer positive tubes into brilliant green bile broth for verification of coliforms.

Since samples from public water supplies with 5 or more sheen colonies must be verified, at least 5 colonies are picked from each positive potable water sample.

The laboratory should make every effort to detect coliforms from samples with excessive non-coliforms on the membrane filter. Any sheen colonies appearing in mixed confluent growth must be verified (see Part III-B).

1.8.2 Fecal Coliforms: Pick at least 10 isolated colonies containing blue to blue-green pigment and transfer to lauryl tryptose broth. Incubate and read. Transfer positive tubes to EC broth where gas production verifies fecal coliform organisms (see Part III-C).

1.8.3 Fecal Streptococci: Pick at least 10 isolated pink to red colonies from MF or pour plates. Transfer to BHI agar or broth. After growth, perform catalase test. If negative (possible fecal streptococci) transfer growth to BHI and 40% bile BHI broth tubes and incubate at 45 C and 35 C respectively. Growth at both temperatures verifies fecal streptococci (see Part III-D).

1.9 MPN Completion of Total Coliform Test

1.9.1 For routine analyses, complete the MPN test on five percent of the positive confirmed samples and a minimum of one sample per test run.

1.9.2 For potable waters, complete the MPN test once each quarter on 10 percent of positive confirmed samples. If insufficient positive tubes result from potable water samples, perform the completed test on positive source waters.

1.10 Measurement of Analyst Precision: If the routine work of the laboratory includes samples from different wastewaters, surface waters, water supplies or finished waters, the following steps should be accomplished for each type.

Step 1

Perform duplicate analyses on the first 15 typical samples with positive responses. Although each set of duplicates must be run by the same analyst, all analysts performing routine analyses should contribute a share of this initial data.

Step 2

Calculate the logarithms of results. If either of a set of duplicate results is zero, add 1 to both values before calculating the logarithms.

Step 3

Calculate the range (R) for each pair of transformed duplicates and the mean ($\overline{R}$) of these ranges.

Step 4

Thereafter, run 10% of routine samples in duplicate. Transform the duplicates as in Step 2 and calculate their range. If this range is greater than 3.27 $\overline{R}$, analyst precision is out of control and all analytical results since the last precision check must be discarded. The analytical problem must be identified and resolved before doing further analyses (1).

Step 5

In order that the criterion used in Step 4 be kept up-to-date, periodically repeat Steps 2 and 3 using the most recent sets of 15 duplicate results.

2. Quality Control in Compliance Monitoring (National Pollution Discharge Elimination System)

2.1 For any legal assurance of non-compliance within a permit, analytical results must exceed the permit limit by a statistically significant amount. This requires allowance of the analytical deviation known to occur at a level equal to the permit limitation.

2.2 It is apparently common for many monitoring agencies to judge compliance with microbiological permit limitations based upon one analysis each for single grab samples taken at about the same time on 3 consecutive days. The largest of these 3 results is then compared to a maximum discharge limitation, while the arithmetic mean is compared to a daily average limitation. Judgment based upon such a sampling program is only valid under certain assumptions; namely that there is no relationship between the discharge level and the time of sampling, and that the variability among analytical results for samples taken simultaneously is either inconsequential or can be properly estimated from previous data.

The first assumption is usually invalid because the discharge level is dependent upon production or processing operations which are not uniform, but rather follow some daily or weekly cycle. To keep such periodic cycles from biasing the results, the grab samples should be taken at randomly-selected times during the study or, if there is some knowledge of the cycle, at times selected systematically to define its extremes.

The second assumption is important in order to make a compliance judgment. As noted in 2.1, this judgment must take into account the estimated analytical standard deviation at a level equal to the permit limit. If an appropriate estimate of this standard deviation is not already available then it must be estimated during the compliance study, either from multiple analyses of each grab sample or from single analyses of multiple grab samples taken simultaneously.

2.3 Whenever the analytical standard deviation must be estimated from the study data the following procedure should be followed for each major type of discharge.

Step 1

Take 3 simultaneous grab samples at each of K (at least 3) randomly or systematically selected sampling times which are expected to represent the range of discharge throughout the study period.

Step 2

Analyze the samples and convert the results to logarithms.

Step 3

Using these logarithms, calculate the following statistics:

$$S = \sqrt{\frac{1}{2K}\left(\sum_{i=1}^{K}\left(\sum_{j=1}^{3} X_{ij}^2 - \left(\sum_{j=1}^{3} X_{ij}\right)^2/3\right)\right)}$$

$$\bar{X}_i = \left(\sum_{j=1}^{3} X_{ij}\right)/3$$

$$\bar{X} = \left(\sum_{i=1}^{K} \bar{X}_i\right)/K$$

where: S = the pooled estimate of the analytical standard deviation

X_{ij} = the analytical result for the jth grab sample taken at sampling time i,

$\bar{X}_i$ = the mean result for the 3 grab samples taken at sampling time i,

$\bar{X}$ = the overall mean for the K sampling times throughout the study period.

Step 4

The maximum discharge limitation (D_{max}) has been violated if for any sampling time i:

$$\bar{X}_i - t_{.95}(2K)(S/\sqrt{3}) > D_{max}$$

where $t_{.95}(2K)$ is the Student's t value at a 95 percent confidence level and 2K degrees of freedom.

Step 5

An average limitation ($\bar{D}$) over the study period (daily, weekly, monthly, etc.) has been violated if:

$$\bar{X} - t_{.95}(2K)(S/\sqrt{3K}) > \bar{D}$$

2.4 If an estimate of the analytical standard deviation (S) is available that has at least 30 degrees of freedom, and if the estimate applies to the same type of waste and was generated from samples at a level near the discharge limitation of the permit under study, then Steps 1–5 can be modified as follows:

Step 1

Take one grab sample at each of K (at least 3) sampling times selected as before.

Step 2

Analyze samples and convert results to logarithms.

Step 3

Calculate: $\bar{X} = (\sum_{i=1}^{K} X_i)/K$

Step 4

D_{max} has been violated if $X_i > D_{max} + 2S$ for any sampling time i.

Step 5

$\bar{D}$ has been violated if $\bar{X} > \bar{D} + 2S/\sqrt{K}$.

3. Comparative Testing of Methodologies

3.1 Protocol for Testing: If a laboratory proposes a new or modified method for acceptance certain minimal testing should be completed.

3.1.1 Collect and test at least 10 samples from each sample source or type for which the method is intended, i.e., fresh surface water, sewage treatment plant influent or effluent, industrial effluents, saline water, etc. Source waters should be chosen which are representative and which span the relevant range of concentrations.

3.1.2 Collect the samples and perform the analyses over 1 week of plant processing, but not less than 5 calendar days.

3.1.3 Take an even number of aliquots (at least 8) from each sample and simultaneously analyze half of them using the accepted method and the other half using the proposed alternative method. Throughout this protocol, the analysis of aliquots from a single sample using one method will be referred to as "replicate analyses" and the resulting data will often be referred to as "replicates."

3.2 Statistical Analyses: Compile the data and perform the following steps for each sample type:

Step 1: Transform the Data

Calculate the logarithms of the basic data and use these transformed data as the basis for all subsequent statistical analyses.

The characteristics of microbiological data have been discussed by many authors. Of particular importance to this procedure is the effect that a logarithmic transformation has on such data (2, 3).

Specifically, a logarithmic transformation normalizes the distribution of results from replicate analyses of the same sample and equalizes the within-sample standard deviations among samples with different microbiological densities. These conditions are required assumptions for many of the statistical procedures that follow.

Step 2: Calculate the Basic Statistics

Calculate the arithmetic mean ($\overline{X}$) and standard deviation (S) for each set of replicates, and tabulate the data and their basic statistics.

Step 3: Test for Suspected Outliers

For each set of n replicates, test for outliers by calculating the Extreme Studentized Deviate as $T = (X_E - \overline{X})/S$, where X_E = the maximum or minimum replicate, whichever is farthest from $\overline{X}$.

Look up the critical T value for the number of replicates on each sample and a .01 (1 percent) significance level in Standard D2777–72, 1975 Book of ASTM Standards, Part 31, Water, page 15. If T is greater than the critical value, reject all data for that sample with 99 percent confidence that X_E is an outlier.

This unfortunate loss of data is necessary because the following statistical procedures have been simplified and only apply to the unique situation where, for each of K sample, there are exactly n replicates by each of the two methods.

Step 4: Test for Equality Among the Within-Sample Standard Deviation Estimates

For each method, test for equality of the standard deviation estimates (S) calculated in Step 2 from the replicates for each sample.

This is done by applying Cochran's Test (4) as follows:

$$C = S_{max}^2 / \sum_{i=1}^{K} S_i^2$$

K = the number of samples

S_i = the standard deviation estimate from the set of replicates for the ith sample,

S_{max} = the largest value for S_i $i = 1, 2, \ldots, K$

Look up the critical C value for n-1 degrees of freedom on each of K standard deviations and a .01 significance level, i.e., $C_{.99}$ (n-1, K) in Dixon and Massey (5).

If $C \leq C_{.99}$ (n-1, K), then pool the standard deviations for the method (M) as follows:

$$S_M^2 = (\sum_{i=1}^{K} S_i^2)/K$$

S_M^2 is the pooled within-sample variance estimate for the method.

If $C > C_{.99}$ (n-1, K), then the within-sample standard deviation of this method is not a constant over the concentration range represented in the data. To correct for this, the data must be stratified, i.e., subdivided, into sets of samples which have a common standard deviation among the results of replicate analyses by this method. Care should be taken to minimize the number of strata to those that are required in order to justify the development of pooled within-sample standard deviation estimates. If both methods require stratification, common strata should be developed which are jointly suitable. Step 4 should be repeated to verify the proposed stratification. Subsequently, Step 5 should be carried out for each stratum independently. If you have any problems with stratifying the data, please seek the assistance of a qualified statistician/data analyst before proceeding.

Step 5: Test for Equality of the Pooled Within-Sample Variance Estimates

Having developed a pooled within-sample variance estimate for each method, within common strata if necessary, test the equality of these estimates by performing the following F test:

$$F = S^2_{M_1} / S^2_{M_2}, \text{ where } S^2_{M_1} > S^2_{M_2}$$

Look up the critical F value for K(n-1) degrees of freedom for each variance and a .05 significance level, i.e., $F_{.95}$ (K(n-1), K(n-1)), in any standard statistical reference. If $F \leq F_{.95}$ (K(n-1), K(n-1)), then the two methods produce results with within-sample variances which are not significantly different. If $F > F_{.95}$ (K(n-1), K(n-1)), then there is a statistically significant difference between the variances.

Step 6: Test for Equality of the Method Means

Finally, calculate the difference between the method means for each of the K samples as:

$$d_i = \bar{X}_{M_1} - \bar{X}_{M_2}, \quad i = 1, 2, \ldots, K.$$

Using the mean ($\bar{d}$) and standard deviation (S_d) of these differences, calculate the Student's statistic as:

$$t = \frac{\bar{d}}{S_d/\sqrt{K}}$$

Refer to a statistical reference for the critical Student's t value at the .05 significance level and K-1 degrees of freedom, i.e, $t_{.95}$(K-1). If $t > t_{.95}$(K-1), then the means for the two methods are significantly different.

3.3 Example of the Statistical Analysis Procedure

Statement of the Problem

Method 1 is the accepted method for determining fecal coliform levels in chlorinated sewage treatment effluents. Recognized problems in Method 1 include: deviation among replicate results is considered excessive, averages show a positive bias of about 40 percent, and the procedure is difficult and time-consuming.

Suppose that Method 2 is quicker and easier than Method 1. Then, if the following statistical analysis shows the performance of Method 2 to be at least equal to Method 1, EPA may designate Method 2 as an alternative to Method 1. If Method 2 offered no operational advantage over Method 1, the statistical analysis would have to show the performance of Method 2 to be significantly superior to Method 1.

Steps 1 and 2

Transform the Data (Table IV-C-1) and Calculate the Basic Statistics:

The logarithmic transformation was found most appropriate for the data in Table IV-C-1, see the results in Table IV-C-2.

Step 3

Test for Suspected Outliers in Table IV-C-2.

$$T = \frac{|X_E - \bar{X}|}{S}$$

Sample 2, method 2, replicate 1:

$X_E = 3.3711$

$$T = \frac{|3.3711 - 3.6773|}{.1738} = 1.76$$

Sample 4, method 2, replicate 1:

$X_E = 2.7497$

$$T = \frac{|2.7497 - 2.5299|}{.1301} = 1.69$$

TABLE IV-C-1

Raw Sample Data from the Analysis of Chlorinated Sewage Treatment Plant Effluents

Sample No.	Date	Method	Replicates (X_R) in counts/100 ml	ΣX_R	$\overline{X}_R$	S_R
1	3/11	1	1850, 1350, 1050, 1800, 1150	7,200	1440	368.1
		2	1200, 850, 650, 500, 1050	4,250	850	285.0
2	3/11	1	7800, 8100, 5900, 5450, 7150	34,400	6880	1163.3
		2	2350, 5300, 5650, 6350, 5450	25,100	5020	1545.8
3	3/17	1	4000, 4400, 2650, 3000, 3650	17,700	3540	715.4
		2	2400, 2150, 2050, 1600, 2200	10,400	2080	297.1
4	3/18	1	625, 512, 512, 462, 700	2,811	562	97.5
		2	562, 350, 300, 275, 275	1,762	352	121.1
5	3/24	1	3870, 2330, 3330, 3670, 2530	15,730	3146	685.2
		2	2730, 3930, 2530, 2200, 1730	13,120	2624	822.3
6	3/25	1	7200, 4400, 7800, 6600, 5200	31,200	6240	1410.0
		2	4400, 4000, 4400, 3600, 6200	22,600	4520	996.0
7	3/26	1	1200, 1000, 1000, 1200, 1600	6,000	1200	245.0
		2	700, 600, 900, 900, 900	4,000	800	141.4

TABLE IV-C-2

Logarithmic Transformation of the Data in Table IV-C-1

Sample No.	Method	Transformed Replicates ($X = \log X_R$)	ΣX	$\overline{X}$	S
1	1	3.2672, 3.1303, 3.0212, 3.2553, 3.0607	15.7347	3.1469	.1115
	2	3.0792, 2.9294, 2.8129, 2.6990, 3.0212	14.5417	3.9083	.1544
2	1	3.8921, 3.9085, 3.7709, 3.7364, 3.8543	19.1622	3.8324	.0756
	2	3.3711, 3.7243, 3.7520, 3.8028, 3.7364	18.3866	3.6773	.1738
3	1	3.6021, 3.6435, 3.4232, 3.4771, 3.5623	17.7082	3.5416	.0903
	2	3.3802, 3.3324, 3.3118, 3.2041, 3.3424	16.5709	3.3142	.0664
4	1	2.7959, 2.7093, 2.7093, 2.6646, 2.8451	13.7242	2.7448	.0735
	2	2.7497, 2.5441, 2.4771, 2.4393, 2.4393	12.6496	2.5299	.1301
5	1	3.5877, 3.3674, 3.5224, 3.5647, 3.4031	17.4453	3.4891	.0984
	2	3.4362, 3.5944, 3.4031. 3.3424, 3.2380	17.0141	3.4028	.1310
6	1	3.8573, 3.6435, 3.8921, 3.8195, 3.7160	18.9284	3.7857	.1033
	2	3.6435, 3.6021, 3.6435, 3.5563, 3.7924	18.2377	3.6475	.0886
7	1	3.0792, 3.0000, 3.0000, 3.0792, 3.2041	15.3625	3.0725	.0836
	2	2.8451, 2.7782, 2.9542, 2.9542, 2.9542	14.4860	2.8972	.0816

Sample 5, method 2, replicate 2:

$$X_E = 3.5944$$

$$T = \frac{|3.5944 - 3.4028|}{.1310} = 1.46$$

Sample 6, method 2, replicate 5:

$$X_E = 3.7924$$

$$T = \frac{|3.7924 - 3.6475|}{.0886} = 1.64$$

The critical T value for sets of 5 replicate and a .01 significance level is 1.76. Since none of the T values is greater than 1.76, we must accept all of the suspected values.

Step 4

Test for Equality Among the Within-Sample Standard Deviation Estimates (S) for Each Method in Table IV-C-2:

Recall that $C = S^2_{max} / \sum_{i=1}^{K} S_i^2$

For Method 1: $C = (.115)^2/.0590 = .2106$

For Method 2: $C = (.1738)^2/.1071$

$= .2822$

The critical Cochran statistic for 7 variances with 4 degrees of freedom each and a .01 significance level equals .5080. Since both the calculated C values are less then the critical value, the within-sample variances are equal and we can proceed to calculated pooled variance and standard deviation estimates for each method:

$$S^2_{M_1} = \frac{.0590}{7} = .0084 \therefore S_{M_1} = .0918$$

$$S^2_{M_2} = \frac{.1071}{7} = .0153 \therefore S_{M_2} = .1237$$

Step 5

Test for Equality of the Pooled Within-Sample Variance Estimates

$$F = S^2_{M_2} / S^2_{M_1} = .0153/.0084 = 1.82$$

The critical F value at .05 significance level, when both variance estimates have 7(4) or 28 degrees of freedom, equals 1.88. Since the critical value is not exceeded, these methods do not produce significantly different within-sample variances. The same statement can be made regarding within-sample standard deviations.

Step 6

Test for Equality of the Method Means

First, for each sample, calculate the difference between the means in Table IV-C-2. These differences (d_i) are shown in Table IV-C-3 along with their mean ($\bar{d}$) and standard deviation (S_d). Then calculate:

$$t = \bar{d}/S_d/\sqrt{n}) = .1765/(.0547/\sqrt{7})$$

$$= .1765/.0207 = 8.54$$

The critical Student's t value at a .05 significance level and 6 degrees of freedom is 2.45. Since the calculated value exceeds the critical value, the methods do produce significantly different means.

To Summarize the Results

(a) The within-sample standard deviations for these methods are not significantly different at the .05 significance level.

(b) The mean for Method 1 is significantly higher than the mean for Method 2 at well over the .05 significance level.

Discussion of the Results

Referring to Table IV-C-3, it can be easily seen that Method 1 produces the larger mean. The difference can be calculated as:

$$\left(\frac{\text{antilog}(3.3733) - \text{antilog}(3.1967)}{\text{antilog}(3.1967)}\right) 100$$

$$= \left(\frac{2362 - 1573}{1573}\right) 100$$

$$= \left(\frac{789}{1573}\right) 100 = 50.2\%$$

Since Method 1 results are known to be about 40% greater than the true value, it can be estimated that Method 2 results are about 7 percent less than the true value. Therefore, besides being quicker and easier, Method 2 offers comparable precision with improved accuracy. Such results would qualify Method 2 for approval as an alternative to Method 1.

4. Method Characterization

The choice of a method of analysis among several considered should be based on comparison of relative performance using these measureable characteristics.

4.1 Specificity is the ability of a method to recover the desired organisms identified by a selective or differential characteristic and verified by additional tests. A method is judged specific if the recovered microorganisms verify as the desired organism, and the colonies designated as "other organisms" do not verify as the desired organism when picked and tested. The acceptable level of specificity for a method cannot be set absolutely, for example as 90%, but rather must be established for standard procedures or for new parameters on a best judgment basis by comparison with the accepted methods.

4.2 Selectivity is the ability of a method to encourage growth of the desired organism while reducing other organisms by some arbitrary degree, for example, three orders of magnitude, compared with growth and recovery on non-selective media.

4.3 Precision is a measure of the deviation among multiple measurements of a single quantity. The most widely used expression of precision is the standard deviation (σ) which is equal to the square root of the variance and indicates the deviation of the values about the mean.

Like accuracy, the true precision of a method must be generated in a collaborative study among at least 15 laboratories.

4.4 Accuracy is a measure of the closeness of observed values to a known true value. The lack of available standards for comparison of microbiological methodology has resulted in methods with known precision, but with limited accuracy information.

In microbiology, accuracy has been determined by applying the test method to a suspension of a pure culture while independently determining the cell number in the suspension using a non-selective medium. Natural water samples are tested with and without the pure culture spike and the recovery determined by difference. The recovery from the spiked sample is compared with the count on the non-selective medium (assumed to be true value) and expressed as a percent of this true value.

True accuracy of a method must be generated in a collaborative study among at least 15 laboratories using known levels of organisms.

4.5 The Section on Sanitary and Health Effects Microbiology in the American Society for Testing and Materials (ASTM) Committee D-19:24 has also proposed a characteristic called Counting Range. The Counting Range is described as the range from the lowest number to the highest number of colonies that can be measured on a single agar plate or membrane filter, without affecting the reliability of the method.

TABLE IV-C-3

Analysis of Difference Between Means

Sample No.	Mean from Table IV-C-2 for Method 1 ($\bar{X}_{M_1}$)	Method 2 ($\bar{X}_{M_2}$)	Difference Between Means ($d_1 = \bar{X}_{M_1} - \bar{X}_{M_2}$)	
1	3.1469	3,9083	.2386	
2	3.8324	3.6773	.1551	
3	3.5416	3.3142	.2274	
4	2.7448	2.5299	.2149	
5	3.4891	3.4028	,0863	
6	3.7857	3.6475	.1382	
7	3.0725	2.8972	.1753	
	$\bar{\bar{X}}_{M_1}$ = 3.3733	$\bar{\bar{X}}_{M_2}$ = 3.1967	$\bar{d}$ = .1765	S_d = .0547

REFERENCES

1. Grant, E. L. and R. S. Leavenworth, 1972. Statistical Quality Control. Fourth Edition. McGraw-Hill, Inc., New York, NY. p. 87.

2. Eisenhart, C. and P. W. Wilson, 1943. Statistical Methods and Control in Bacteriology. Bacteriological Reviews 7:57.

3. Velz, C. J., 1951. Graphical Approach to Statistics, Part 4: Evaluation of Bacterial Density. Water and Sewage Works, 98:66.

4. Dixon, W. J. and F. J. Massey, Jr., 1969. Introduction to Statistical Analyses, Third Edition. McGraw-Hill, Inc., New York, NY. p. 310.

5. ibid. p. 537.

PART V LABORATORY MANAGEMENT

Part V addresses those laboratory activities which supplement the analytical methodology and are primarily the responsibility of the laboratory manager.

PART V. LABORATORY MANAGEMENT

Section A Development of a Quality Control Program

1. Intralaboratory Quality Control

1.1 To insure a viable quality assurance effort, management must recognize the need for a formal program and require its development. Management must commit itself to the program by setting aside 15% of the laboratory man-years for quality control activity. It must meet with supervisors and staff to establish levels of responsibility for management, supervisors and analysts. Laboratory personnel should participate in the planning and structuring of the QA program.

1.2 Once the QA program is functioning, supervisors review laboratory operations and quality control with analysts on a frequent (weekly) basis. Supervisors use the results of the regular meetings with laboratory personnel to inform management on a regular (monthly) basis of the status of the QA program. These meetings are important to identify the problems at the laboratory level and to get the backing of management in the actions necessary to correct problems.

2. Documentation of the Intralaboratory Quality Control Program

Unless a record is made of the quality control checks and procedures described in Part IV of this Manual, there is no proof of performance, no evidence for future reference and for practical purposes, no quality control program in operation. The following documentation should be made and maintained.

2.1 An Operating Manual is prepared which describes the sampling techniques, analytical methods, laboratory operations, maintenance and quality control procedures. Specific details are given on all procedures and quality control checks made on materials, supplies, equipment, instrumentation and facilities. The frequency of the checks, the person responsible for each check (with necesssary back-up assignments), the review mechanism in the QC program to be followed, the frequency of the review and the corrective actions to be taken are specifed. A copy is provided to each analyst.

2.2 A Sample Log is maintained which records chronologically information on sample identification and origin, the necessary chain of custody information, and analyses performed.

2.3 A Written Record is also maintained of all analytical QC checks: positive and negative culture controls, sterility checks, replicate analyses by an analyst, comparative data between analysts, use-test results of media, membrane filters and laboratory pure water, replicate analyses done to establish precision of analysts, or of methodology used to determine non-compliance with bacterial limits and water quality standards.

3. Interlaboratory Quality Control

An interlaboratory quality control program consists of: 1) formal collaborative method studies to establish precision and accuracy of selected methodology, 2) specific minimal standards for personnel, sampling and sample preservation procedures, analytical methodology, equipment, instrumentation, facilities and within-laboratory quality control programs, 3) verification of acceptable standards through annual on-site inspections, 4) periodic performance tests of analytical capabilities using unknown samples, 5) follow up on problems detected in onsite inspections and performance evaluations.

EPA has established such a interlaboratory quality control program in response to the FWPCA Amendments of 1972 and the Safe Drinking Water Act of 1974.

PART V. LABORATORY MANAGEMENT

Section B Manpower and Analytical Costs

Laboratories planning to begin or increase microbiological activities have difficulty in determining their added manpower, equipment and supply needs. This Section provides estimates of added costs:

1. **Time Expenditures**

2. **Specialized Equipment and Supplies**

1. Time Expenditures for Microbiological Analyses

The following estimates of time required for membrane filter (MF) and most probable number (MPN) tests were prepared in response to information requests from Regional Offices and States planning new or additional microbiological work. The estimates are presented as guidelines only.

The Microbiological Methods Section, BMB and the Quality Assurance Branch, both of EMSL-Cincinnati prepared the estimates based on average performance of one technician qualified by short-course or on-the-job training and experience in the specific techniques. The procedures are those described in This Manual.

1.1. MF Analyses

If only fecal coliform bacteria are being tested, thirty (30) samples (estimating 3 dilutions/sample) can be prepared by one analyst on day one. Two hours are required to complete the counting, calculation of results and verification on day two.

If fecal and total coliform tests are performed using 3 dilutions, twenty (20) samples can be analyzed in one day by one analyst. On the second day, counting of plates takes an estimated 2 1/2 hours. An estimated four hours are needed for preparation of media, dilution water, dishes and pipettes for the test period if analyses are performed over five days. The time estimates include 10% devoted to quality control procedures.

1.2 MPN Analyses

After preparation of tubed media, one worker can process 15 samples for MPN analyses for total and fecal coliforms in an 8-hour day. The procedures include 5 tube × 5 or more dilutions to assure the positive and negative tubes result in 3 significant dilutions needed to obtain the optimum MPN index. If 15 samples are analyzed each day for a 5 day week, preparation and clean-up of tubed media and supplies for the test period would require an estimated 8 hours. Estimates include 10% of time devoted to quality control procedures (Part IV of this Manual).

Since the MPN requires reading and transfer of growth from positive tubes of lauryl tryptose broth to brilliant green bile broth and EC broth at 24 and 48 hours, the time span for one day's samples may cover four days. These times are indicated as 8, 4, 2 and 1 hours on the four successive days. New samples tested on the 2, 3, 4th day, etc., add to each day's workload. This add-on effect results in 8 hours + 4 hours + 2 hours + 1 hour for a workload of 15 hours from the 4th day onward. Either the sample load must be reduced or one must provide a second technician. The accumulative load is shown in Table V-B-1.

2. Specialized Equipment and Supplies

The following tables of materials and equipment are provided as guidance to laboratories. Tables V-B-2, 3 and 4 are designed for a minimal program of 2 samples/day, 7 days a week. Estimates of expendable supplies are based on those used in 1 year. Tables V-B-5, 6, 7 and 8 are designed for laboratories planning a program of 15 samples per day, 5 days per week. Estimates for expendable supplies are based on those used in 1 week.

The following assumptions are made:

The basic laboratory facility and equipment are available (space, benches, lighting, utilities, chemicals and standard glassware).

The laboratory will do membrane filtrations as single analyses at three dilution levels or MPN's as five tube, five dilution tests.

The amounts of supplies or numbers of items are based on needs, then adjusted to take advantage of quantity discounts by the dozen, box, 100's, etc. For both laboratory efforts, the expendable items are based on 2 weeks usage, while consumable media are based on usage/week in the high effort laboratory and usage/year in the low effort laboratory.

Laboratories doing primarily drinking water analyses may require fewer items; other laboratories analyzing only polluted waters may require more items.

Costs are current figures provided to give rough estimates only. The necessary characteristics of equipment and materials are cited to assist the purchaser in his choices.

TABLE V-B-1

Estimated Time Required for Fifteen MPN Analyses/Day*

Day 1	Day 2	Day 3	Day 4	Day 5	Day 6	Day 7	Day 8
First Series Samples Start 15 MPNS 8 hours	4 hours	2 hours	1 hour				
	Second Series 15 MPNs 8 hours	4 hours	2 hours	1 hour			
		Third Series 15 MPNs 8 hours	4 hours	2 hours	1 hour		
			Fourth Series 15 MPNs 8 hours	4 hours	2 hours	1 hour	
				Fifth Series 15 MPNs 8 hours	4 hours	2 hours	1 hour

*The time required per MPN analysis is not 1/15 of the time estimated because there is a time savings in preparing larger numbers of samples.

TABLE V-B-2

General Equipment and Supplies
Minimum Program

Item	Quantity	Cost of Quantity*
Incubator, gravity convection, 35 ± 0.5 C, DWH, 18 × 19 × 28", (46 × 48 × 71 cm)	1	$360.00
Incubator, waterbath, 44.5 ± 0.2 C for fecal coliforms, LWH 18 × 12 × 7½", (46 × 31 × 19 cm)	1	380.00
Sterilizer, steam, bench top, electric heat with temperature and pressure controls and gauges I.D.: 9" (23 cm) diameter chamber, 16" (41 cm) deep	1	1500.00
pH meter, analog, 0–14 pH range, accuracy ± 0.1 pH units, temperature compensated, with electrode	1	225.00
Oven, double wall, gravity connection RT to 225 C, automatic temperature control, I.D.: LHD 19 × 18 × 16", (48 × 46 × 41 cm)	1	575.00
Refrigerator, 3 cubic ft. (.19 M^3) with freezer compartment	1	425.00
Thermometer, mercury, range 0–50 C graduated in 0.1 C. Meets NBS specs.	2	40.00
Cylinder, graduated, 100 ml, 1 ml graduations	12	66.00
Cylinder, graduated, 50 ml, 1 ml graduations	18	85.00
Bottles, dilution, 99 ml mark, screw-cap, Pyrex glass	48	50.00
Bottle, sample, polypropylene, wide mouth, screw-capped autoclavable,		
125 ml	12	5.90
250 ml	12	9.00
500 ml	12	13.00
Bottle, sample, glass, wide mouth, screw-cap, autoclavable		
130 ml	12	6.80
210 ml	12	8.32
Beaker, stainless steel, 4 qt. waterbath for media preparation	1	15.00
Pump, vacuum, polypropylene, water powered, 11.5 liters/min capacity	1	2.92

*1978 prices.

TABLE V-B-2

General Equipment and Supplies
Minimum Program (continued)

Item	Quantity	Cost of Quantity*
Burner, Bunsen, utility	1	5.25
Needle, inoculating, in holder	2	2.50
Tongs, flask	1 pr.	5.25
Pen, ink, felt-tip, waterproof	12	11.00
Balance, torsion, two pan, 200 g capacity, accurate to 10 mg, with 100 g weight	1	310.00
Hot Plate/Magnetic Stirrer, variable speed and heat, 6 × 6", (15 × 15 cm) top	1	125.00
Pipettor, automatic, volume of 5–50 ml, speed of 10–60 deliveries/minute with glass syringe	1	485.00

*1978 prices.

TABLE V-B-3

Equipment and Supplies for MF Analyses
Minimum Program

Item	Quantity	Cost of Quantity*
Filters, membrane, white, gridded 47 mm, 0.45 μm or equivalent pore size	2100 MFs in pkgs of 100	$400.00
Pads absorbent, 47 mm (optional)	2100 MFs pads in pkgs of 100	32.00
Dishes, petri, plastic, tight-lid, 50 mm × 12 mm	2100 dishes in boxes of 500	210.00
Bag, plastic, waterproof for 44.5 C waterbath incubator	1000 in boxes of 500	60.00
Pipet, glass Mohr or bacteriological, 10 ml in 0.1 ml increments	24	45.00
Pipet, glass Mohr or bacteriological, 2.0 or 2.2 ml, in 0.1 increments	18	28.00
Pipet, glass Mohr or bacteriological, 1.0 or 1.2 ml, in 0.1 increments	18	25.00
Can, pipet, stainless steel, 2.5 × 2.5 × 16" 6.5 × 6.5 × 41 cm	3	42.00
Jar, polypropylene, for disinfection of pipets, 6.5 × 20" (16.5 × 50 cm)	1	20.00
Flask, vacuum, pyrex glass, 1 liter	1	6.00
Water Trap, glass bottle, stoppered with glass tube inlet/outlet	1	1.00
Tubing, rubber, vacuum, 3/6" O.D. & 3/32" I.D. (1.3 cm O.D. & .24 cm I.D.)	48	9.00
Pinchclamp, flat jaw	2	2.00
Forceps, blunt with smooth tip	2	3.00
Microscope, dissecting, binocular, 15 power	1	200.00
Illuminator, microscope, fluorescent, fits round tube microscope	1	62.00

*1978 prices.

TABLE V-B-3

Equipment and Supplies for MF Analyses
Minimum Program
(continued)

Item	Quantity	Cost[1]
M Endo MF broth (optional)	8 × ¼#/yr	$60.00
LES Endo agar (optional)	8 × ¼#/yr	72.00
M-FC broth	8 × ¼#/yr	64.00
Rosolic Acid (for M-FC broth)	54 × 1 g/yr	90.00
Methanol, 95%, in small vial for forceps disinfection	1 pt/yr	1.00
Ethanol, 95%, unadulterated, for M-Endo, MF broth and LES Endo agar	500 ml/yr	5.00

Equipment and Supplies for Verification of MF Analyses
Minimum Program

Item	Quantity	Cost[1]
Tubes, Culture, w.o. lip, borosilicate glass, 150 × 20 mm, reusable.[2]	8 × 24	$30.00
Tube, fermentation, borosilicate, w.o. lip, 75 × 10 mm, reusable.[2]	8 × 24	15.00
Rack, wire, for 10 × 4 culture tubes	6	48.00
Basket, wire, rectangular galvanized, 10 × 6 × 6", (25 × 15 × 15 cm)	6	35.00
Caps, culture tube, aluminum 22 mm I.D.	16 × 12	40.00
Lauryl Tryptose broth	1#/yr	13.80
Brilliant Green Bile broth	1#/yr	17.90
EC broth	1#/yr	16.45

[1]1978 prices.
[2]Available as disposables.

TABLE V-B-4

Equipment and Supplies for MPN Analyses
Minimum Program

Item	Amount	Cost[1]
Tubes, Culture, w.o. lip, borosilicate glass, 150 × 20 mm, reusable.[2]	2 × 576	$170.00
Tubes, Culture, w.o. lip, borosilicate glass, 150 × 25 mm, reusable.[2]	288	70.00
Tube, fermentation, borosilicate, w.o. lip, 75 × 10 mm, reusable.[2]	2 × 720	105.00
Rack, wire, for 10 × 5 culture tubes	15 × 1	225.00
Basket, wire, rectangular galvanized, 10 × 6 × 6", (25 × 15 × 15 cm)	10 × 1	80.00
Caps, culture tube, aluminum 22 mm I.D.	60 × 12	150.00
Caps, culture tube, aluminum 27 mm I.D.	24 × 12	60.00
Lauryl Tryptose broth	14 × 1#/yr	200.00
Brilliant Green Bile broth	16 × 1#/yr	300.00
EC broth	15 × 1#/yr	250.00

[1]1978 prices.
[2]Available as disposables.

TABLE V-B-5

General Equipment and Supplies
Full Program in Microbiology

Item	Quantity	Cost of Quantity*
Pipets, glass, Mohr or bacteriological T.D., 10 ml in 0.1 ml graduations	96	70.00
Pipets, glass, Mohr or bacteriological T.D., 2 or 2.2 ml in 0.1 ml graduations	96	60.00
Pipets, glass, Mohr or bacteriological T.D., 1 ml in 0.1 ml graduations	96	55.00
Can, pipet, stainless steel, 2.5 × 2.5 × 16" (6.5 × 6.5 × 41 cm)	15	210.00
Jar, polypropylene, for disinfection of pipets, 6.5 × 20" (16.5 × 50 cm)	2	40.00
Cylinder, graduated, 100 ml	24	132.00
Cylinder, graduated, 50 ml	36	170.00
Bottle, dilution, pyrex glass, marked at 99 ml, screw-cap	96	100.00
Bottle, sample, glass, wide mouth, screw-cap, autoclavable		
125 ml	72	42.00
250 ml	72	50.00
Bottle, sample, polypropylene, wide mouth, screw-cap, autoclavable		
125 ml	96	72.00
250 ml	96	108.00
500 ml	48	52.00
1000 ml	24	38.00
Thermometer, mercury, 0–50 C, graduated in 0.1 C, meets NBS specifications	2	40.00
Beaker, stainless steel, 4 quart, waterbath for media preparation	1	15.00
Flask, vacuum, aspirator, pyrex glass, 1 liter	3	18.00
Water-trap, glass (glass bottle, stoppered, with glass tube inlet and outlet)	1	1.00

*1978 prices.

TABLE V-B-5

General Equipment and Supplies
Full Program in Microbiology (continued)

Item	Quantity	Cost of Quantity*
Tubing, rubber, vacuum, 3/16′ I.D. × 3/32′ O.D. (1.3 cm I.D. × 2.4 cm O.D.)	48′	9.00
Needle, inoculating, with holder	4	6.00
Tongs, flask	1 pr	5.25
Pen, ink, felt tip, waterproof	12	11.00
Balance, single pan, electric, top loader, 200 g capacity, sensitivity 0.02 g for weighings of 2–200 grams	1	700.00
pH meter, analog, 0–14 pH range accuracy ± 0.1 pH units, temperature compensated line voltage, with comb. electrode	1	225.00
or		
pH meter, same as above, with accuracy ± 0.05 pH units	1	500.00
Burner, gas, Bunsen	2	10.50
Hot plate-magnetic stirrer combination, variable speed and heat, 15 × 15 cm. top	1	125.00
Incubator, 35 C ± 0.5 C, water jacketed radiant heat, 18 × 38 × 27″, (46 × 96 × 68 cm)	2	3380.00
or		
Incubator, 35 C ± 0.5 C, forced air, large, single chamber I.D.: DWH 18 × 36 × 27″, (46 × 96 × 68 cm)	2	2300.00
Incubator, 44.5 C ± 0.2 C water bath circulating, I.D., LWD 36 × 18 × 9½″, (91 × 46 × 24 cm)	1	720.00
Sterilizer, large rectangular, double-wall steam, electric automatic, I.D., LWH 38 × 20 × 20″ (97 × 51 × 51 cm)	1	14000.00
Oven, double wall, gravity convection RT to 225 C I.D., LWH 24 × 20 × 20″ (61 × 51 × 51 cm)	1	750.00
Refrigerator, 13 cubic foot, with freezer compartment, automatic defrost	1	500.00

*1978 prices.

TABLE V-B-6

Equipment and Supplies for MF Analyses
Full Program in Microbiology

Item	Quantity	Cost of Quantity*
Membrane filtration assembly, stainless steel, for 47 mm filters	3	$360.00
Membrane filtration assembly, plastic, autoclavable, for 47 mm filters	3	24.00
Membrane filtration assembly, pyrex, for 47 mm filters	3	120.00
UV sterilizer unit for sterilizing MF filtration assemblies	1	550.00
Manifold, PVC, 3 place, for multiple filtrations	1	200.00
Microscope, binocular, dissecting type, 15 power	1	200.00
Lamp, fluorescent, microscope illuminator, fits round tube microscope	1	62.00
Filters, membrane, white, 47 mm 0.45 μm or equiv. pore size, gridded	300	57.00
Pads, absorbent, 47 mm	300	4.50
Dishes, petri, plastic, 50 × 12 mm, tight lid	500	50.00
Bags, plastic, waterproof for submersion of M-FC dishes in waterbath	500	30.00
Bottle, rinse water, round, wide-mouth, screw-cap, polypropylene, 1000 ml	24	38.00
Forceps, blunt with smooth tip	3	4.00
Counter, mechanical, hand type	1	9.00
Pump, vacuum, polypropylene, water powered, 11.5 liters/min capacity	1	3.00
or		
Pump, vacuum/pressure, electric, portable, produces 20" mercury vacuum	1	130.00
Pinchclamp, flatjaw	12	6.00

*1978 prices.

TABLE V-B-7

Equipment and Supplies for MPN Analyses
Full Program in Microbiology

Item	Quantity	Cost of Quantity[1]
Basket, wire, rectangular galvanized, 10 × 6 × 6" (25.4 × 15 × 15 cm)	190	$1520.00
Rack, wire, for 10 × 5 culture tubes, 150 × 25 mm	150	2250.00
Tubes, culture w.o. lip, borosilicate glass, 150 × 20 mm, reusable.[2]	30 × 576	2560.00
Tubes, culture w.o. lip, borosilicate glass, 150 × 25 mm, reusable.[2]	17 × 288	1200.00
Tubes, fermentation w.o. lip, borosilicate glass, 75 × 10 mm, reusable.[2]	21 × 720	1100.00
Caps, culture tube, aluminum, 22 mm I.D.	1250 × 12	3125.00
Caps, culture, tube, aluminum, 27 mm I.D.	408 × 12	1020.00

[1]1978 prices.
[2]Available as disposables.

TABLE V-B-8

Media for Full Program in Microbiology Laboratory
Usage for each Week/100 Samples

Medium	Usage/Week	Cost/lb[1]
MF Analyses:		
M Endo MF Broth	0.1 lb/week	$17.90
M-FC Broth	0.07 lb/week	24.85
LES Endo Agar	0.17 lb/week	20.95
Rosolic Acid (for M-FC Broth)	1 g/week	1.70/g
MPN Analyses:		
Lauryl Tryptose Broth	1.96 lb/wk[2]	13.80
Brilliant Green Bile Broth	2.2 lb/wk	17.90
EC Broth	2.04 lb/wk	16.45
EMB Agar	variable	18.70

MF Verification Costs on 5 Samples/Week
20 Colonies/Sample

Medium	Usage/Week	Cost/lb[1]
Lauryl Tryptose Broth	.07 lb/week	13.80
Brilliant Green Bile Broth	.07 lb/week	17.90
EC Broth	.07 lb/week	16.45

[1] 1978 prices.
[2] For single strength medium.

PART V. LABORATORY MANAGEMENT

Section C Laboratory and Field Safety

Introduction

This Section has been compiled from the best available sources. The procedures given for general laboratory safety follow the OSHA regulations (1, 2). Specific recommendations for microbiology were selected from literature of the Center for Disease Control, the National Institute for Occupational Safety and Health of the Public Health Service, the U.S. Environmental Protection Agency and other sources (3–12).

Safety procedures should be performed as an integral part of the analytical methods and should be included in program planning on a day to day basis.

The objectives of a laboratory and field safety program are 1) to protect the laboratory worker, the laboratory environment and the surrounding community from microbial agents studied and 2) to protect the integrity of the microbiological studies. The program is discussed as follows:

1. **Administrative Considerations**
2. **Sources of Hazard**
3. **Field Guidelines**
4. **Laboratory Guidelines**
5. **Biohazard Control**
6. **Safety Check List**

1. Administrative Considerations

1.1 Development of a Safety Program

1.1.1 If a laboratory safety program is to be effective, management must know the causes of infections in order to develop and upgrade safety procedures, equipment and rules, and to reduce incidence of infection. The lack of information on sources of laboratory infection (Table V-C-1 and 2) prevents improvement in safety programs and emphasizes the need for reporting all accidents. A preponderance of evidence in the literature indicates that if the known causes of laboratory infections were eliminated, the remaining infections could be considered to be caused by airborne transmission.

Some common microbiological procedures shown to produce aerosols include pipetting into a petri plates or flasks, opening lyophilized culture ampuls, opening culture containers, inserting a hot loop into a culture container and removing the cover from standard blender after mixing a sample.

1.1.2 Job attitudes can be the cause of laboratory accidents: overly rigid work habits, failure to recognize dangerous situations, work at excessive speed and deliberate violations of rules. These attitudes can only be overcome by the development of safe work habits through continued education and training.

TABLE V-C-1

Laboratory-acquired Infections Related to Personnel and Work*

Distribution	Total # of Infections	Type of Infective Agent Bacteria	Viruses	Rickettsia	Parasites	Fungi
No. of cases	2262	1303	519	293	62	85
No. of deaths	96	53	31	8	2	2
Type of Personnel (where known)						
Trained Scientific Personnel**	1534	856	358	206	57	57
Students	82	80	—	1	1	—
Animal Caretakers	221	137	39	40	1	4
Clerks, occasional visitors, maintenance	87	48	—	15	1	23
Types of Work (where known)						
Diagnostic	525	393	96	3	16	17
Research	726	289	230	155	31	21
Biological Reagent Production	51	19	31	1	—	—
Classwork	37	37	—	—	—	—
Combination of Activities	599	335	81	133	11	39

*Reference 3.

**Includes research assistants, professional and technical workers and graduate students.

TABLE V-C-2

Sources of Laboratory-acquired Infections*

Source of Infection	Type of Infective Agent		
	Bacteria	Parasites	Fungi
Work Situation but No Known Incident			
Clinical specimens	103	17	—
Autopsy, including known accidents	95	—	—
Aerosols	135	1	39
Handled infected animals and ectoparasites	126	5	—
Work with agent	412	16	27
Discarded glassware	21	—	1
Known accidents			
Needle and syringe	66	3	2
Pipetting	66	1	—
Spilling and splattering	41	5	2
Injury with broken glass, etc.	44	—	1
Bite of animal or ectoparasite	28	8	2
Centrifuge	2	—	3

*Reference 3.

1.1.3 Safety training of laboratory and field personnel, formulation of safety regulations, and the establishment of mechanisms for reporting and investigating accidents are prime responsibilities of the laboratory and field supervisors and higher management. First aid courses should be provided to the laboratory supervisor and at least one other permanent employee.

Each employee should have a copy of the safety program. Joint supervisor-employee safety committees should identify potential laboratory hazards and formulate workable safety regulations. Laboratory safety regulations should stress the protection of the laboratory personnel, janitorial and maintenance staff and others who might come in contact with the laboratory and its personnel.

1.2 Reporting Laboratory Infections and Accidents

1.2.1 Because many laboratory accidents are not reported, most laboratory infections are never traced to a specific cause. To improve safety, it is important to maintain good records of laboratory accidents and infections. It is necessary to know the pathogens involved and the circumstances under which the infection or accident occurred. It is recommended that each laboratory set up a formal system for reporting accidents as they occur so that appropriate prophylatic measures may be instituted. A record of accidents with after effects, particularly those resulting in infection, can be of considerable value.

1.2.2 In EPA, laboratory-acquired infections or accidents must be reported to the immediate supervisor. EPA Form 1440, Supervisor's Report of Accident, and Form CA 1, "Federal Employee's Notice of Traumatic Injury and Claim for Continuation of Pay/Compensation," must be completed by the employee and the supervisor and filed within two working days. Form CA 2, "Federal Employee's Notice of Occupational Disease and Claim for Compensation" is to be completed by the employee and submitted within 30 days.

It is recommended that the safety officer at each installation conduct a quarterly safety inspection of the facilities to identify and correct dangerous conditions or procedures. He should make full use of the safety check lists developed for the laboratory. An example of a safety check list is given at the end of this Section.

2. Sources of Hazard

2.1 Causative Agents

Only a small percentage of microorganisms are capable of producing disease in man; they include bacteria, fungi, yeasts, protozoa, actinomycetes, animal and human viruses, and rickettsiae (3, 4). Because innocuous and infectious microorganisms cannot be differentiated in natural materials and because microorganisms considered harmless may produce disease in man under favorable conditions, it is good practice to treat all microorganisms and materials as if they are pathogens or disease carriers.

2.2 Sources in Sampling

Laboratory and field personnel collecting samples and isolating cultures from natural sources must be made aware that pathogens are present in environmental samples. Disease-causing organisms are found in natural waters, municipal effluents and sludges, industrial wastewaters from packing plants, in soils and runoff from feedlots and from septic tank systems. These pathogens have also been found in inadequately-treated finished water systems and in ground water supplies. Before working with environmental samples, field personnel should have thorough training in aseptic technique and handling pathogens.

2.3 Sources in the Laboratory

Table V-C-1 and 2 present data on laboratory-acquired infections in the United States which were collected in a survey of 5000 laboratories over a 20-year period (3). Table V-C-1 shows the distribution of laboratory-acquired infections according to personnel and their work. It is significant that trained scientific personnel had 1534 of the 1924 infections identified by type of personnel. Among laboratory workers, researchers led with 726 of 1938 infections. These data suggest that the majority of infections occur because of carelessness of trained and knowledgeable workers and not because of ignorance. However, new technicians should be made aware that culturing microorganisms from natural sources develops extremely large numbers of cells which could cause disease if not properly handled.

Table V-C-2 presents data on the sources of laboratory-acquired infections. This table shows that the five most common accidents causing laboratory infection were:

1. Accidental inoculation with syringes and needles.

2. Accidental oral aspiration of infectious material through a pipet.

3. Cuts or scratches from contaminated glassware.

4. Spilling or splattering of pathogenic cultures on floors, table tops, and other surfaces.

5. Bites of animals or ectoparasites.

It is significant that in about 80% of the laboratory infections studied in Table V-C-2 the mechanism was not known. It was only known that the individual had worked with the agent or had tended infected animals.

3. Field Safety Guidelines

The sample collector or investigator must also consider safety in his work. The potential for accidents in field work is much greater than in the laboratory. The following rules on field safety were extracted from a comprehensive safety manual developed by EPA's National Enforcement and Investigations Center at Denver (5). They are intended as guidelines to assist the laboratory in developing its own protocol.

3.1 Automotive Safety

3.1.1 The driver should make certain that he has valid state and agency driver licenses on his person before operating a vehicle.

3.1.2 If the driver observes a questionable or unsafe condition when first operating a vehicle, he should return it directly to the carpool regardless of work demands.

3.1.3 Continuous driving in excess of ten hours, in any 24-hour period, is not recommended.

3.1.4 Occupants of vehicles should wear seat belts and shoulder harnesses, where provided, whenever vehicles are in motion. The driver should carry a kit provided by the laboratory that includes fire extinguisher, flares, reflectors, and a first-aid kit.

3.1.5 Safety screens should be installed in carryall and van-type vehicles to separate the cargo and passenger compartments. If safety screens are not installed, cargo should not be stacked higher than the back of the seat.

3.1.6 Employees required to tow a trailer should be instructed in the proper handling of the equipment involved.

3.1.7 Vehicles used to tow any kind of trailer should be equipped with west-coast type mirrors and the necessary connections for trailer signal, tail lights, brakes, and safety chains.

3.1.8 If a boat is transported in a pickup truck, it should not obstruct the vision of the driver or extend over the vehicle cab.

3.1.9 A driver backing a vehicle with a trailer in tow should have someone outside the vehicle to direct him.

3.2 Boat Safety

3.2.1 Only qualified employees should operate watercraft. Boat operators must have completed advanced emergency first-aid training.

3.2.2 The boat operator must not operate the boat without a second person on board.

3.2.3 The boat operator is responsible for the safety of persons and equipment on board. He should provide a a boat safety briefing to occupants before embarking.

3.2.4 Occupants must wear life jackets onboard. No exceptions are permitted. Soft-soled, non-skid shoes are recommended.

3.2.5 Flare gun, fire extinguishers and first aid kit should be kept on each boat.

3.2.6 Boats operated in estuaries or open seas should be equipped with depth-finding instruments, navigational aids and two-way radios adequate to communicate with at least one shore station. Boats with marine radios should monitor distress frequency except when transmitting.

3.2.7 Boats should not be operated in high winds, storms, heavy rain, fog, etc. Boats should not be operated more than one half mile from launch point on estuaries, large lakes, and large rivers until acquiring reliable weather forecasts.

3.2.8 The operator should attach a red pennant to the radio mast when operating at slow speeds (e.g., sampling, dredging, towing).

3.2.9 The operator must install and use lights according to established practice for night operations.

3.3 General Rules for Sampling

3.3.1 Require two people on each gauging or night sampling crew or on other hazardous projects.

3.3.2 Wear safety glasses, safety shoes, hard hats, respirators, gas masks, and ear-protection devices, as appropriate in hazardous areas.

3.3.3 Wear fluorescent vests or jackets while sampling from roadways and bridges. During sampling, post a yellow flasher on approaches at each end of the bridge. On heavily traveled roads post flagmen or warning devices at each end of bridge that lacks 24-inch walkways. Such sampling points should be avoided where possible.

3.3.4 Wear rubber gloves while handling samples that might be toxic or corrosive. Wear work gloves while handling sampling equipment. During collection and transport, the field worker should store containers to prevent spilling or splashing of samples. Disinfect hands immediatcly after handling sewage samples and equipment for sampling sewage.

3.3.5 Do not sample from railroad bridges unless there is an adequate walkway or the railroad dispatcher has been contacted and it has been positively determined that no trains will run during sampling period.

3.3.6 Equip vehicles for sampling and associated work with rotary amber caution lights. Operate such lights whenever vehicles are driven slowly on roadways or are parked near roadways. Do not park vehicles on bridges.

3.3.7 Inform employees of the safety rules in force within industrial sites. Employees must conform to rules promulgated by the industry while on-site.

3.3.8 Properly ground electrical apparatus employed in field operations and use battery straps to handle or move wet-cell batteries.

3.4 Sampling from Manholes

3.4.1 Erect barricades around manholes where samples are being collected. Do not leave manholes uncovered while unattended or unbarricaded.

3.4.2 Do not enter a manhole until it is cleared by using a blower for at least five minutes. Following the five-minute ventilation, use a lead acetate swab to check for H_2S. If H_2S is present, wear a respirator when entering manhole. Substitute a respirator for blower ventilation where explosive gases are present (i.e., in storm sewer or domestic sewer).

3.4.3 A sample collector entering manholes must wear safety lines handled by two persons outside of manholes. Keep safety lines taut at all times. Fifteen minutes is the maximum time allowed in manhole. Keep a vehicle at hand in case of emergency.

3.4.4 Do not enter sewer lines for any reason.

3.5 Sampling Channels and Streams

3.5.1 Sample collector must work from behind a barricade or wear a safety line attached to a secure object when sampling fast-moving channels or streams from shore, walkway, etc. The same rule applies to any open channel when footing is questionable, i.e., snow, steep bank, etc.

3.5.2 Wade only to knee-depth in swift waterstreams, or to hip-depth in placid water. When wading in fast moving water secure safety lines to shore and have at least two other persons in attendance.

3.5.3 Attach lines from sampling devices to a secure object but never to sampling personnel.

3.6 Sampling Under Ice

3.6.1 Two people are a minimum crew for operations involving ice cover. One person must remain on solid footing until thickness of ice is known.

3.6.2 Do not sample on ice if the ice thickness is less than four inches.

3.6.3 Wear life preservers and secure safety lines to an object on shore when sampling on ice-covered water.

4. Laboratory Safety Guidelines

The following safety rules are intended as guidelines. They were developed from the available safety literature (3–9) and have considered the Occupational Safety and Health Administration (OSHA) regulations (1, 2). Using such source materials, the laboratory director, laboratory supervisor or senior professional should develop rules that are specific for the laboratory program and the organisms involved.

4.1 Personal Conduct and Clothing

4.1.1 Store coats, hats, jackets, and other items of personal clothing outside of the microbiology laboratory. Do not mix laboratory and street clothes in the same locker.

4.1.2 Wear a non-flammable laboratory gown or coat in the laboratory. If clothing becomes contaminated, autoclave before laundering. Laboratory clothing should not be worn in clean areas or outside the building. Open-toed shoes, or extreme shoe styles should not be worn, since they provide little protection or are unstable.

4.1.3 Wear goggles or safety glasses to protect eyes from UV irradition.

4.1.4 Wash hands carefully after laboratory and field duties, using a germicidal soap.

4.1.5 Use forceps or rubber gloves when there is a significant danger of contamination such as during the clean-up of pathogenic material.

4.1.6 Do not touch one's face, lick labels or put pencils and other materials in one's mouth.

4.1.7 Don't smoke, eat, drink or chew gum in the laboratory or while sampling. Do not keep food or drinks in the lab refrigerator or cold room. Do not brew coffee or tea in the laboratory area.

4.1.8 Keep conversation to an absolute minimum during bench work to prevent self-infection or loss of analytical data.

4.1.9 Keep reading matter, surplus materials and equipment out of the laboratory area.

4.1.10 Laboratory and field personnel handling polluted samples should be vaccinated against typhoid, tetanus and polio.

4.2 Laboratory Equipment

4.2.1 Limit traffic through the work areas.

4.2.2 Treat all cultures and samples as if they are potentially pathogenic. The degree of risk is increased greatly in culture work because the microorganisms are produced in very large numbers.

4.2.3 Do not mouth-pipet polluted water, wastewater or other potentially infectious or toxic fluids; use a bulb or other mechanical device. See Part II-B, 1.8.2.

4.2.4 For potable waters, plug pipets with non-absorbent cotton. Do not use pipets with wet plugs.

4.2.5 Use a hooded bunsen burner or shielded electric incinerator to protect against splattering during culture work.

4.2.6 Maintain benches in a clear and uncluttered condition for maximum efficiency and safety.

4.2.7 Perform all culture work in a biohazard hood to protect cultures and workers.

4.2.8 Do not use the kitchen type blender for mixing materials containing infectious agents. Safety blenders are available in which infectious materials may be mixed without dissemination of infectious aerosols.

4.2.9 When a vacuum line is used, interpose suitable traps or filters to insure that infectious agents do not enter the system.

4.2.10 Lyophilization procedures can be a source of laboratory infection. When vacuum is applied during lyophilization, the contaminated air is withdrawn from the ampuls through the pump and into the room. Use biological air filters or air decontamination procedures to reduce hazard. Aerosols are also often created by opening lyophilized ampuls. Reduce this hazard by wrapping the ampul in a disinfectant-soaked pledget of cotton before breaking.

4.2.11 Read II-C-6 for instructions on proper packing of cultures for mail shipment before sending any isolates to a central laboratory for confirmation.

4.2.12 Periodically clean out freezers, ice chests and refrigerators to remove any broken ampuls, tubes, etc., containing infectious materials. If units contain pathogenic cultures, use rubber gloves during this cleaning. Use respiratory protection if actinomycetes, fungi or other easily disseminated agents are involved.

4.3 Disinfection/Sterilization

4.3.1 Disinfect table tops and work carts before and after laboratory work. A bottle of disinfectant and gauze squares or towelling for washing and wiping purposes should be available in laboratory for routine and emergency use.

4.3.2 Use a disinfectant which specifies germicidal activity against the organisms most often encountered in the laboratory. Organo-iodine complexes, quaternary ammonium compounds, phenolics and alcohols which are effective against vegetative bacteria and viruses are recommended for general use. However, these disinfectants are not sporocidal. If spore-forming bacteria are encountered, formaldehyde or formaldehyde/alcohol solution is recommended. See Table V-C-3.

Mercury salts, chlorine-containing compounds or home-use products are not recommended for the laboratory.

TABLE V-C-3

Normal Use Concentration of Disinfectants

Compound	Use Concentration mg/liter
Organo-Iodine Complexes	100–150
Quaternary Ammonium Compounds	700–800
Phenolics	½-1%
Alcohol, 70% w/v	water solution
Formaldehyde	8%
Formaldehyde in 70% Alcohol Solution	8%

4.3.3 If a culture or infective material is spilled, notify the laboratory supervisor at once, then disinfect and clean up the area.

4.3.4 Never pour viable cultures or contaminated materials in the sink. Never leave infectious material or equipment unattended during use.

4.3.5 Immediately after use, place contaminated pipets in a disinfectant container which allows complete immersion; place cultures and contaminated materials in color-coded biohazard bags and seal. Disinfectant containers of pipets and sealed bags of materials are autoclaved as units.

4.3.6 Place used glassware in special cans marked for autoclaving. Keep broken glassware in a separate container. Place plastic items in separate cans to prevent fusing of plastic around glass items.

4.3.7 Mark contaminated items as Contaminated before removal from the laboratory for autoclaving. Use temperature-sensitive tapes which indicate exposure to heat. Preprinted tapes or tags simplify this task.

4.3.8 Check autoclaves with the use of spore strips or spore suspensions of *B. stearothermophilus* and maximum-minimum recording thermometers. Ideally autoclaves are equipped with temperature recording devices so that a permanent record may be maintained.

Check hot air ovens and gas sterilizers periodically with spore strips or the indicator, *B. subtilis* var. *niger*.

4.3.9 Wet-mop floors weekly, using water containing a disinfectant. Dry or wet pickup vacuum cleaners with high-efficiency exhaust air filters are recommended. Wax floors with bacteriostatic floor waxes if available.

4.4 Chemicals and Gases

4.4.1 Label containers plainly and permanently. Dispose of material in unlabelled containers carefully. Wipe or rinse residual material from the external surfaces of reagent containers after use.

4.4.2 Store flammable solvents in an approved solvent storage cabinet or a well-ventilated area.

4.4.3 When opening bottles which may be under pressure i.e., hydrochloric acid, ammonium hydroxide, cover the bottle with a towel to divert chemical spray.

4.4.4 Use bottle carriers to transport bottles containing hazardous chemicals (acids, corrosives, flammable liquids). Large cylinders are transported only by means of a wheeled cart to which the cylinder is secured. Store and transport compressed gas cylinders with shipping caps on, in an upright position, always securely clamped or chained to a firm support and away from heat.

4.4.5 Reagents and chemical which might react in water drains or be dangerous to the environment must be disposed of in other ways. Examples are 1) sodium azide which reacts with metal drains to produce very explosive lead or copper azides and 2) mercury and its salts which should not be returned to the environment. Consult reference texts to determine the proper disposal procedure for each chemical (8, 9).

4.5 Handling Glassware

4.5.1 Discard broken, chipped or badly scratched glassware. Use gloves or sweep up broken glass, do not use bare hands. Pick up fine glass particles with wet paper towelling.

4.5.2 Fire polish tubing and rods.

4.5.3 Protect hands with gloves, towel, or tubing holder when inserting tubing into stoppers. Lubricate the tubing with water or glycerine. Handle tubing close to the stopper and out of line with end of the tube.

4.5.4 Use asbestos-centered wire gauze when heating glass vessels over a burner.

4.5.5 Do not attempt to catch falling glassware.

4.6 Electrical Equipment

4.6.1 Keep materials, tools and hands dry while handling electrical equipment.

4.6.2 Use grounded outlets only.

4.6.3 Do not use electrical equipment near flammable solvents.

4.6.4 Use only carbon dioxide or dry powder fire extinguishers in case of fire in or near any electrical equipment.

4.7 Emergency Precautions

4.7.1 Install and maintain both foam and carbon dioxide fire extinguishers within easy access of the laboratory.

4.7.2 Fire exits should be clearly marked and accessible.

4.7.3 Install and maintain a complete first aid kit and an oxygen respiration unit in the laboratory.

5. Biohazard Control

5.1 Safety Cabinets

5.1.1 The safety cabinet is the most important primary barrier available to the microbiologist for isolation and containment of microorganisms and for protecting the laboratory environment, and the surrounding area from contamination. Transfers of cultures especially pathogenic fungi, actinomycetes and yeasts should be conducted in the safety cabinets.

5.1.2 UV lamps are commonly used in biohazard hoods to maintain sterility of the work area. Goggles should be worn to protect the worker and cultures should be protected from undesirable exposure (see Part IV-A, 4 in this Manual).

5.1.3 There are several types of ventilated cabinets available for use (10, 13):

(a) Partial Barrier Cabinet

The open or closed front cabinet is usually referred to as a partial barrier ventilated cabinet. This cabinet can be used with the glove panel removed, depending upon an inward flow of air of at least 100 linear ft. per min. to prevent escape of airborne particles. It can also be used with the glove panel in place and arm-length gloves attached, in which case it will be maintained under a reduced air pressure of about one inch of water gauge. When operated closed, the partial barrier needs an attached air lock for movement of materials. A third mode of operation consists of using a cabinet with glove panel attached, but with gloves removed.

(b) Absolute Barrier Cabinet

The second type of ventilated cabinet is the gas-tight cabinet system, referred to as an absolute barrier cabinet. Absolute barrier cabinets are connected to form a modular cabinet system with enclosed refrigerators, incubators, etc. Air is drawn into the cabinet system through ultrahigh efficiency filters and is exhausted through ultrahigh efficiency filters.

(c) Vented Laminar Flow Cabinet

This third type of safety cabinet is not gas-tight. It relies on high efficiency filters to protect the worker and the environment. In the Figure V-C-1, the blower (1) set into the bottom of the cabinet pushes air up through a rear duct into the top of the cabinet (2) and causes a slight negative pressure in the work area (3). In the top of the cabinet (2), part of the air is forced out the exhaust through a High Efficiency, Particulate, Air (HEPA) filter (4). However, the major part of the air is forced down through a large HEPA filter (5), in a vertical flow into the work area (6). This vertical flow (laminar flow) combines with the negative pressure from the blower to draw in enough make-up from the room to replace that exhausted above. The make-up air (7) is pulled through vents at the edge of the hood opening and is drawn down to the blower without contacting the work area. It combines with the filtered air and is recirculated through the HEPA filters. The laminar flow draws off any contaminating particles emanating from the work area. HEPA filters are 99.99% efficient in removing particles 0.3 µm diameter or larger, by the DOP test.

5.1.4 Selection of a Cabinet

The partial barrier cabinet with open front gives some protection to the worker and the laboratory environment but does not protect the cultures. With the glove panel in place, it protects the worker and adjacent laboratory area. The absolute barrier and laminar flow cabinets provide the greatest protection to the worker and the cultures.

For routine water bacteriology and limited work with pathogens, a partial barrier cabinet with glove panel can be used. If a significant portion of the workload involves pathogenic microorganisms, a gas-tight absolute barrier cabinet with glove ports or laminar flow cabinet is recommended. These provide protection of cultures as well as the worker and the work environment. Because laminar flow cabinets do not require glove openings, yet do protect personnel and culture work and are easy to use, they are recommended for all but the most hazardous microbiological operations.

5.2 Biohazard Identification

5.2.1 The revised Federal Occupational Safety and Health Act of 1972 requires biological hazard signs and tags to signify the actual or potential biological hazard (1, 10). The hazards are defined as infectious agents that present a risk to human well-being. These signs and tags are used to identify equipment, containers, rooms, materials, experimental animals, or combinations of the above, which contain or are contaminated with viable hazardous agents.

5.2.2 The biological or biohazard symbol design is shown in Figure V-C-2. It has a fluorescent orange or orange-red color, and may contain appropriate wording to indicate the nature or identity of the hazard, the name of the individual responsible for its control, precautionary information, etc. The wording must not be superimposed on the symbol. Background color is optional, but enough contrast must be provided so that the symbol can be clearly defined.

6. Safety Check List

The laboratory safety check list that follows is provided as a guide for a laboratory to incorporate wholly or in part into its own safety program.

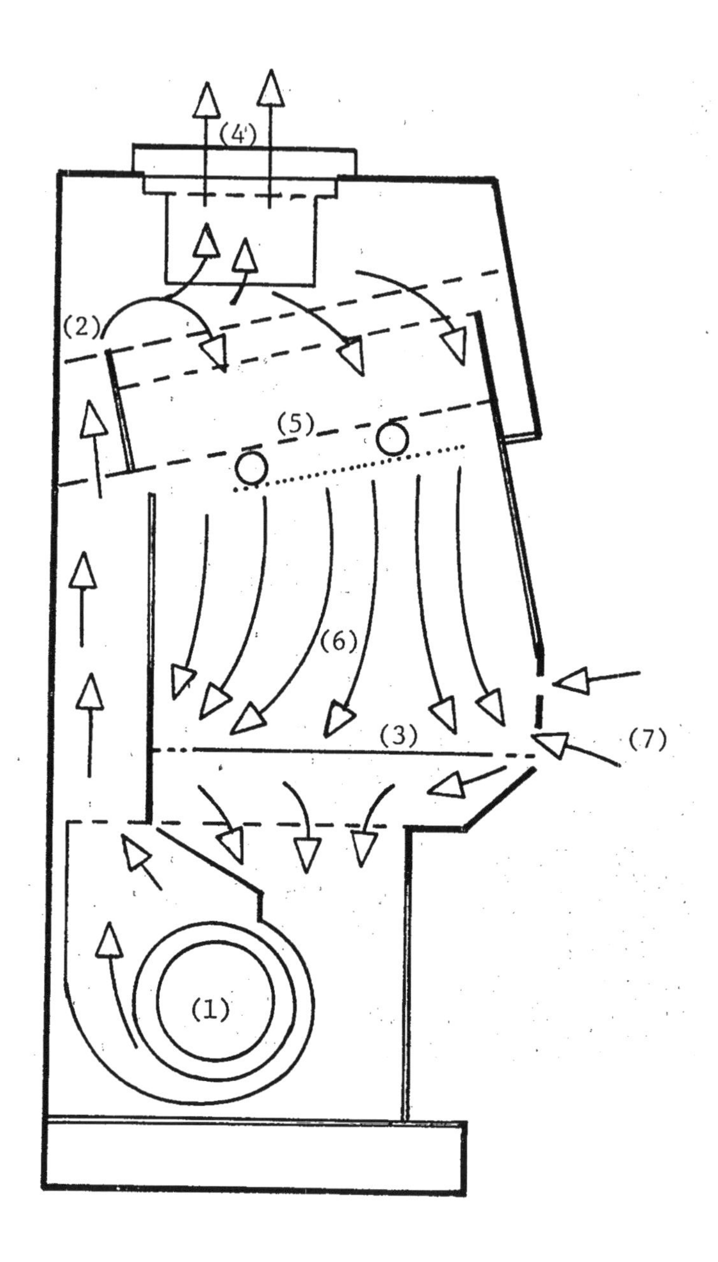

FIGURE V-C-1. Laminar Flow Cabinet.

FIGURE V-C-2. Example of Biohazard Sign.

Safety Check List
for Microbiological Water Laboratories

Survey By:
Laboratory:
Location:
Date:
Code: S=Satisfactory, U=Unsatisfactory

1. Administrative Considerations

(a) Laboratory has a formal documented safety program. ________

(b) Each worker has a copy of the safety program. ________

(c) Employees are aware of procedures for reporting accidents and unsafe conditions. ________

(d) New employees are instructed on laboratory safety. ________

(e) Joint supervisor-employee safety committee has been established to identify potential laboratory hazards. ________

(f) Records are maintained of accidents and consequences. ________

(g) Name and phone number of the supervisor and an alternate are posted at door of the laboratories so he may be contacted in case of an emergency. ________

(h) Laboratory supervisor and at least one other permanent employee have attended appropriate first aid courses. If so, when: ____________________. ________
(date)

(i) Emergency telephone numbers for fire, ambulance, health centers, and poison control center are placed in a conspicuous location near the telephone. ________

(j) Employees know the location of first aid supplies. ________

(k) Emergency first aid charts, and hazardous agents charts are posted in the laboratory. ________

(1) Fire evacuation plan is established for the laboratory and is posted in a conspicuous location. ________

2. Personal Conduct

(a) Personal clothing is stored outside of the microbiology laboratory. ________

(b) Lab coats and street clothes are kept in separate lockers. ________

(c) Laboratory coats are worn at all times in the laboratory. ________

(d) Germicidal soap or medicated surgical sponges are available for employees' use. ________

(e) Preparing, eating or drinking food and beverages are not permitted in the laboratory. ________

(f) Smoking or chewing gum are not permitted in the laboratory. ________

(g) Food or drink are not stored in laboratory refrigerators. ________

(h) Reading materials are not kept in the laboratory. ________

(i) Laboratory coats are not worn outside the lab. ________

(j) Employees who have cuts, abrasions, etc. on face, hands, arms, etc. do not work with infectious agents. ________

3. Laboratory Equipment

(a) Bulb or mechanical device is used to pipet polluted water, wastewater or other potentially infectious or toxic fluids. ________

(b) Pipets are immersed in disinfectant after use. ________

(c) Benches are maintained in clear and uncluttered condition. ________

(d) Centrifuge cups and rubber cushions are in good condition. ________

(e) A suitable disinfectant is available for immediate use. ________

(f) Blender is used with sealed container assembly. ________

(g) Microscopes, colony counters, etc. are kept out of the work area. ________

(h) Water baths are clean and free of growth and deposits. ________

(i) Employees are instructed in the operation of the autoclave and operating instructions are posted near the autoclave. ________

(j) Autoclaves, hot air sterilizing ovens, water distilling equipment, and centrifuges are checked routinely for safe operation. ________

Give frequency and last date

Autoclave ____________________ ________

Water still ____________________ ________

Centrifuge ____________________ ________

Hot Air Oven ____________________ ________

(k) No broken, chipped or scratched glassware are in use. ________

(1) Broken glass is discarded in designated containers. ________

(m) Electrical circuits are protected against overload with circuit breakers or ground-fault breakers. ________

(n) Power cords, control switches and thermostats are in good working order. ________

(o) Water taps are protected against back-siphoning. ________

4. Disinfection/Sterilization

(a) Proper disinfectant is used routinely to disinfect table tops and carts before and after laboratory work. ________

(b) Receptacles of contaminated items are marked. ________

(c) Performance checks of autoclaves, gas sterilizers and hot air ovens are conducted with the use of spore strips, spore ampuls, indicators, etc. ________

Item	Frequency	Last Date
________	________	________
________	________	________
________	________	________

(d) Safety glasses are provided to employees. ________

(e) Safety glasses are used with toxic or corrosive agents and during exposure to UV irradiation. ________

5. Biohazard Control

(a) Biohazard tags or signs are posted in hazardous areas. ________

(b) Safety cabinets of the appropriate type and class are provided. ________

(c) Lab personnel are vaccinated for typhoid fever, tetanus and polio. ________

(d) Floors are wet-mopped weekly. with a disinfectant solution. ________

(e) Personnel are trained in the proper procedures for handling lyophilized cultures where used. ________

6. General Handling and Storage of Chemicals and Gases

(a) Containers of reagents and chemicals are labelled properly. ________

(b) Flammable solvents are stored in an approved storage cabinet or well-ventilated area away from oil burners, hot plates, etc. ________

(c) Bottle carriers are provided for hazardous substances. ________

(d) Gas cylinders are securely clamped to a firm support. ________

(e) Toxic chemicals are clearly marked poison or toxic. ________

7. Emergency Precautions

(a) Foam and carbon dioxide fire extinguishers are installed within easy access to laboratory and are properly maintained. Frequency ________________. ________

(b) Eye wash stations ______, showers ______, oxygen respirators ______, and fire blankets ______ are available within easy access. ________

(c) Fire exits are marked clearly. ________

(d) First aid kits are available and in good condition. ________

(e) At least one full-time employee is trained in first aid. ________

(f) Source of medical assistance is available and known to employees. ________

8. Suggested Areas of Improvement:

9. General Comments:

______________________________ __________

(Signature of Installation Officer) (date)

REFERENCES

1. 29 Code of Federal Regulations (CFR) Part 1910, "Occupational Safety and Health Standards" and Amendments.

2. Occupational Health and Safety Administration, 1974. Provisions for Federal Worker Safety and Health, OSHA (October, 1974), Reference File 41:6241

3. American Public Health Association, 1963. Diagnostic Procedures and Reagents, (4th ed.), APHA, Inc. pp. 89.

4. Pike, R. M., 1976. Laboratory-associated infections: summary and analysis of 3921 cases. Health Laboratory Science, 13:105.

5. National Field Investigation Center, 1973. NFIC-Denver Safety Manual, NFIC-Denver, EPA, Denver, CO. EPA 330/9–74–002 (In revision).

6. Wedum, A. G., 1961. Control of laboratory airborne infection, Bacterial Reviews 25.

7. Reitman, M. and A. G. Wedum, 1971. Infectious hazards of common microbiological techniques. *In:* Handbook of Laboratory Safety. (N. V. Steere, ed.), The Chemical Rubber Co., Cleveland, OH pp. 633.

8. Shapton, D. A. and R. G. Board, Ed. 1972. *In:* Safety in Microbiology. Academic Press, New York, NY.

9. Steere, N.V., editor, 1971. Handbook of Laboratory Safety. The Chemical Rubber Co., Cleveland, OH.

10. Manufacturing Chemists Association, 1975. Laboratory Waste Disposal Manual, Washington, DC.

11. US Public Health Service, 1974. National Institutes of Health, Biohazards Safety Guide, USPHS, DHEW, GPO 1740–00383.

12. U.S. Environmental Protection Agency, 1977. Occupational Health and Safety Manual, TN5 (9–12–77).

13. Runkle, K. S. and G. B. Phillips, Ed. 1969. Microbial Contamination Control Facilities. Van Nostrand Reinhold Co., New York, NY.

PART V. LABORATORY MANAGEMENT

SECTION D LEGAL CONSIDERATIONS

This Section is intended to guide practicing microbiologists in assessment of their responsibilities and role under the three federal laws on water quality. It is not intended as formal legal guidance or as representative of official legal position. It is based on the three Federal laws on water quality, (1, 2, and 3), on A Primer on the Law, Evidence and Management of Federal Water Pollution Control Cases, Legal Support Division, US EPA May 1972 (4) and on "Enforcement Activities," David I. Shedroff, in Proceedings of the First Microbiology Seminar on Standardization of Methods, US EPA, March 1973 (5). Lawyers should be consulted for the exact interpretation of the laws and their applications.

The Section describes the portions of the Federal laws on water quality that are relevant to microbiologists and relates analytical methods and record-keeping to these laws. It also shows how the analytical results become evidence in administrative or court proceedings and will help the analyst to understand his role as a witness. A brief outline of the contents of the Section follows:

1. Enabling Legislation

1.1 Scope and Application: To understand the role of microbiology in environmental and compliance monitoring, it is necessary to briefly describe EPA's responsibilities for development of methodology, assistance to the States, promulgation of criteria and guidelines, establishment of compliance with permits and conduct of enforcement actions. Much of the work of the microbiologist in EPA will involve generation of data to determine compliance with the Federal laws on water quality. These laws with related regulations limit the choice of analytical technique and sometimes require more documentation than the analyst might otherwise provide.

1.2 Federal Laws Dealing with Water Quality (6)

Congress has passed three principal laws on water quality which concern the microbiologist: The Federal Water Pollution Control Act, as amended; the Marine Protection, Research and Sanctuaries Act, commonly known as the Ocean Dumping Law; and the Safe Drinking Water Act. These acts have a common theme:

1.2.1 Legislation is passed by Congress providing the general framework of Federal interest and control of an area of the environment. EPA promulgates general rules describing, requiring and/or limiting the qualities of wastewater discharges or drinking water.

1.2.2 EPA or appropriate state agencies issue permits placing specific limitations on discharges, establish Maximum Contaminant Levels in drinking and ambient water, and provide general rules controlling underground injections.

1.2.3 Permittees or others subject to the particular act may be required to self-monitor their discharges, and report findings to the State and/or EPA.

1.2.4 Provisions are made for enforcement actions when permit, variance, or abatement schedules are violated.

1.3 Federal Water Pollution Control Act Amendments of 1972, Public Law 92-500 (1)

(See Appendix A for listing and summary of pertinent sections of the law, and the related microbiological activities).

1.3.1 Background and Summary (6): This is the most comprehensive program ever enacted to prevent, reduce, and eventually eliminate water pollution. The two general goals of the Act are: To achieve wherever possible by July 1, 1983, water that is clean enough for recreational uses and for the protection of aquatic life; and by 1985 to have no discharges of pollutants into the Nation's waters. The law extends the Federal program to all U.S. surface waters, not just interstate waters. The States have submitted water quality standards for intrastate waters to EPA for approval or revision. While the States retain primary responsibility to prevent, reduce, and eliminate water pollution they must now do so within the framework of a new national program. The law sets forth guidelines for the control of industrial and municipal water pollution, expands water quality standards, establishes a new system of permits for discharges into the Nation's waters, and creates stringent enforcement machinery and heavier penalties for violations.

1.3.2 The Regulatory Scheme (6): Under this act the States establish the minimum water quality standards for streams and these standards are approved by EPA. The Administrator determines minimum acceptable effluent limits for municipal treatment plants and for specific industries based on current technology. These limits become more restrictive over time. Using the more stringent water quality or treatment limitations, the Administrator or State determines specific limits for a discharge. These limits set forth in a permit for a direct discharge are enforceable by civil penalty, civil or criminal process, or revocation of permit.

A microbiologist may be called as a witness to prove the violation of the permit by direct discharge or to prove the violation of specific limitations placed on industrial firms discharging to municipal plants. Generally these microbiological limitations will relate to discharges from municipal treatment plants or from industries such as food processors.

Other enforcement activities for the microbiologist under the FWPCA include a permit program covering sewage sludge discharges and standards for discharges from marine sanitation devices and their performance (40 CFR/40 Amendments, 41 No. 20, January 29, 1976). If such permits or standards contain microbiological limitations, the microbiologist may be called on to prove or disprove a violation (Section 405). Standards have been set for discharges from marine sanitation devices and their performance (Federal Register 41, January, 1976).

1.3.3 Analytical Guidelines and Criminal Sanctions (6)

Analytical guidelines and criminal sanctions for improper analysis or furnishing false results are included in the Act. Section 304 (g) calls for the Administrator to promulgate "...guidelines establishing test procedures for the analysis of pollutants...". The initial guidelines for monitoring the National Pollution Discharge Elimination System (NPDES) (7) contain test procedures for total coliforms, fecal coliforms, and fecal streptococci using methods referenced from the 14th Edition of *Standard Methods*(8). The guidelines defining secondary treatment originally included fecal coliform limitations, and permits issued to municipal plants until July, 1976 contain such limitations. These regulations were amended in July, 1976 and no longer contain microbiological limitations. Such limitations may still be required in permits issued after that date if required for compliance with water quality standards or if such parameters are required in order to comply with State law. Guidelines for certain food related industries include fecal coliforms as a limiting parameter. The amended guidelines for municipal and industrial wastewaters place restrictions on the measurement of fecal coliforms in chlorinated or toxic wastewaters. For these wastewaters the membrane filter or most probable number (MPN) methods may be used, but the MPN is the method of choice when the results may be involved in controversy. (Refer to the amendments to 40 CFR Part 136 (9)). The methods are described in Part III, B, C and D.

Microbiologists performing analyses required under this Act should be aware of the specialized enforcement procedures in Section 309 (c) of FWPCA relating to analyses and reports of results: "Any person who knowingly makes any false statement, representation, or certification in any application, record, report, plan, or other document filed or required to be maintained under this Act or who falsifies, tampers with, or knowingly renders inaccurate any monitoring device or method required to be maintained under this Act, shall upon conviction, be punished by a fine of not more than $10,000, or by imprisonment for not more than six months, or by both."

1.3.4 Alternative test procedures are permitted for use in NPDES, 40 CFR Part 136 and in the Drinking Water regulations under 40 CFR Part 141. Information on application for use of alternative test procedures is given in these issues of CFR. The details of the comparative testing which may be required are given in this Manual, under Quality Control, Part IV-C.

1.4 The Marine Protection, Research and Sanctuaries Act of 1972, (Ocean-Dumping) Public Law 92–532 (2)

(See Appendix A for listing and summary of pertinent sections of the law, and the related microbiological activities).

1.4.1 The Regulatory Scheme: The Marine Protection, Research and Sanctuaries Act of 1972 regulates the dumping into ocean waters of all types of materials which would adversely affect human health and welfare, the marine environment, ecological systems or economic development.

This Act bans dumping of radiological, chemical or biological warfare agents and high-level radioactive wastes. With one exception, permits are required for transporting materials for ocean dumping and for the dumping itself. The Corps of Engineers issues permits for dredge spoils; EPA issues permits for all other materials (see Appendix A). The exception is fish-processing wastes. Since they are a natural ocean waste product, no permit is required unless harbors or other protected waters are involved as the receiving waters, or unless the EPA Administrator finds that such deposits in certain offshore areas could endanger health, the environment or ecological systems.

In evaluating permit applications, EPA and the Corps of Engineers must consider:

The need for the proposed dumping.

The effect on human health and welfare, including economic, aesthetic and recreational values.

The effect on fisheries, resources, plankton, fish, shellfish, wildlife, shorelines, beaches and marine ecosystems.

The effect of dumping particular volumes and concentrations of materials and the persistence of the effect.

The effect on other uses such as scientific study, fishing and other resource exploitation.

Appropriate locations and methods of disposal or recycling, including land-based alternatives and the probable impact of requiring the use of such alternate locations or methods.

The law charges the Secretary of Commerce with responsibility for a comprehensive and continuing research program involving the possible long-range effects of pollution, overfishing or man-induced changes in ocean ecosystems. Research efforts are to be coordinated with EPA and the Coast Guard.

The basic research objective of the law is to find ways to minimize or to end all ocean dumping within five years. It will cover the effects of dumping materials into ocean or coastal waters and into the Great Lakes or their connecting waters.

1.4.2 Civil and Criminal Sanctions: The law provides for both civil and criminal penalties for violations but there is no penalty for dumping materials from a vessel as emergency action to safeguard life at sea. Any individual may initiate a civil suit to enjoin any person, including Federal, State and local government or agency, who is alleged to be violating any prohibition, limitations, criterion or permit established or issued under this law.

1.5 The Safe Drinking Water Act of 1974, Public Law 93–523 (3)

(See Appendix A for listing and summary of pertinent sections of the law, and the related microbiological activities).

1.5.1 National Objectives: This Act has as its main objective the establishment and enforcement of primary drinking water standards. These standards, which are to be enforced by the States, will apply to public water systems and specify maximum levels of: 1) Those contaminants which may have adverse health effects (primary standards) and, 2) those contaminants which should be limited to protect the public welfare (secondary standards). The protection of underground drinking water sources by regulation of State underground injection control programs and the appointment of a 15-member National Drinking Water Advisory Council are also provided for under this Law.

1.5.2 The Regulatory Scheme

Interim Primary Drinking Water Regulations: The Administrator of EPA proposed national interim primary drinking water regulations 90 days after enactment of the Law. The interim regulations were promulgated in December, 1975 (10).

Secondary Drinking Water Regulations: Following the enactment date of this Law, proposed secondary regulations will address the aesthetic characteristics of water such as taste, appearance, etc. Ninety days later these secondary regulations will be promulgated, but will not be enforceable.

Study of Maximum Contaminant Levels in Drinking Water: The National Academy of Sciences conducted a study to determine the Maximum Contaminant Levels which should be recommended to protect human health. The study investigated contaminants that might have adverse health effects but the levels of which could not be determined in drinking water. The results of the above study are reported to Congress. EPA then publishes proposals in the Federal Register for recommended maximum levels, which will subsequently be promulgated.

Revised Primary Drinking Water Regulations: Following the National Academy of Sciences study, EPA proposes revised primary drinking water regulations to be adopted in

180 days and to be effective 18 months after promulgation. These primary drinking water regulations specify a Maximum Contaminant Level or require the use of treatment techniques for each contaminant in lieu of Maximum Contaminant Levels.

Regulation of State Underground Injection Control Programs: The Act provides for regulations which contain minimum requirements for effective (State) programs to prevent underground injection which endangers drinking sources. To be approved, a State program must prohibit underground injection without a state permit within three years after the enactment date of the Law. Applicants for underground injection permits must satisfy the State that the injection will not endanger drinking water sources. No regulations are promulgated that allow underground injection which endangers drinking water sources.

1.5.3 Quality Control Requirement

In the Safe Drinking Water Act, the Administrator can specify analytical methodology which includes quality control and testing procedures.

1.5.4 Enforcement

(a) State Primary Enforcement Responsibility: A State has the primary enforcement responsibility for protection of drinking water if it has a program acceptable to EPA. The EPA Administrator must determine that the State has drinking water regulations no less stringent than Federal regulations. The State may permit variances and exemptions as prescribed in the Law, and must have an adequate plan for providing safe drinking water under emergency circumstances. In addition, the State must have monitoring programs that comply with Federal requirements and must possess sufficient enforcement authority.

Approval by the EPA Administrator of the State's underground injection program gives to the State the primary enforcement responsibility until such time as the Administrator determines that the State no longer meets its requirements under the Act.

(b) Federal Enforcement: If a State fails to assure enforcement of drinking water regulations, the EPA Administrator notifies the State concerning the violation, and provides advice and technical assistance to the State and to the public water system that is in noncompliance, to bring the system into compliance.

1.5.5 Civil and Criminal Sanctions: The EPA Adiministrator may bring a civil action to require compliance with either the national primary drinking water regulations or with any requirement of an applicable underground injection control program. A maximum penalty of $5,000 per day may be imposed by the court for each day in which a violation occurs.

2. Application of the Laws to Microbiology

2.1 Gathering and Preserving Evidence

2.1.1 Stream Standards: To establish a standard violation, it is necessary to show the navigable waterway is below approved water quality standards. Water quality criteria specify permissible levels of chemical and biological constituents for receiving waters. It must be demonstrated that the defendant's discharge caused or contributed to a reduction in receiving water quality below one or more applicable standards. Samples should be collected 1) upstream of the discharge, 2) at the point of discharge and 3) downstream of the discharge at a point after a reasonable mixing zone.

Although all State water quality standards include criteria for the same basic parameters, there are differences among the states as to the sampling and test procedures which must be followed in order to establish a standards violation. It is thus imperative that only the testing method specified be used in order to show that a particular state water quality standard has or has not been violated.

2.1.2 Effluent Standards: The Agency has not provided specific guidance in sampling and test protocol to microbiologists working with the effluent standards and compliance monitoring. A protocol must vary with the way in which the permits are written. The standards may set maximum values and average values by the day, week, month, and year without defining the number and kinds of sample required to establish compliance with a given standard. Specifications of numbers and types of samples are needed to provide weekly, monthly or yearly maxima or averages.

Each Regional Office of EPA has developed an approach to compliance monitoring. Uniformity of sampling and testing schemes is desirable for data validation and for comparison of data between different laboratories. The need for uniformity is even more important with the transfer of the responsibility to the states for issuance of permits, as legally permitted. Some states are following EPA-Regional guidance while others are developing their own plans for compliance monitoring.

The sampling plan should select the sampling techniques, volumes, frequency, replication, etc., to meet the standard being challenged. All sampling and analyses to verify compliance with a particular standard should be performed in the same fashion and with the same frequency. Three rules for a good sampling plan are: 1) fit the design to meet the effluent standard, i.e., use a reasonable means to obtain statistical validity 2) apply the plan uniformly and 3) document the plan, indicate its source and record its use.

2.1.3 Drinking Water Standards: The Safe Drinking Act has established Maximum Contaminant Levels (MCL's) on an interim basis for community and non-community systems. The MCL's are based primarily on the 1962 Public Health Service Standards. For microbiology, the minimal sampling frequency per month is specified in Title 40 Part 141. Samples shall be taken at regular time intervals and in numbers proportional to the populations served. Samples shall also be taken at points representative of the conditions in the distribution systems.

2.1.4 Constitutional Protections: Sample evidence taken from the defendant's (individual or corporate) property without his consent cannot be introduced into evidence in either a civil or a criminal case because of the Fourth Amendment guarantee against unreasonable searches and seizures. Consent need not be obtained to take samples on the public portions of a waterway, usually up to the ordinary high water level (11). Almost all Fourth Amendment objections can be prevented by sending an advance, written notification of the time, scope, and purpose of any proposed EPA inspection, or sampling visit and by obtaining the written consent of the party to be inspected (12). If a search warrant has not been obtained, unannounced investigatory inspections may be made only if the voluntary consent of a person in authority is secured (13).

2.2 Admissibility of Evidence

2.2.1 Types of Legal Action: Violations of Public Laws 92–500, 92–532, or 93–523 can result in civil penalties assessed by the Administrator, a hearing board, or a court; criminal sanctions by a court; or a court order requiring a discharge source to take or cease a particular action. Except for civil penalties imposed by the Administrator, formal hearing will be required.

2.2.2 Authentication of Testimony: Authentication as a condition precedent to admissibility may require testimony under oath (14). Witnesses are subject to cross-examination. The form and admissibility of evidence presented in Federal courts are clearly defined in the Federal Rules of Civil Procedure (15) and the Federal Rules of Evidence (14). These formal rules of evidence may be partially annulled for administrative hearings. Laboratory records may not be acceptable evidence without proof of authenticity.

The hearsay rule is one of the most directive statutes. It states that generally persons may only testify to what they know personally and that they must be subject to cross-examination. However, some exceptions to the hearsay rule are allowed. Evidence that would normally be hearsay is admissible in the administrative hearing, if it has probative value in the opinion of the hearing officer. Also, evidence may be presented in written form in a hearing, but this is more likely when the preceeding is not a full-fledged trial-type hearing.

Additional exceptions to the hearsay rule are cited in the recently-enacted Federal Rules of Evidence (14). These rules state that under certain circumstances a witness does not have to be present for his statement to be admissible. In addition, records of regularly-conducted business activities, public reports, reports prepared by law enforcement personnel and factual findings resulting from legal investigations may be admitted without the testimony of the person or persons involved.

2.2.3 Admissibility of Records: Under Rule 803 (6) of the Federal Rules of Evidence (14), written records made in the regular course of any business (i.e., laboratory operation) may also be introduced into evidence in civil actions without the testimony of the person(s) who made the record. Prior to enactment of the Federal Rules of Evidence, this authority was contained in the Federal Business Records Act, 29 U.S. Code, Section 1732A.

Although preferable, it is not always possible to have the individuals who collected, kept, and analyzed samples testify in court. In addition, if the opposing party does not intend to contest the integrity of the sample or testing evidence, admission under the Business Records Act can save much trial time. For these reasons, it is important that the procedures followed in evidence, sample collections and analyses be standardized and described in an instruction manual which can be offered as evidence of the standard operating procedure followed by the laboratory.

2.2.4 Limitations on the Admissibility of Records: Although the statutes do not specifically cover the point, it is clear from the examination of cases that one of the requirements for admissibility is that the document has inherent probability of trustworthiness. Thus, a trial judge has discretion in allowing or not allowing a document into evidence if there is doubt as to its trustworthiness. One criterion for the judge to consider is whether the particular analysis was done as a routine matter or whether it was specifically done in anticipation of litigation. This caution in admitting evidence is an indication of distrust of the situation, not of the individuals involved.

2.2.5 Contacts with Parties to Adjudicatory or Adversary Proceedings: The following statements are quoted directly from the May 5, 1975, memorandum of the Acting Assistant Administrator for Enforcement, US EPA (16):

> As we are now becoming involved in more and more adjudicatory hearings on NPDES (National Pollution Discharge Elimination System) permits and in enforcement actions, both through Administrative Orders and in the Courts, it is very important that our staffs clearly understand that contacts and discussions with parties to these proceedings be carefully controlled.
>
> We have recently had inquiries about cases in which requests for adjudicatory hearing had been granted and in which EPA technical staff members, without the knowledge of either the attorney assigned or of the Enforcement Director, met with company representatives to discuss the pending case. In each case, the merits of EPA's position as compared to that of the company were discussed, as were possible areas of compromise with respect to EPA's position.
>
> Case preparation and decisions on strategy for adjudicatory hearings are the responsibility of our regional attorneys with assistance from Headquarters Counsel for Adjudicatory Hearings. Accordingly, I would appreciate your instructing your staff members not to discuss permit questions or technical issues applicable to a particular industrial facility which is the subject of an adjudicatory hearing or an enforcement action until the appropriate Enforcement Division attorney, either in the Regional Office or in Headquarters is notified.
>
> The foregoing is not to be construed as discouraging settlement discussions in pending cases, but is only intended to

provide for orderly resolution of matters which can be negotiated.

2.3 Preparation for Testimony

2.3.1 Gathering and Preserving Evidence in Water Pollution Enforcement Actions: In every water pollution suit, expert testimony will be of primary importance. To meet its burden of proof, the Government may have to present expert testimony on sampling, laboratory analyses, test results and the harmful effect attributable to the defendant's discharge. If the Government's expert witnesses do not testify effectively, the lawsuit may be jeopardized.

2.3.2 Testimony on Sampling: In the order of proof in a trial concerning pollution there will be testimony by witnesses who have taken samples. The samples may be effluents, receiving waters, potable waters, sludges or sediments. These witnesses will explain how, where and when the samples were taken. The choice of sampling location and what to sample depends to a large extent on the type of legal action contemplated.

2.3.3 Documentation of Procedures: In anticipation of possible court presentation of evidence, laboratories must maintain an orderly, complete and permanent record-keeping and filing system. A laboratory operating manual should be used in all laboratories. The manual formalizes the operation of the laboratory by describing in detail the sampling procedures, the line of technical responsibility, specific analytic methods followed, data handling procedures, the continuous quality control program established for daily operations, participation in interlaboratory and intralaboratory quality control programs and safety guidelines. Since routines change, personnel should sign dated receipts that indicate they have received the operating instructions and modifications when issued.

Complete records of samples received must be kept in a separate log and official chain of custody requirements must be observed. The laboratory data records, analytical results and computations should be written, preferably in a bound book or on bench cards, that can be incorporated into a permanent record log. Provision should be made for the signatures of sample collectors, analysts and direct line supervisors in the sample log and in the data log so that the laboratory data are authenticated. As described in Part V-A of this manual, a quality control log book should be maintained on a day-to-day basis. It should record quality control checks on: media and supplies, equipment and instrumentation, the actual analyses, data handling and storage. Training of analysts should include familiarization with the quality control book and identification of their responsibilities in the program. Each analyst should have a personal copy of the manual as a guide.

2.3.4 Pre-Trial Discovery: Whenever an agency is a party to any federal court litigation, it will be subject, under the Federal Rules of Civil Procedure, to pre-trial discovery. The agency will be required to answer the opposing party's questions and to produce requested documents. Technical personnel responding to a motion to produce documents should deliver related documents to the agency attorney handling the case. Documents should not be withheld because they appear to be damaging to the government's case. The responsible government attorneys will determine, on the basis of the law of discovery, which documents must be submitted to the opposing party.

A sensible filing system should be set up and followed. The objective of a filing system is to store information so that it can be found quickly. Information which is known but is not reflected in the file is of no use and will not be available when needed. However, the files should be examined regularly and outdated or superfluous information discarded to maintain manageability. Critical or outspoken comments on notes, route slips or in margins, should not be retained unless the originator and recipient are prepared to defend them in court.

2.3.5 Testimony on Methodology: A witness may be required to provide testimony on methods of analyses and test results. For acceptance in court, the witness must be able to testify that the analytical method or procedure employed has wide use in the microbiological community. For example, the procedures outlined in *Standard Methods* (8) and in This Manual are recognized and accepted. In court cases on record, results obtained using *Standard Methods* have been admitted into evidence while deviations from *Standard Methods* have had to be explained and justified.

It may be necessary to present testimony on parameters that are not included in these publications or on special types of samples to which the methods described are not applicable. In such cases effective testimony may be based upon the best methodology currently available, utilizing as substantiating evidence published reports, other method manuals, etc. to demonstrate that the methods do have recognition in the scientific community.

The specific test methods to be used in the application of the Federal Water Pollution Control Laws may be identified in the Code of Federal Regulations (CFR). For example, the procedures required for Section 304 (g) of the Federal Water Pollution Control Amendments of 1972 appear in 40 CFR, Part 136. These guidelines establish the methodology to be used for compliance monitoring and the methods become those that are acceptable as standards in court. Part 136 of 40 CFR also provides a mechanism and rule for obtaining approval for any alternate procedure that may be proposed when the recommended method is not appropriate.

2.3.6 Testimony by Expert: A court may require that an expert witness' opinion be based on studies and tests conducted or supervised by him personally. However, experts are frequently permitted to offer testimony in the form of an opinion in the area of competence or based on someone else's work. Such testimony can be developed through the use of hypothetical questions and objections tend to add weight to the expert's testimony rather than to cast doubt on the witness' competency.

2.4 Testimony in Court

2.4.1 General Instructions for a Witness: The following suggestions are made for prospective witnesses to lessen the apprehensions everyone feels when first testifying before a board, commission, hearing officer, or in court. Even veteran witnesses often experience some anxiety. However, if a witness is properly prepared on the subject matter of his testimony and his conduct on the witness stand, he is much more confident about testifying. The witness will be required to take an oath to tell nothing but the truth. The important point is that there are two ways to tell the truth—one is in a halting hesitant manner, which makes the board member, hearing officer, judge or jury doubt that the witness is telling all the facts in a truthful way, and the other is in a confident straight forward manner, which gives credence to the witness' words.

If a scientist is a witness in a case involving testimony concerning the appearance of an object, place or condition, he should refresh his recollection by inspecting the object, place or condition, etc., before the hearing or trial. Later he should try to picture the item and recall the important points of his testimony. He should repeat this procedure until he has thoroughly familiarized himself with the points that will be made in the testimony.

Before testifying, the witness should visit a court trial or board hearing and listen to other witnesses testifying. This will familiarize him with such surroundings and help him to understand court protocol and the problem of testimony. The scientist should arrive at the hearing in time to listen to other witnesses testify before taking the witness chair himself.

A good witness listens to the question and then answers it calmly and directly in a sincere manner. He knows the facts and can communicate them. He testifies in this manner on cross-examination as well as on direct examination.

The witness should wear neat, clean clothes when he testifies and should dress conservatively. He should speak clearly and not chew gum while testifying.

2.4.2 Direct Examination

(a) In a discussion on administrative procedures, E. Barrett Prettyman, Retired Chief Judge, U.S. Court of Appeals for the District of Columbia, gave the following advice (4):

> The best form of oral testimony is a series of short, accurate, and complete statements of fact. It is to be emphasized that the testimony will be read by the finder of the facts, and that he will draw his findings from what he reads...confused, discursive, incomplete statements of fact do not yield satisfactory findings.

(b) The witness should stand upright when taking the oath, pay attention, say "I do" clearly, and not slouch in the witness chair. If the witness has prepared answers to possible questions, he should not memorize them. It is, however, very important that he familiarize himself as much as possible with the facts about which he will be called to testify.

(c) During direct examination, the witness may elaborate and respond more fully than is advisable on cross-examination. However, when volunteering information, he should not ramble or stray from the main point raised in his lawyer's question. Testimony is a dialogue, not a monologue. If testimony concerns a specialized technical area, the court or hearing board will find it easier to understand if it is presented in the form of short answers to a logical progression of questions. In addition, by letting his lawyer control the direction of his testimony, the witness will avoid making remarks which are legally objectionable or tactically unwise.

(d) The witness should be serious at all times and avoid laughing or talking about the case in the building where the hearing or trial is being held.

(e) While testifying, the witness should talk to the board member, hearing officer or jury, looking at him or them most of the time, and speaking frankly and openly as if to a friend or neighbor. He should speak clearly and loudly enough so that anyone in the hearing room or courtroom can hear him easily. The witness makes certain that the reporter taking the verbatim record of his testimony is able to hear him and record what he says. The case will be decided entirely on the words that are reported as the testimony given at the hearing or trial. The witness must give complete statements in sentence form; half statements or incomplete sentences may convey the thought in the context of the hearing, but be unintelligible when read from the cold record months later.

2.4.3 Cross-Examination

(a) Concerning cross-examination, the following advice is given to prospective witnesses (4):

> Don't argue. Don't fence. Don't guess. Don't make wisecracks. Don't take sides. Don't get irritated. Think first, then speak. If you do not know the answer to a question, say so. If you do not know the answer but have an opinion or belief on the subject based on information, say exactly that and let the hearing officer decide whether you shall or shall not give such information as you have. If a 'yes or no' answer to a question is demanded but you think that a qualification should be made to any such answer, give the 'yes or no' and at once request permission to explain your answer. Don't worry about the effect an answer may have. Don't worry about being bulldozed or embarrassed; counsel will protect you. If you know the answer to a question, state it as precisely and succinctly as you can. The best protection against extensive cross-examination is to be brief, accurate and calm.

The hearing officer, board member or jury wants only the facts, not hearsay, conclusions, or opinions. The witness usually will not be allowed to testify about what someone else has told him.

(b) The witness must be polite, even to the attorney for the opposing part. He should not be a cocky witness. This will lose him the respect and objectivity of the trier of the facts in the case. He should not exaggerate or embroider his testimony.

(c) The witness should stop instantly when the judge, hearing officer or board member interrupts, or when the other attorney objects to what is said. He must not try to sneak the answer in or nod his head for a "yes" or "no" answer. The reporter has to hear an answer to record it. If the question is about distances or time and the answer is only an estimate, he must say that it is only an estimate.

(c) The witness should listen carefully to the questions asked. No matter how friendly the other attorney may seem on cross-examination, he may be trying to damage the testimony. He must understand the question completely and should have it repeated if necessary, then give a thoughtful answer. He must not give a snap answer. He cannot be rushed into answering, yet taking too much time would make the board member, hearing officer or jury think the witness is making up the answers.

(d) The witness must answer the question that is asked—not the question that he thinks the examiner (particularly the cross-examiner) intended to ask. The printed record shows only the question asked, not what was in the examiner's mind and a non-responsive answer may be very detrimental to the case. This situation exists when the witness thinks "I know what he is after but he hasn't asked for it." Answer only what is asked. The witness must explain his answers if necessary.

(e) If by chance one's answer is wrong, correct it immediately; if the answer was not clear, clarify it immediately. The witness is sworn to tell the truth. Every material truth should be readily admitted, even if not to the advantage of the party for whom he is testifying. He must not stop to figure out whether the answer will help or hurt his side.

(f) The witness must give positive, definite answers when at all possible and avoid saying "I think", "I believe", "in my opinion." If he does not know, he must say so and not make up an answer. One can be positive about the important things which he naturally would remember. If asked about little details which a person naturally would not remember it is best to say that one does not remember, but he must not let the cross-examiner place him in the trap of answering question after question with "I don't know."

(g) The witness must not act nervous. He should avoid mannerisms which will make him appear frightened, not telling the truth, or not telling all that he knows. Above all, it is most important that the witness not lose his temper. Testifying at length is fatiguing. Fatigue will be recognized by crossness, nervousness, anger, careless answers and a willingness to say anything or answer any questions in order to leave the witness stand. When the witness feels these symptoms, he must recognize them and strive to overcome these feelings. Some attorneys on cross-examination try to wear out the witness so he will lose his temper and say things that are not correct, or that will hurt the testimony. The witness must not let this happen.

(h) If the witness does not want to answer a question, he should not ask the judge, hearing officer or board member whether he must answer it. If it is an improper question, his attorney will object for him. One must not ask the presiding officer, judge or board member for advice or help in answering a question. The witness is on his own. If the question is an improper one, his attorney will object. If the judge, hearing officer, or board member then directs the witness to answer it, he must do so. He cannot hedge or argue with the opposing attorney.

(i) There are trick questions which may be asked and which, if answered, signify "yes" or "no", and will damage the credibility of the testimony. Two examples follow:

(1) "Have you talked to anybody about this matter?" If you say "no", the hearing officer or board member, or a seasoned jury, will know that is not correct because good lawyers always talk to the witnesses before they testify. If one says "yes", the lawyer may try to infer

that you were told what to say. The best thing to say is that you have talked to Mr. ________, your lawyer, to the appellant, etc., and that you were just asked what the facts were. All that is wanted is the truth.

(2) "Are you getting paid to testify in this appeal?" The lawyer asking this hopes your answer will be "yes", thereby inferring that you are being paid to say what your side wants you to say. Your answer should be something like "No, I am not getting paid to testify, I am only getting compensation for my time off from work, and my expenses incurred in being here." A witness should never be paid a contingency fee as it indicates strongly that since his compensation depends upon the results, he may be inclined to overstate the case.

REFERENCES

1. Federal Water Pollution Control Act Amendments of 1972, Public Law 92–500, October 18, 1972, 86 Stat. 816, 33 United States Code (USC) Sec. 1151.

2. Marine Protection, Research and Sanctuaries Act of 1972, Public Law 92–532, October 23, 1972, 86 Stat. 1052.

3. Safe Drinking Water Act, Public Law 93–523, December 16, 1974, 88 Stat. 1660, 42 United States Code (USC) 300f.

4. U.S. Environmental Protection Agency, Legal Support Division, 1972. A Primer on the Law, Evidence, and Management of Federal Water Pollution Control Cases, Washington, D.C. pp. 43–52, 54–58.

5. Shedroff, D. I., 1973. Enforcement activities. In: Proceedings of the First Microbiology Seminar on Standardization of Methods, EPA-R4-73-022, Office of Research and Monitoring, U.S. Environmental Protection Agency, Washington, D.C., pp. 1–11.

6. Shedroff, D. I., 1976. Personal Communication. Office of Enforcement, US EPA, Washington, DC.

7. Guidelines Establishing Test Procedures for Analysis of Pollutants, 40 Code of Federal Regulations (CFR) Part 136, Published in Federal Register, 38, p. 28758, October 16, 1973.

8. American Public Health Association. 1976. Standard Methods for the Examination of Water and Wastewater (14th ed.) American Public Health Association, Inc., Washington, DC. p. 874

9. Guidelines for Establishing Test Procedures, 40 Code of Federal Regulations (CFR) Part 136, Published in Federal Register, 40, 52780, Dec. 1, 1976.

10. National Interim Primary Drinking Water Regulations, 40 Code of Federal Regulations (CFR) Part 141, Published in Federal Register, 40, 59566, December 24, 1975.

11. Borough of Ford City vs. United States, 345 F. 2d 645 (3rd Cir. 1965).

12. Camara vs. Municipal Court, 387 U.S. 523 (1967); See vs. Seattle, 387 U.S. 547 (1967).

13. United States vs. Hammond Milling Co., 413 F. 2d 608 (5th Cir. 1969), cert. den. 396 U.S. 1002 (1970); and United States vs. Thriftimart, Inc. 429 F. 2d 1006 (9th Cir. 1970) cert. den. 400 U.S. 926 (1970).

14. The Federal Rules of Evidence, Public Law 93–595, January 2, 1975, 88 Stat. 1926, 28 United States Code (USC) App.

15. Federal Rules of Civil Procedure, Rule 43, adopted by the U.S. Supreme Court pursuant to Title 28, U.S. Code (USC) Section 2072, as amended effected July, 1975.

16. Johnson, R. H., Acting Assistant Administrator for Enforcement, EPA Office of Enforcement. May 5, 1975. "Contacts with Parties to Adjudicatory or Adversary Proceeding", Memorandum to EPA Assistant and Regional Administrators.

APPENDIX A

TABLE-1

FEDERAL WATER POLLUTION CONTROL ACT AMENDMENTS OF 1972, PUBLIC LAW 92-500
Microbiological Activities Under Relevant Sections of the Law

Sections of Law	Summary of Sections	Microbiological Activity
104(a)(5) Water Quality Surveillance System	The Administrator is required to establish and maintain a water quality surveillance system with States and other Federal Agencies. Agencies in the system will collect and disseminate basic data on the chemical, physical and biological effects of varying water quality. They are to develop new methods for identifying and measuring the effects of pollution on the chemical, physical and biological integrity of the water.	Conduct research, develop methodology and technology, complete necessary analyses, perform surveys, and provide expertise in microbiology.
106 (c) Grants for State Pollution Control Program	EPA is to provide assistance and guidance to the States on the development and operation of procedures and systems to monitor water quality, including biological monitoring.	Provide necessary assistance in microbiological expertise and consultation to the States.
108 (a) Pollution Control in the Great Lakes	EPA is to conduct projects in cooperation with other agencies for demonstrating new methods and developing plans for their use in controlling pollution on the Great Lakes.	Analyze water samples to support method and development plans for Great Lakes pollution control.
301 (b) and 402 Permits for Publicly Owned (Municipal) Treatment Works	Municipal treatment plants must attain ". . . secondary treatment", as defined by Administrator, or treatment necessary to meet water quality standards, whichever is more stringent, by 1977, and ". . . best practicable waste treatment technology over the life of the works" by 1983.	Analyses to determine compliance or non-compliance with microbiological portions of permit requirements.

TABLE-1
(Continued)

FEDERAL WATER POLLUTION CONTROL ACT AMENDMENTS OF 1972, PUBLIC LAW 92-500
Microbiological Activities Under Relevant Sections of the Law

Sections of Law	Summary of Sections	Microbiological Activity
301 (b) and 402 Permits for Non-Publicly Owned (Industrial) Treatment Works	(1) Existing Plants: Must attain ". . . best practicable control technology currently available", or water quality standards, whichever is more stringent, by 1977, and ". . . best available technology economically achievable" by 1983. (2) New Plants: Must comply with ". . . National Industrial Standards of Performance" which for a particular industry reflect ". . . the greatest degree of effluent reduction . . . achievable through the application of the best available control technology, processes, operating methods, or other alternatives."	Analyses to determine compliance or non-compliance with microbiological portions of permit requirements.
301 (b) and 307 (b) Pretreatment Standards for Discharges by Non-Publicly Owned Enterprises into Publicly-Owned Plants	Private industry discharging into public treatment plants must demonstrate compliance with pretreatment standards which are determined by the type of waste source and whether the plant is already existing or is new since the passage of the Act. Standards for both new and old plants are designed to prevent the discharge through publicly-owned treatment works of pollutants which ". . . interfere with, pass through, or (are) otherwise incompatible with such works." Existing sources must comply by three years after promulgation of applicable pretreatment standards.	Analyses to determine compliance or non-compliance with microbiological portion of the pretreatment standard.
304(a) (b) and (g) Information and Guidelines	EPA must develop water quality criteria which reflect knowledge of the effects on plankton, fish, shellfish, wildlife, plant life, esthetics and recreation which may be expected from presence of pollutants in any body of water or in ground water. Information must be developed on what factors are needed to restore and maintain the chemical, physical and biological integrity of navigable waters, ground waters, coastal waters and oceans. The Administrator is also required to issue guidelines for identifying and evaluating the nature and extent of nonpoint sources of pollutants.	Develop microbial water quality criteria based on the analyses of all navigable, ground and coastal waters and the ocean.

TABLE-1
(Continued)

FEDERAL WATER POLLUTION CONTROL ACT AMENDMENTS OF 1972, PUBLIC LAW 92–500
Microbiological Activities Under Relevant Sections of the Law

Sections of Law	Summary of Sections	Microbiological Activity
304 (b) Publication of Effluent Limitation Guidelines	The Administrator shall publish regulations providing guidelines for effluent limitations.	Provide advice, technical assistance and analyses required to establish effluent limitations.
304 (g) Guidelines for Test Procedures	The Administrator shall promulgate guidelines establishing test procedures for the analyses of pollutants.	Provide advice, technical assistance and analyses required to establish microbiological procedures.
307 (a) Toxic Pollutants	Discharge limitations are established or materials are prohibited that are designated by the Administrator as toxic, taking into account ". . . toxicity . . . persistence . . . degradability . . . presence of the affected organisms and the nature and extent of the effect of the toxic pollutant on such organisms."	Identification and quantification of pollutants, including viruses, designated as toxic by the Administrator.
308 Inspections, Monitoring and Entry	Owners and operators of pollution point sources shall establish and maintain records; make reports; install, use, and maintain monitoring equipment or methods, including biological monitoring methods; and sample effluents at locations, intervals, and with methods prescribed by the Administrator.	Inspection of microbiological portions of records, microbiological equipment or methods used by owner or operator of pollution point source; sampling and analysis of effluents required to be sampled by the owner or operator of the pollution point source.
309 Federal Enforcement	On the basis of any information available that indicates non-compliance of the requirements of a permit issued by a State, the Administrator may notify the person in alleged violation and the State of such findings. If after the thirtieth day after notification the State has not commenced appropriate enforcement action, the Administrator may issue a compliance order or bring civil action to enforce the permit conditions or limitations.	Provide advice and technical assistance to the State and persons in non-compliance to bring them into compliance; provide analytical data, expertise and testimony as required to establish EPA's case.

TABLE-1
(Continued)

FEDERAL WATER POLLUTION CONTROL ACT AMENDMENTS OF 1972, PUBLIC LAW 92-500
Microbiological Activities Under Relevant Sections of the Law

Sections of Law	Summary of Sections	Microbiological Activity
310 International Pollution Abatement	The Administrator may call a hearing when he has reason to believe pollution is occurring from U. S. sources ". . . which endangers the health or welfare of persons in a foreign country."	Microbiological surveys to determine if domestic pollution is adversely affecting a foreign country; survey results provide the Administrator with data to assist him in deciding whether to call a hearing.
311 Oil and Hazardous Substance Liability	This section bans the discharge of oil and any other ". . . elements and compounds which, when discharged in any quantity into . . . waters . . . present an imminent and substantial danger to the public health or welfare, including, but not limited to, fish, shellfish, wildlife, shorelines, and beaches." The ban applies to any substance which fits this description and that the Administrator designates as "hazardous."	Performance of degradability tests and experiments under actual or simulated conditions.
312 (b) Marine Sanitation Devices	Vessel sanitation devices must conform to performance standards issued by the Administrator. New vessels must comply within two years of promulgation; existing vessels have five years in which to comply.	Analyses to confirm compliance with microbiological portions of performance standards. Analyses to determine if the device operates in conformity with the standards; the Coast Guard is responsible for such testing of the devices.
403 (c) Ocean Discharge Criteria	The Administrator is required to promulgate guidelines for determining the degradation of territorial waters, coastal waters and oceans.	Conduct required analyses to establish guidelines for monitoring the degradation of territorial and coastal waters and the oceans.

TABLE-1
(Continued)

FEDERAL WATER POLLUTION CONTROL ACT AMENDMENTS OF 1972, PUBLIC LAW 92–500
Microbiological Activities Under Relevant Sections of the Law

Sections of Law	Summary of Sections	Microbiological Activity
405 Permits for Disposal of Sewage Sludge	Permits are required for disposal of sewage sludge (including removal of in-place sewage sludge from one location and its deposit in another location) where disposal ". . . would result in any pollutant . . . entering . . . waters."	Analysis of sludges at time of transport and at disposal site to determine compliance or non-compliance with permit requirements.
504 Emergency Powers	An injunction prohibiting discharge by a particular source may be issued on proof of ". . . imminent and substantial endangerment to the health of persons or to the welfare of persons where such endangerment is to the livelihood of such persons."	Detection of pathogens in water and from their sources; and enumeration of indicators authorizing closure of shellfish beds and identification of pollutant sources.

APPENDIX A

TABLE-2

MARINE PROTECTION, RESEARCH AND SANCTUARIES ACT OF 1972, PUBLIC LAW 92-532
Microbiological Activities Under Enforcement and Compliance Monitoring Sections

Sections of Law	Summary of Sections	Microbiological Activity
102 EPA Permits for Ocean Dumping	Establishment of a program for the issuance of EPA permits based on criteria which consider the effects of ocean dumping on human welfare, shellfish and fisheries resources, plant and animal life, shorelines, beaches, and marine ecosystems.	Analyses to determine compliance or non-compliance with microbiological portions of permit requirements. Possible conduct of microbiological analyses to determine that the proposed dumping ". . . will not unreasonably degrade or endanger human health, welfare, or amenities, or the marine environment, ecological systems, or economic potentialities." (Regulations in 40 CFR 227 set forth criteria for evaluation of permit applications for materials containing living organisms. See especially 227.36.)
103 Corps of Engineer Permits	The responsibility for issuing permits (based on the criteria in Section 102 above) for the ocean dumping of dredged materials is under the jurisdiction of the Army Corps of Engineers.	Analyses to determine compliance or non-compliance with microbiological portions of permit requirements.

APPENDIX A

TABLE-3

SAFE DRINKING WATER ACT OF 1974, PUBLIC LAW 93–523
Microbiological Activities Under Enforcement and Compliance Sections

Sections of Law	Summary of Sections	Microbiological Activity
1412 (a) (1) Establish Primary Interim Standards	EPA has the responsibility for establishing national interim primary drinking water regulations which will protect human health by using the technology which is generally available.	Provide the expertise and advice to assist in the establishment of the required safe interim primary standards for drinking water.
1412 (c) Proposed National Secondary Drinking Water Regulations	The Agency promulgates National Secondary Drinking Water regulations which are mostly related to the aesthetic characteristics of drinking water.	Provide the expertise and advice to EPA for establishing National Secondary Standards relating to microbiology.
1412 (e) Study by Independent Organizations	The Administrator shall arrange studies with the National Academy of Sciences or other independent scientific organization to determine maximum contaminant levels of known or anticipated contaminants, and to identify those contaminants in drinking water which are at levels too low to measure.	Provide input as needed for establishment, performance and evaluation of studies by the National Academy of Sciences or other independent scientific organization.
1413 (b) Recommended Maximum Contaminant Levels	The Agency must also establish for each contaminant a maximum contaminant level which will produce no known adverse effects and allows an adequate margin of safety.	Conduct research and monitoring analyses to establish acceptable levels and maximum levels for bacterial indicators, pathogens and viruses.
1414 (a) (1) (A) Failure by State to Assure Enforcement of Standards	If the Administrator finds that a State with primary enforcement responsibility has not maintained compliance in its public water systems, he shall so notify the State and provide assistance in achieving compliance.	Provide advice and technical assistance to the State and water systems to bring them into compliance.

TABLE-3
(Continued)

SAFE DRINKING WATER ACT OF 1974, PUBLIC LAW 93-523
Microbiological Activities Under Enforcement and Compliance Sections

Sections of Law	Summary of Sections	Microbiological Activity
1414 (a) (1) (B) Civil Action	If the Administrator determines that the State abused its discretion in carrying out its primary enforcement responsibility, the Administrator may commence a civil action to enforce the standards.	Provide analytical data, expertise and testimony as required to establish EPA's case.
1414 (f) Non-Compliance/Public Hearings	If the Administrator finds non-compliance by a public water system in a State with primary enforcement responsibility, he may hold hearings to gather information from technical and other experts and may issue recommendations on actions which will achieve compliance.	Provide confirming analytical data, recommendations and expert advice on microbiological aspects demonstrating non-compliance.
1421 Underground Injection Control Systems	The Administrator shall publish and promulgate regulations for State underground injection control systems after public hearing. The States will issue permits for underground injection which will not endanger drinking water sources; and will inspect, monitor and keep records of the permitted underground injection wells.	Determine the feasibility of microbiological criteria and perform analyses of injected wastewaters and of ground waters if required.
1431 (a) Emergency Powers	EPA may obtain an injunction against a non-complying water system on proof of the presence of a contaminant which presents an imminent and substantial endangerment to human health and if the appropriate State or local authority has not acted.	Proof of presence of bacterial indicators, pathogens or viruses in sufficient numbers to pose a danger to human health.
410 Amendments to the Bottled Drinking Water Standards (Section 4. Chapter IV of the Federal Food, Drug, and Cosmetic Act)	After the promulgation of drinking water regulations, the Food and Drug Administration must either promulgate amendments to the bottled drinking water standards or publish reasons for not making amendments.	Conduct analyses to determine if the quality of bottled drinking water meets drinking water regulations.

APPENDIX B

FROM: *Manual for the Interim Certification of Laboratories Involved in Analyzing Public Drinking Water Supplies,* EPA 600/8-78-008, May, 1978. OMTS, Office of REsearch and Development, U.S. Environmental Protection Agency, Washington, DC 20460

Chapter V

MICROBIOLOGY: CRITERIA AND PROCEDURES FOR INTERIM CERTIFICATION OF LABORATORIES INVOLVED IN ANALYSIS OF PUBLIC WATER SUPPLIES

The criteria and procedures described herein, shown in **bold**, are minimum requirements considered essential for laboratories seeking certification for microbiological analysis of public water supplies. The requirements include laboratory equipment and supplies, laboratory practices, methodology, sample collection, and certain quality control measures. The other items, involving personnel, facilities, additional quality control procedures, data reporting, and action response, are optional. For a commercial laboratory to qualify for certification in microbiology, it must process a minimum of 20 potable water samples per month using either the multiple tube procedure or membrane filter test.

Until National Revised Primary Drinking Water Regulations require certification of water supply laboratories, all specifications will be considered as guidelines to be used by certification officials. At that time, minimal requirements will be essential to certification of laboratories involved in analysis of public water supplies.

The minimum requirements must be in compliance, or action must be taken to correct deficiencies prior to certification. A laboratory that exceeds these minimum requirements is encouraged to maintain and improve those higher standards for facilities, equipment, methodology, and quality control, as well as to continue the upgrading of personnel through training efforts to ensure routine production of reliable data.

The required methods of analyses are referenced in "Standard Methods for the Examination of Water and Wastewater," 13th edition; however, some criteria in this document are more specific and permit fewer variations than "Standard Methods."

The guidelines for quality assurance procedures are those in EPA's quality assurance program as cited in the EPA Manual, "Microbiological Methods for Monitoring the Environment" (EMSL EPA Cincinnati). A valuable source of further detail and background information for the laboratory evaluator is available in EPA's "Handbook for Evaluating Water Bacteriological Laboratories" (EPA-670/9-75-006, August 1976).

Minimum requirements are shown throughout in **bold.**

PERSONNEL[1] (OPTIONAL REQUIREMENTS)

Analyst

The analyst performs microbiological tests with minimal supervision in those specialties for which he is qualified by education and/or training and experience.

[1] Exceptions will be made for those persons employed by the laboratory and currently doing the required analyses prior to promulgation of the interim regulations provided that within 2 years after June 24, 1977, they receive a minimum of 2 weeks of additional training in water microbiology.

- *Academic training:* Minimum of high school diploma in academic or laboratory-oriented vocational courses.
- *Job training:* Minimum of 30 days on-the-job training plus one week of supplementary training acceptable to the Federal and State regulatory agency or agency responsible for primacy. Personnel should take advantage of courses available to Federal and State regulatory agencies.
- *Supervision:* Supervision by an experienced professional scientist. In the small water plant laboratory consisting of a single analyst, the services of a State-approved outside consultant must be available.

Supervisor/Consultant

The supervisor directs technical personnel in the proper performance of laboratory procedures and the reporting of results. If no technical supervisor is available, a consultant should be available.

- *Academic training:* Minimum of a bachelor's degree in microbiology, biology, chemistry, or a closely related field. Exceptions will be made for employees of laboratories that serve communities with populations of 50,000 or less if they receive at least 2 weeks of additional training in water microbiology from a Federal agency, State agency, or university.
- *Job training:* Technical training in water microbiology for a minimum of 2 weeks from a Federal agency, State agency, or university in the parameter to be tested. Consultant must have 1 year of bench experience, approved by the State, in total coliform analysis. State laboratory expertise would be the most desirable source of outside consultation.
- *Experience:* One year of bench experience in sanitary (water, milk, or food) microbiology.

LABORATORY FACILITIES (OPTIONAL REQUIREMENTS)

Laboratory space should be adequate (200 ft^2 and 6 linear ft of bench space per analyst) to accommodate periods of peak work load. Working space requirements should include sufficient bench-top area for processing samples; storage space for media, glassware, and portable equipment items; floor space for stationary equipment (incubators, waterbaths, refrigerators, etc.); and associated area for cleaning glassware and sterilizing materials. The space required for both laboratory work and materials preparation in small water plant laboratories may be consolidated into one room, with the various functions allocated to different parts of the room.

Facilities should be clean, air-conditioned, and with adequate lighting at bench top (100 ft-candles).

Laboratory safety, which must be an integral and conscious effort in laboratory operations, should provide safeguards to avoid electric shock, prevent fire, prevent accidental chemical spills, and minimize microbiological dangers, facility deficiencies, and equipment failures. While safety is not an aspect of laboratory certification, the evaluation should point out on an informal basis, potential safety problems observed during an on-site visit.

LABORATORY EQUIPMENT, SUPPLIES, AND MATERIALS (MINIMUM REQUIREMENTS)

The laboratory must have available or access to the items required for the total coliform membrane filter or most probable number procedures as listed below.

- *pH Meter:* Accuracy must be ±0.1 units.
- *Balances–top loader or pan:* Balance must be clean, not corroded, and be provided with appropriate weights of good quality. Balance must tare out and detect 50-mg weight accurately: this sensitivity is required for use in general media preparation of 2g or larger quantities.
- *Temperature-monitoring devices:*

  –Glass or metal thermometers must be graduated in 0.5°C increments.
  –Continuous temperature recording devices must be sensitive to within 0.5°C.
  –Liquid column of glass thermometers must have no separation.
  –A certified thermometer or one of equivalent accuracy must be available.

- *Air (or water jacketed) incubator/incubator rooms/waterbaths/aluminum block incubators:*

  –Unit must maintain internal temperature of 35.0° ± 0.5°C in area of use at maximum loading.
  –When aluminum block incubators are used, culture dishes and tubes must be snug-fitting in block.

- *Autoclave:*

  –Autoclave must be in good operating condition when observed during operational cycle or when time-temperature charts are read. Vertical autoclaves are not recommended. For most efficient operation, a double-walled autoclave constructed of stainless steel is suggested (optional).
  –Autoclave must have pressure and temperature gauges on exhaust side and an operating safety valve.
  –Autoclave must reach sterilization temperature (121°C) and be maintained during sterilization cycle: no more than 45 minutes is required for a complete cycle.
  –Depressurization must not produce air bubbles in fermentation media.

- *Hot-air oven:* Oven must be constructed to ensure a stable sterilization temperature. Its use is optional for sterilization of glass pipets, bottles, flasks, culture dishes, etc. (optional).
- *Refrigerator:* Refrigerator must hold temperature at 1° to 4.4°C (34° to 40°F).
- *Optical/counting/lighting equipment:* Low power magnification device (preferably binocular microscope with 10 to 15x) with fluorescent light source must be available for counting MF colonies. A mechanical hand tally can be used for counting colonies (optional).
- *Inoculation equipment:*

  –Loop diameter must be at least 3 mm and of 22 to 24 gauge Nichrome, chromel, or platinum-iridium wire. Single-service metal loops, disposable dry heat-sterilized hardwood applicator sticks, pre-sterilized plastic, or metal loops may be used (optional).

- *Membrane filtration equipment:*

  –Units must be made of stainless steel, glass, or autoclavable plastic. Equipment must not leak and must be uncorroded.
  –Field equipment is acceptable for coliform detection only when standard laboratory MF procedures are followed.

- *Membrane filters and pads:*

  –Membrane filters must be manufactured from cellulose ester materials, white, grid-marked, 47-mm diameter, 0.45 μm pore size. Another pore size may be used if the manufacturer gives performance data equal to or better than the 0.45-μm membrane filter.
  –Membranes and pads must be autoclavable or presterilized.

- *Laboratory glassware, plastic ware, and metal utensils:*

  –Except for disposable plastic ware, items must be resistant to effects of corrosion, high temperature, and vigorous cleaning operations. Metal utensils made of stainless steel are preferred (optional).
  –Flasks, beakers, pipets, dilution bottles, culture dishes, culture tubes, and other glassware must be of borosilicate glass and free of chips, cracks, or excessive etching. Volumetric glassware should be Class A, denoting that it meets Federal specifications and need not be calibrated before use.
  –Plastic items must be of clear, inert, nontoxic material and must retain accurate calibration marks after repeated autoclaving.

- *Culture dishes:*

  –Sterile tight or loose-lid plastic culture dishes or loose-lid glass culture dishes must be used.
  –For loose-lid culture dishes, relative humidity in the incubator must be at least 90 percent.
  –Culture dish containers must be aluminum or stainless steel; or dishes may be wrapped in heavy aluminum foil or char-resistant paper.
  –Open packs of disposable sterile culture dishes must be resealed between uses.

- *Culture tubes and closures:*

  –Culture tubes must be made of borosilicate glass or other corrosion resistant glass and must be of a sufficient size to contain the culture medium, as well as the sample portions employed, without being more than 3/4 full. It is desirable that the fermentation vial extend above the medium (optional).
  –Caps must be snug-fitting stainless steel or plastic; loose-fitting aluminum caps or screw caps are also acceptable.

- *Measuring equipment:*

  –Sterile, glass or plastic pipets must be used for measuring 10 ml or less.
  –Pipets must deliver the required volume quickly and accurately within a 2.5 percent tolerance.
  –Pipets must not be badly etched; mouthpiece or delivery tips must not be chipped; graduation marks must be legible.
  –Open packs of disposable sterile pipets must be resealed between uses.
  –Pipet containers must be aluminum or stainless steel.
  –Graduated cylinders must be used for samples larger than 10 ml; calibrated membrane filter funnel markings are permissible provided accuracy is within a 2.5 percent tolerance.

GENERAL LABORATORY PRACTICES (MINIMUM REQUIREMENTS)

Sterilization Procedures

- The following times and temperatures must be used for autoclaving materials:

Material	*Temperature/Minimum Time*
Membrane filters and pads	121°C/10 min.
Carbohydrate-containing media (lauryl tryptose, brilliant green lactose bile broth, etc.)	121°C/12-15 min.
Contaminated materials and discarded tests	121°C/30 min.
Membrane filter assemblies (wrapped), sample collection bottles (empty), individual glassware items	121°C/30 min.
Rinse water volumes of 500 ml to 1,000 ml	121°C/45 min.
Rinse water in excess of 1,000 ml	121°C/time adjusted for volume; check for sterility
Dilution water blank	121°C/30 min.

Membrane filter assembles must be sterilized between sample filtration series. A filtration series ends when 30 minutes or longer elapse between sample filtrations. At least 2 minutes of UV light or boiling water may be used on membrane filter assembly to prevent bacterial carry-over between filtrations (optional).

Dried glassware must be sterilized at a minimum of 170°C for 2 hours.

Laboratory Pure Water (Distilled, Deionized, or Other Processed Waters)

- An analyst must test the quality of the laboratory pure water or have it tested by the State or by a State-authorized laboratory.

- Only water determined as laboratory pure water (see quality control section) can be used for performing bacteriological analyses.

Although processed water may be acceptable for routine chemistry, there is a good chance that it contains enough of some constituent to be toxic or stimulatory to microorganisms (optional).

Rinse and Dilution Water

Stock buffer solution must be prepared according to "Standard Methods" using laboratory pure water adjusted to pH 7.2. Stock buffer must be autoclaved or filter-sterilized, labeled, dated, and stored at 1° to 4.4°C. The stored buffer solution must be free of turbidity.

Rinse and dilution water must be prepared by adding 1.25 ml of stock buffer solution per liter of laboratory pure water. Final pH must be 7.2 ± 0.1.

Media Preparation and Storage

The following are minimum requirements for storing and preparing media:

- Laboratories must use commercial dehydrated media for routine bacteriological procedures as quality control measures.
- Lauryl tryptose and brilliant green lactose bile broths must be prepared according to "Standard Methods"; lactose broth is not permitted.
- Dehydrated media containers must be kept tightly closed and stored in a cool, dry location. Discolored or caked dehydrated media cannot be used.
- Laboratory pure water must be used; dissolution of the media must be completed before dispensing to culture tubes or bottles.
- The membrane filter broth and agar media must be heated in a boiling water bath until completely dissolved.
- Membrane filter (MF) broths must be stored and refrigerated no longer than 96 hours. MF agar media must be stored, refrigerated and used within 2 weeks.
- Most probable number (MPN) media prepared in tubes with loose-fitting caps must be used within 1 week. If MPN media are refrigerated after sterilization, they must be incubated overnight at 35°C to confirm usability. Tubes showing growth or gas bubbles must be discarded.
- Media in screw cap containers may be held up to 3 months, provided the media are stored in the dark and evaporation is not excessive (0.5 ml per 10 ml total volume). Commercially prepared liquid and agar media supplies may be used (optional).
- Ampouled media must be stored at 1° to 4.4°C (34° to 40°F); time must be limited to manufacturer's expiration date.

METHODOLOGY (MINIMUM REQUIREMENTS)

The required procedures, which are mandatory, are described in the 13th edition of "Standard Methods": standard coliform MPN tests (p. 664-668), single step or enrichment standard total coliform membrane filter procedure (p. 679-683). Tentative methods are not acceptable. All other procedures are considered alternative analytical techniques as described in section 141.27

of the National Interim Primary Drinking Water Regulations. Application for the use of alternative methods may require acceptable comparability data.

The membrane filter procedure is preferred because it permits analysis of large sample volumes in reduced analysis time. The membranes should show good colony development over the entire surface. The golden green metallic sheen colonies should be counted and recorded as the coliform density per 100 ml of water sample. The following rules for reporting any problem with MF results must be observed:

- *Confluent growth:* Growth (with or without discrete sheen colonies) covering the entire filtration area of the membrane. Results are reported as "confluent growth per 100 ml, with (or without) coliforms," and a new-sample requested.
- *TNTC (Too numerous to count):* The total number of bacterial colonies on the membrane is too numerous (usually greater than 200 total colonies), not sufficiently distinct, or both. An accurate count cannot be made. Results are reported as "TNTC per 100 ml, with (or without) coliforms," and a new sample requested.
- *Confluent growth and TNTC:* A new sample must be requested, and the sample volumes filtered must be adjusted to apply the MF procedure; otherwise the MPN procedure must be used.
- *Confirmed MPN test on problem supplies:* If the laboratory has elected to use the MPN test on water supplies that have a continued history of confluent growth or TNTC with the MF procedure, all presumptive tubes with heavy growth without gas production should be submitted to the confirmed MPN test to check for the suppression of coliforms. A count is adjusted based upon confirmation and a new sample requested. This procedure should be carried out on one sample from each problem water supply once every 3 months.

SAMPLE COLLECTION, HANDLING, AND PRESERVATION (MINIMUM REQUIREMENTS)

When the laboratory has been delegated responsibility for sample collecting, handling, and preservation, there must be strict adherence to correct sampling procedures, complete identification of the sample, and prompt transfer of the sample to the laboratory as described in "Standard Methods," 13th edition, section 450, p. 657-660.

The sample must be representative of the potable water system. The sampling program must include examination of the finished water at selected sites that systematically cover the distribution network.

Minimum sample frequency must be that specified in the National Interim Primary Drinking Water Regulations, 40 CFR 141.21.

The collector must be trained in sampling procedures and approved by the State regulatory authority or its delegated representative.

The water tap must be sampled after maintaining a steady flow for 2 or 3 minutes to clear service line. The tap is free of aerator, strainer, hose attachment, or water purification devices.

The sample volume must be a minimum of 100 ml. The sample bottle must be filled only to the shoulder to provide space for mixing.

The sample report form must be completed immediately after collection with location, date and time of collection, chlorine residual, collector's name, and remarks.

Sample bottles must be of at least 120 ml-capacity, sterile plastic or hard glass, wide mouthed with stopper or plastic screw cap, and capable of withstanding repeated sterilization. Sodium thiosulfate (100 mg/l) is added to all sample bottles during preparation. As an example, 0.1 ml of a 10 percent solution is required in a 4-oz (120-ml) bottle.

Date and time of sample arrival must be added to the sample report form when sample is received in the laboratory.

State regulations relating to chain-of-custody, if required, must be followed in the field and in the laboratory.

Samples delivered by collectors to the laboratory must be analyzed on the day of collection.

Where it is necessary to send water samples by mail, bus, United Parcel Service, courier service, or private shipping, holding/transit time between sampling and analyses must not exceed 30 hours. When possible, samples are refrigerated during transit and during storage in the laboratory (optional).

If the laboratory is required by State regulation to examine samples after 30 hours and up to 48 hours, the laboratory must indicate that the data may be invalid because of excessive delay before sample processing. Samples arriving after 48 hours shall be refused without exception and a new sample requested. (The problem of holding time is under investigation by EPA.)

QUALITY CONTROL PROGRAM

Minimum Requirement

A written description for current laboratory quality control program must be available for review. Management, supervisors, and analysts participate in setting up the quality control program. Each participant should have a copy of the quality control program and a detailed guide of his own portion. A record on analytical quality control tests and quality control checks on media, materials, and equipment must be prepared and retained for 3 years.

Analytical Quality Control Tests for General Laboratory Practices and Methodology

Minimum and optional requirements for analytical quality control tests for general practices and methodology are:

- *Minimum requirements:*

  —At least five sheen or borderline sheen colonies must be verified from each membrane containing five or more such colonies. Counts must be adjusted based on verification. The verification procedure must be conducted by transferring growth from colonies into lauryl tryptose broth (LTB) tubes and then transferring growth from gas-positive LTB cultures to brilliant green lactose bile (BGLB) tubes. Colonies must not be transferred exclusively to BGLB because of the lower recovery of stressed coliforms in this more selective medium. However, colonies may be transferred to LTB and BGLB simultaneously. Negative LTB tubes must be reincubated a second day and confirmed if gas is produced. It is desirable to verify all sheen and borderline sheen colonies (optional).

  —A start and finish MF control test (rinse water, medium, and supplies) must be conducted for each filtration series. If sterile controls indicate contamination, all data on samples affected must be rejected and a request made for immediate resampling of those waters involved in the laboratory error.

—The MPN test must be carried to completion, except for gram staining, on 10 percent of positive confirmed samples. If no positive tubes result from potable water samples, the completed test except for gram staining must be performed quarterly on at least one positive source water.

—Laboratory pure water must be analyzed annually by the test for bactericidal properties for distilled water ("Standard Methods," 13th edition, p. 646). Only satisfactorily tested water is permissible in preparing media, reagents, rinse, and dilution water. If the tests do not meet the requirements, corrective action must be taken and the water retested.

—Laboratory pure water must be analyzed monthly for conductance, pH, chlorine residual, and standard plate count. If tests exceed requirements, corrective action must be taken and the water retested.

—Laboratory pure water must not be in contact with heavy metals. It must be analyzed initially and annually thereafter for trace metals (especially Pb, Cd, Cr, Cu, Ni, and Zn). If tests do not meet the requirements, corrective action must be taken and the water retested.

—Standard plate count procedure must be performed as described in "Standard Methods," 13th edition, p. 660-662. Plates must be incubated at 35° ± 0.5°C for 48 hours.

—Requirements for laboratory pure water:

pH	**5.5 - 7.5**
Conductivity	**Greater than 0.2 megohm as resistivity or less than 5.0 micromhos/cm at 25°C**
Trace metals:	
A single metal	**Not greater than 0.05 mg/l**
Total metals	**Equal to or less than 1.0 mg/l**
Test for bactericidal properties of distilled water ("Standard Methods," 13th edition, p. 646)	**0.8 - 3.0**
Free chlorine residual	**0.0**
Standard plate count	**Less than 10,000/ml**

—Laboratory must analyze one quality control sample per year (when available) for parameter(s) measured.

—Laboratory must satisfactorily analyze one unknown performance sample per year (when available) for parameter(s) measured.

- *Optional requirements:*

—Duplicate analyses should be run on known positive samples at a minimum frequency of one per month. The duplicates may be run as a split sample by more than one analyst, with each split being a 50-ml sample.

—Water plant laboratories should examine a minimum of one polluted water source per month in addition to the required number of distribution samples.

—If there is more than one analyst in laboratory, at least once per month each analyst should count the sheen colonies on a membrane from a polluted water source. Colonies on the membrane should be verified and the analysts' counts compared to the verified count.

—A minimum number of the official water supply samples required for each system should be analyzed by the State laboratory. For example, systems that are required to have less than 10 samples examined per month should submit one additional sample to a State authorized laboratory. Water systems with sample requirements above 10 per month would submit two additional samples to a State authorized laboratory.

Quality Control Checks of Laboratory Media, Equipment, and Supplies

Minimum and optional requirements for quality control checks of laboratory media, equipment, and supplies are:

- *Minimum requirements:*

—pH meter must be clean and calibrated each use period with pH 7.0 standard buffer. Buffer aliquot must be used only once. Commercial buffer solutions must be dated on initial use.

—Balances (top loader or pan) must be calibrated annually.

—Glass thermometers or continuous recording devices for incubators must be checked yearly and metal thermometers quarterly (or at more frequent intervals when necessary) against a certified thermometer or one of equivalent accuracy.

—Temperature in air (or water jacketed) incubator/incubator room/waterbaths/aluminum block incubators must be recorded continuously or recorded daily from in-place thermometer(s) immersed in liquid and placed on shelves in use.

—Date, time, and temperature must be recorded continuously or recorded for each sterilization cycle of the autoclave.

—Hot air oven must be equipped with a thermometer calibrated in the range of 170°C or with a temperature recording device. Records must be maintained showing date, time, and temperature of each sterilization cycle. It is desirable to place the thermometer bulb in sand and to avoid overcrowding (optional).

—Membrane filters used must be those recommended by the manufacturer for water analysis. The recommendation must be based on data relating to ink toxicity, recovery, retention, and absence of growth-promoting substances.

—Washing processes must provide clean glassware with no stains or spotting. With initial use of a detergent or washing product and whenever a different washing product is used, the rinsing process must demonstrate that it provides glassware free of toxic material by the inhibitory residue test ("Standard Methods," 13th edition, p. 643).

—At least one bottle per batch of sterilized sample bottles must be checked by adding approximately 25 ml of sterile LTB broth to each bottle. It must be incubated at 35 ± 0.5°C for 24 hours and checked for growth.

—Service contracts or approved internal protocols must be maintained on balances, autoclave, water still, etc., and the service records entered in a log book.

–Records must be available for inspection on batches of sterilized media showing lot numbers, date, sterilization time-temperature, final pH, and technician's name.

- *Optional requirements:*

–Positive and negative cultures should be used, and testing should be carried out to determine recovery and performance compared to a previous acceptable lot of medium.
–Media should be ordered on a basis of 12-month needs. Bottles should be dated on receipt and when opened initially. Except for large volume uses, media should be purchased in 1/4-lb bottles. Bottles of media should be used within 6 months after opening; however, in no case should opened media be used after one year. Opened bottles should be stored in a desiccator to extend storage time beyond 6 months. Shelf life of unopened bottles is 2 years.
–Testing should be carried out in media and membranes to determine recovery and performance compared to previous acceptable lot.
–Lot number of membrane filters and date of receipt should be recorded.
–Heat sensitive tapes and spore strips or ampoules should be used during sterilization. Maximum registering thermometer is recommended.

DATA REPORTING (MINIMUM REQUIREMENTS)

Where the laboratory has the responsibility for sample collections, the sample collector should complete a sample report form immediately after each sample is taken. The information on the form includes sample identification number, sample collector's name, time and date of collection, arrival time and date in the laboratory, direct count, MF verified count, MPN completed count, analyst's name, and other special information.

Results should be calculated and entered on the sample report form to be forwarded. A careful check should be made to verify that each result was entered correctly from the bench sheet and initialed by the analyst.

All results should be reported immediately to the proper authority.

Positive results are reported as preliminary without waiting for MF verification or MPN completion. After MF verification and/or MPN completion, the adjusted counts should be reported.

A copy of the sample report form should be retained either by the laboratory or State program for 3 years. If results are entered into a computer storage system, a printout of the data should be returned to the laboratory for verification with bench sheets.

ACTION RESPONSE TO LABORATORY RESULTS (MINIMUM REQUIREMENT)

When action response is a designated laboratory responsibility, the proper authorities should be promptly notified on unsatisfactory sample results, and a request should be made for resampling from the same sampling point.

SAMPLE FORMS FOR ON-SITE EVALUATION OF LABORATORIES INVOLVED IN ANALYSIS OF PUBLIC WATER SUPPLIES–MICROBIOLOGY

LABORATORY: ______________________________

STREET: ______________________________

CITY: ______________________________ STATE: ______________________________

TELEPHONE NUMBER: ______________________________

SURVEY BY: ______________________________

AFFILIATION: ______________________________

DATE: ______________________________

CODES FOR MARKING ON-SITE EVALUATION FORMS:

S–Satisfactory
U–Unsatisfactory
NA–Not Applicable

Laboratory ______________________ Evaluator ______________________

Location ______________________ Date ______________________

PERSONNEL

The form dealing with personnel can be found on the following page.

LABORATORY FACILITIES

Space in laboratory and preparation room is adequate for needs during peak work periods (200 ft^2 and 6 linear ft of usable bench space per analyst).

Facilities are clean, with adequate lighting (100 ft-candles) and air conditioning.

NOTE: Material on pages 53-65, except where indicated, are minimum requirements.

PERSONNEL

Laboratory ____________________

Location ____________________

Date ____________

Evaluator ____________________

Position/title	Name	Academic training				Present specialty	Experience (years/area)
		HS	BA/BS	MA/MS	Ph.D.		
Laboratory director							
Supervisor/consultant							
Professionals (note discipline)							
Technician/analyst							

Laboratory ______________________ Evaluator ______________

Location ______________________ Date ______________

LABORATORY EQUIPMENT, SUPPLIES, AND MATERIALS

1. pH Meter

Manufacturer ______________________ Model ______________

Clean, calibrated to 0.1 pH units each use period; record maintained . . . . ____

Aliquot of standard pH 7.0 buffer used only once . . . . ____

2. Balance—Top Loader or Pan

Manufacturer ______________________ Model ______________

Clean. Detects a 50-mg weight accurately (for a general media preparation of >2-g quantities) . . . . ____

Good quality weights in clean condition . . . . ____

3. Temperature-Monitoring Devices

Accuracy checked annually against a certified thermometer or one of equivalent accuracy . . . . ____

Legible graduations in 0.5°C-increments . . . . ____

No separation in liquid column . . . . ____

4. Incubator or Incubator Room

Manufacturer ______________________ Model ______________

Sufficient size for daily work load . . . . ____

Uniform temperature maintained on shelves in all areas used (35.0° ± 0.5°C) . . . . . . ____

Calibrated thermometer with bulb immersed in liquid and located on shelves in use. . ____

Temperature recorded daily or recording thermometer sensitive to ± 0.5°C . . . . ____

5. Autoclave

Manufacturer ______________________ Model ______________

Reaches sterilization temperature (121°C), maintains 121°C during sterilization cycle, and requires no more than 45 min for a complete cycle. . . . . ____

Pressure and temperature gauges on exhaust side and an operating safety valve . . . . . ____

No air bubbles produced in fermentation vials during depressurization. . . . . ____

Record maintained on time and temperature for each sterilization cycle . . . . ____

6. Hot-Air Oven

Manufacturer ______________________ Model ______________

Operates at a minimum of 170°C . . . . ____
Thermometer inserted or oven equipped with temperature-recording thermometer device. . . . . ____
Time and temperature record maintained for each sterilization cycle . . . . ____
Thermometer bulb in sand (optional)

7. Refrigerator

Temperature maintained at 1° to 4.4°C (34° to 40°F) . . . . ____

8. Optical Equipment

Low power magnification device (preferably binocular microscope with 10 to 15X) with fluorescent light source for counting MF colonies . . . . ____
Colonies counted with a mechanical hand tally (optional).

9. Inoculation Equipment

Sterilized loops of at least 3-mm diameter, 22 to 24 gauge Nichrome, Chromel, or platinum-iridium wire. . . . . ____

or

Disposable dry heat-sterilized hardwood applicator sticks or presterilized loops . . . . . ____

10. Membrane Filtration Equipment

Manufacturer ______________________ Model ______________

Made of stainless steel, glass, or autoclavable plastic . . . . ____
Nonleaking and uncorroded . . . . ____

11. Membrane Filters and Pads

Manufacturer ______________________ Model ______________

Filters recommended by manufacturer for water analyses. . . . . ____
Filters and pads presterilized or autoclavable . . . . ____

12. Glass, Plastic, and Metal Utensils for Media Preparation

Washing process provides glassware free of toxic residue as demonstrated by the inhibitory residue test . . . . ____
Glass items of borosilicate, free of chips and cracks . . . . ____
Utensils clean and free from foreign residues or dried medium . . . . ____
Plastic items clear with visible graduations . . . . ____

13. Sample Bottles

Wide-mouth hard glass bottles; stoppered or plastic screw-capped; capacity at least 120 ml . . . . ——
Glass-stoppered bottles with tops covered with aluminum foil or kraft paper . . . . ——
Screw-caps have leakproof nontoxic liners that can withstand repeated sterilization (30 min at 121°C) . . . . ——
Sterile sample bottles contain 10 mg of dechlorinating agent per 100 ml of sample . . ——

14. Pipets

Brand ______________________ Type ______________

Sterile; glass or plastic; with a 2.5 percent tolerance . . . . ——
Tips unbroken; graduations distinctly marked. . . . . ——

15. Pipet Containers

Aluminum, stainless steel . . . . ——
Pipets wrapped in quality kraft paper (char-resistant) . . . . ——
Open packs of disposable sterile pipets resealed between uses. . . . . ——

16. Culture Dishes

Brand ______________________ Type ______________

Sterile plastic or glass . . . . ——
Open packs of disposable sterile plastic dishes resealed between uses . . . . ——
Dishes are in containers of aluminum or stainless-steel with covers or are wrapped with heavy aluminum foil or char-resistant paper. . . . . ——

17. Culture Tubes and Closures

Sufficient size to contain sterile medium and sample without danger of spillage. . . . . ——
Metal or plastic caps; plastic plugs. . . . . ——
Borosilicate glass or other corrosion-resistant glass . . . . ——

Laboratory ______________________________ Evaluator ______________

Location ______________________________ Date ______________

GENERAL LABORATORY PRACTICES

1. Sterilization Procedures

Satisfactory sterilization procedures and/or records . . . . ____
Tube broth media and reagents sterilized at 121°C 12 to 15 min . . . . ____
Tubes and flasks packed loosely in baskets or racks for uniform heating and cooling . ____
MF presterilized or autoclaved at 121°C for 10 min with fast exhaust . . . . ____
MF assemblies and empty sample bottles sterilized at 121°C for 30 min. . . . . ____
MF assemblies sterilized between sample filtration series . . . . ____
Rinse water volumes of 500 to 1,000 ml sterilized at 121°C
for 45 min . . . . ____
Dilution water blanks autoclaved at 121°C for 30 min . . . . ____
Wire loops, needles, and forceps sterilized. . . . . ____
Total exposure of MPN media to heat not over 45 min. . . . . ____
Timing for sterilization begins when autoclave reaches 121°C. . . . . ____
Individual glassware items autoclaved at 121°C for 30 min. ____
Individual dry glassware items sterilized 2 hours at 170°C (dry heat) . . . . ____
Pipets, culture dishes, and inoculating loops in boxes sterilized at 170°C for 2 hours . ____
MPN media removed and cooled as soon as possible after sterilization and stored in
cool dark place (optional)
UV light or boiling water for at least 2 min may be used on membrane filter assemblies to reduce bacterial carry-over between each filtration (optional)

2. Laboratory Pure Water

Only laboratory pure water used in preparing media, reagents, rinse water, and dilution water. . . . . ____
Laboratory pure water not in contact with heavy metals . . . . ____

Source: Laboratory-prepared ______________ Purchased ______________

If laboratory-prepared:
Still manufacturer ______________________________
Deionizer manufacturer ______________________________
Record of recharge frequency ______________________________

Production rate and quality adequate for laboratory needs. ____
Inspected, repaired, cleaned by service contract or in-house service . . . . ____

a. Chemical quality control

Record of satisfactory annual analyses for trace metals
A single metal not greater than 0.05 mg/l . . . . ____

Total metals: equal to or less than 1.0 mg/1. . . . . ——

Record of monthly analyses of laboratory pure water

Conductance: >0.2 megohm resistivity or <5.0 microhmos/cm . . . . ——
pH: 5.5-7.5 . . . . ——
Standard plate count: <10,000/ml. . . . . ——
Free chlorine residual: 0.0 . . . . ——

b. Microbiological quality control

Test for bactericidal properties of distilled water (0.8-3.0) performed at least annually . . . . ——

Testing laboratory ____________ Date ____________ Ratio ______________

3. Rinse and Dilution Water

Stock buffer solution prepared according to "Standard Methods," 13th edition. . . . . ——
Stock buffer solution adjusted to pH 7.2 . . . . ——
Stock buffer autoclaved at 121°C, stored at 1° to 4.4°C (34° to 40°F) or filter sterilized. . . . . ——
Stock buffer labeled and dated . . . . ——
Stock potassium phosphate buffer solution (1.25 ml) added per liter distilled water for rinse and dilution water . . . . ——
Final pH 7.2 ± 0.1. . . . . ——

4. Media

Dehydrated media bottle kept tightly closed and protected from dust and excessive humidity in storage areas . . . . ——
Dehydrated media not used if discolored or caked . . . . ——
Laboratory pure water used in media preparation. . . . . ——
Dissolution of media complete before dispensing to culture tubes or bottles . . . . ——
MPN tube media with loose-fitting caps used in less than 1 week . . . . ——
Tube media in screw-capped tubes held no longer than 3 months . . . . ——
Ampouled media stored at 1° to 4.4°C and time limited to manufacturer's expiration date. . . . . ——
Media stored at low temperatures are incubated overnight prior to use and tubes with air bubbles discarded . . . . ——
Media protected from sunlight . . . . ——
MF media stored in refrigerator; broth medium used within 96 hours, agar within two weeks if prepared in tight-fitting dishes. . . . . ——

5. Lauryl Tryptose Broth

Manufacturer ______________________ Lot No. ______________

Single strength composition, 35.6 g per liter pure water . . . . ____
Single strength pH 6.8 ± 0.2; double strength pH 6.7 ± 0.2. ____
Not less than 10 ml per tube . . . . ____
Media made to result in single strength after addition of sample portions . . . . ____

6. Brilliant Green Lactose Bile Broth

Manufacturer ______________________ Lot No. ______________

Medium composition 40 g per liter pure water . . . . ____
Final pH 7.2 ± 0.2 . . . . ____

7. M-Endo Media

Manufacturer ______________________ Lot No. ______________

Medium composition 48.0 g per liter pure water optionally 15 g agar added/l . . . . . . ____
Reconstituted in laboratory pure water containing 2 percent ethanol (not denatured) . . . . ____
Final pH 7.2 ± 0.2 . . . . ____
Medium held in boiling water bath until completely dissolved . . . . ____

8. Standard Plate Count Agar

Correct composition, sterile and pH 7.0 ± 0.2 . . . . ____
Sterile medium not remelted a second time after sterilization. . . . . ____
Culture dishes incubated 48 hours at 35° ± 0.5°C. . . . . ____
No more than 1.0 ml or less than 0.1 ml sample plated (sample or dilution). . . . . ____
Liquefied agar, 10 ml or more; medium temperature between 44° to 46°C. . . . . ____
Melted medium stored no longer than 3 hours before use . . . . ____
Only plates with between 30 to 300 colonies counted; when 1 ml of undiluted sample is plated, colony density may be less than 30 . . . . ____
Only two significant figures recorded and calculated as standard plate count/ 1.0 ml . . . . ____

Laboratory ____________________________ Evaluator ____________________

Location ______________________________ Date ________________________

METHODOLOGY

Methodology specified in "Standard Methods," 13th edition, or EPA manual . . . . ____

M-Endo broth, M-Endo agar, or Les Endo agar used in a single step procedure . . . . ____

In two-step Les M-Endo procedure, MF incubated on lauryl-tryptose-broth-saturated absorbent pad for 1.5 to 2 hours at $3.5^\circ \pm 0.5^\circ$C; then on M-Endo broth or Les Endo agar for 20 to 22 hours at $35^\circ \pm 0.5^\circ$C . . . . ____

1. Total Coliform Membrane Filter Procedure

Samples containing excessive bacterial populations (greater than 200), confluency, or turbidity retested by the MPN procedure. . . . . ____

Filtration assembly sterile at start of each series . . . . ____

Absorbent pads saturated with medium, excess discarded; or 4.0 ml of agar medium can be used per culture dish instead of a pad . . . . ____

Sample shaken vigorously immediately before test . . . . ____

Test sample portions measured and not less than 100 ml . . . . ____

Funnel rinsed at least twice with 20- to 30-ml portions of sterile buffered water . . . . ____

MF removed with sterile forceps grasping ~~area~~ outside effective filtering area. . . . . ____

MF rolled onto medium pad or agar so air bubbles are not trapped. . . . . ____

2. Incubation of Membrane Filter Cultures

Total incubation time 22 to 24 hours at $35^\circ \pm 0.5^\circ$C . . . . ____

Incubated in tight-fitting culture dishes or loose-fitting dishes incubated in high relative humidity chambers . . . . ____

3. Membrane Filter Colony Counting

Samples repeated when coliforms are "TNTC" or colony growth is confluent, possibly obscuring coliform development and/or detection . . . . ____

Total coliform count calculated in density per 100 ml . . . . ____

Samples containing five or more coliforms per 100 ml are resampled and tested . . . . ____

Low power magnification device with fluorescent light positioned for maximum sheen visibility . . . . ____

4. Verification of Total Coliform Colonies

All typical coliform (sheen) colonies or at least five randomly selected sheen colonies verified in lauryl tryptose broth and BGLB. . . . . ____

Counts adjusted based on verification . . . . ____

All atypical coliform (borderline sheen) colonies or at least five randomly-selected colonies verified in LTB and BGLB . . . . ____

Counts adjusted based on verification . . . . ____

5. MF Field Equipment

Manufacturer ______________________________ Model ______________

Only standard laboratory MF procedures adapted to field application . . . . ____

6. Total Coliform Most Probable Number Procedure

a. Presumptive Test

Five standard portions, either 10 or 100 ml . . . . ____
Sample shaken vigorously immediately before test . . . . ____
Tubes incubated at 35° ± 0.5°C for 24 ± 2 hours . . . . ____
Examined for gas (any gas bubble indicates positive test) . . . . ____
Tubes that are gas-positive within 24 hours submitted promptly to confirm test . ____
Negative tubes returned to incubator and examined for gas within 48 ± 3 hours; positives submitted to confirm test . . . . ____
Public water supply samples with heavy growth and no gas production confirmed for presence of suppressed coliforms. . . . . ____
Adjusted count reported based upon confirmation . . . . ____
Adequate test labeling and tube dilution coding (optional)

b. Confirmed Test

Presumptive positive tube gently shaken or mixed by rotating . . . . ____
One loopful or one dip of applicator transferred from presumptive tube to BGLB. . . . . ____
Incubated at 35°C ± 0.5°; checked within 24 hours ± 2 hours for gas production. ____
Positive confirmed tube results recorded; negative tubes reincubated and read within 48 ± 3 hours . . . . ____
Unsatisfactory sample defined as three or more positive confirmed tubes. . . . . . . ____

c. Completed Test

Applied to 10 percent of all positive samples each quarter . . . . ____
Applied to all positive confirmed tubes in each test completed. . . . . ____
Positive confirmed tubes streaked on EMB plates for colony isolation . . . . ____
Plates adequately streaked to obtain discrete colonies. . . . . ____
Incubated at 35° ± 0.5°C for 24 ± 2 hours . . . . ____
Typical nucleated colonies, with or without sheen, on EMB plates selected for completed test identification . . . . ____
If typical colonies absent, atypical colonies selected for completed test identification . . . . ____
If no colonies or only colorless colonies appear, confirmed test for that particular tube considered negative . . . . ____
An isolated typical colony or two atypical colonies transferred to lauryl tryptose broth . . . . ____
Incubated at 35° ± 0.5°C; checked for gas within 48 ± 3 hours. . . . . ____
Cultures producing gas in lauryl tryptose broth within 48 ± 3 hours are considered coliforms . . . . ____

Laboratory ______________________ Evaluator ______________

Location ______________________ Date ______________

SAMPLE COLLECTION, HANDLING, AND PRESERVATION

Representative samples of potable water distribution system . . . . ____
Minimal sampling frequency as specified in the National Interim Primary Drinking Water Regulations . . . . ____
Sample collector trained and approved as required by State regulatory authority or its delegated representative . . . . ____

1. Sample Bottles

Sterile sample bottles of at least 120 ml; able to withstand repeated sterilization . . . . ____
Ample air space remains after sample collected to allow for adequate mixing. . . . . . . ____
Sodium thiosulfate, 100 mg/l, added to sample bottle before sterilization . . . . ____

2. Sampling

Sample collected after maintaining a steady flow for 2 to 3 min to clear service line. . . . . ____
Tap free of aerator, strainer, hose attachment, water purification, or other devices . . ____
Samples refrigerated when possible during transit and storage periods in the laboratory (optional)

3. Sample Identification

Sample identified immediately after collection . . . . ____
Identification includes, water source, location, time and date of collection, and collector's name; insufficiently identified samples discarded. . . . . ____
Chlorine residual where applicable . . . . ____

4. Sample Transit Time

Transit time for potable water samples sent by mail or commercial transportation, not in excess of 30 hours . . . . ____
No sample processed after 48-hour transit/storage . . . . ____
Samples delivered to laboratory by collectors examined the day of collection . . . . . . ____
Data marked as questionable on samples analyzed after 30 hours . . . . ____

5. Sample Receipt in Laboratory

Sample logged in when received in laboratory, including date and time of arrival and analysis. . . . . ____
Chain-of-custody procedures required by State regulations followed . . . . ____

Laboratory________________________ Evaluator________________

Location__________________________ Date___________________

QUALITY CONTROL

A written laboratory quality control program is available for review. . . . . ____

1. Analytical Quality Control

A record containing results of analytical control tests available for review . . . . ____

a. Verification of MF Colonies

At least five coliforms verified from each positive sample . . . . ____
Sheen colonies in mixed confluent growth reported and verified (optional)

b. Negative Coliform Controls

A start and finish MF control test (rinse water, medium, and supplies) run with each filtration series . . . . ____
When controls indicate contamination occurred, all data on affected samples rejected and resampling requested . . . . ____

c. Total Coliform Confirmed Test

Presumptive tubes with heavy growth but no gas production submitted to confirmed test to check for suppression of coliforms. Confirmation procedure carried out every 3 months on one sample from each problem water supply . . . ____

d. Duplicate analyses (optional)

Duplicate analyses run on positive polluted samples not to exceed 10 percent but a minimum of one per month (optional) . . . . ____

e. Positive Control Samples (optional)

One positive control sample (polluted water) run each month (optional)

f. Colony Counting (If More Than One Analyst in Laboratory) (optional)

Two or more analysts count sheen colonies; all colonies are verified; analysts' counts compared to verified counts; procedure is carried out at least once per month (optional)

g. Check Analyses by State Laboratories (optional)

A minimum of samples, proportional to the local laboratory work load, processed by State laboratory (see criteria for recommendations) (optional)

2. Quality Control of Equipment, Supplies and Media

a. Records

Satisfactory records containing complete quality control checks on equipment, supplies, and media available for inspection . . . . ———

b. Equipment and Supplies

Service contracts or approved internal protocol maintained on balance, autoclave, water still, etc.; service records entered in a log book . . . . ———

Glass thermometers calibrated annually against a certified thermometer; metal thermometers checked quarterly . . . . ———

pH Meters standardized with pH 7.0 buffer. . . . . ———

Laboratory pure water analyzed as described in criteria . . . . ———

Lot numbers and dates of receipt of membrane filters recorded (optional)

Heat-sensitive tapes and/or spare strips/ampoules used during sterilization (optional)

c. Media Quality Control

Laboratory chemicals of Analytical Reagent Grade . . . . ———

Dyes certified for bacteriological use. . . . . ———

pH checked and recorded on each batch of medium after preparation and sterilization . . . . ———

Causes for deviations beyond ± 0.2 pH units specified . . . . ———

Media ordered on a basis of 12-month need; purchased in 1/4-lb. quantities, except those used in large amounts (optional)

Bottles dated on receipt and when opened (optional)

Opened bottles of routinely used media discarded within 6 months (if stored in desiccator storage may be extended) (optional)

Shelf life of unopened bottles not in excess of 2 years (optional)

New lots of media quality tested against satisfactory lot using natural water samples (optional)

Laboratory___________________________________ Evaluator____________________

Location_____________________________________ Date________________________

DATA REPORTING

Sample information and laboratory data fully recorded . . . . ____
Direct MF counts and/or confirmed MPN results reported promptly . . . . ____
After MF verification and/or MPN completion, adjusted counts reported . . . . ____
One copy of report form retained in laboratory or by State program for 3 years . . . . ____
Test results assembled and available for inspection (optional)

Laboratory______________________________ Evaluator______________

Location________________________________ Date_________________

ACTION RESPONSE TO LABORATORY RESULTS

Unsatisfactory test results given action response and resampled as defined in National Interim Primary Drinking Water Regulations . . . . ____

State and responsible local authority notified within 48 hours after check samples confirm coliform occurrence . . . . ____

All data reported to State and local authorities within 40 days. . . . . ____

APPENDIX C

BIBLIOGRAPHY

1. Kittrell, R. W., 1969. A Practical Guide to Water Quality Standards of Streams, Publ. No. CWR-5. U.S. Department of the Interior, FWPCA, U.S. Government Printing Office, Washington, DC.

2. American Public Health Association, 1975. Standard Methods for the Examination of Water and Wastewater, (14th ed.) Washington, DC.

3. Society of American Bacteriologists, 1957. Manual of Microbiological Methods, McGraw-Hill Book Company, Inc., New York, NY.

4. Geldreich, E. E., 1975. Handbook for Evaluating Water Bacteriological Laboratories, (2nd ed.) U.S. Environmental Protection Agency, Cincinnati, OH. EPA-670/9–75–006.

5. Scarpino, P. V., 1971. Bacterial and viral analysis of water and wastewater. Chapter 12. *In:* Water and Water Pollution Handbook. Volume 2. (L. L. Ciaccio, ed.) Marcel Dekker, Inc., New York, NY. 2:639.

6. Mitchell, Ralph (ed.), 1972. Water Pollution Microbiology. John Wiley and Sons, Inc., New York, NY.

7. Current Practices in Water Microbiology, 1973. U.S. Environmental Protection Agency, Office of Water Program Operations, National Training and Operational Technology Center, Cincinnati, OH.

8. Greeson, P.E., T. A. Ehlke, G. A. Irwin, B. W. Lium, and K. V. Slack, 1977. Techniques of Water Resources Investigations of the United States Geological Survey. Chapter A4, Methods for Collection and Analysis of Aquatic Biological and Microbiological Samples. Book 5, Laboratory Analysis. U.S. Dept., of the Interior, Superintendent of Documents, U.S. Government Printing Office, Washington, D.C. 20402.

9. Bordner, R. H., C. F. Frith and J. A. Winter, Eds., 1977. Proceedings of the Symposium on Recovery of Indicator Organisms Employing Membrane Filters. U.S. Environmental Protection Agency, EPA-60019–77–024, EMSL-Cincinnati, Cincinnati, OH 45268.

INDEX

TECHNICAL REPORT DATA
(Please read Instructions on the reverse before completing)

1. REPORT NO. EPA-600/8-78-017	2.	3. RECIPIENT'S ACCESSION NO.
4. TITLE AND SUBTITLE MICROBIOLOGICAL METHODS FOR MONITORING THE ENVIRONMENT Water and Wastes		5. REPORT DATE December 1978
		6. PERFORMING ORGANIZATION CODE
7. AUTHOR(S) Editors: Robert H. Bordner and John A. Winter, EMSL-Cincinnati; Pasquale Scarpino, University of Cincinnati		8. PERFORMING ORGANIZATION REPORT NO.
9. PERFORMING ORGANIZATION NAME AND ADDRESS SAME AS BELOW		10. PROGRAM ELEMENT NO. 1HD 621
		11. CONTRACT/GRANT NO. 68-03-0431
12. SPONSORING AGENCY NAME AND ADDRESS Environmental Monitoring and Support Lab. - Cinn, OH Office of Research and Development U.S. Environmental Protection Agency Cincinnati, OH 45268		13. TYPE OF REPORT AND PERIOD COVERED Final
		14. SPONSORING AGENCY CODE EPA/600/06

15. SUPPLEMENTARY NOTES
Project Officer: John Winter, EMSL, Cincinnati

16. ABSTRACT

This first EPA manual contains uniform laboratory and field methods **for** microbiological analyses of waters and wastewaters, and is recommended in enforcement, monitoring and research activities. The procedures are prepared in detailed, stepwise form for the bench worker. The manual covers coliform, fecal coliform, fecal streptococci, Salmonella, actinomycetes and Standard Plate Count organisms with the necessary support sections on sampling, equipment, media, basic techniques, safety, and quality assurance.

17. KEY WORDS AND DOCUMENT ANALYSIS

a. DESCRIPTORS		b. IDENTIFIERS/OPEN ENDED TERMS	c. COSATI Field/Group
Aquatic microbiology	Methodology	Analytical procedures	
Coliform bacteria	Microbiology	Standard plate count	
Enterobacteriaceae	Surface waters	Total coliforms	
Potable water	Statistics	Fecal coliforms	06/M
Public law	Water analysis	Pathogens	
Quality assurance	Water pollution	Indicator organisms	
Safety	Water quality	Fecal streptococci	
	Waste water		

18. DISTRIBUTION STATEMENT RELEASE TO PUBLIC	19. SECURITY CLASS *(This Report)* Unclassified	21. NO. OF PAGES 354
	20. SECURITY CLASS *(This page)* Unclassified	22. PRICE

EPA Form 2220-1 (9-73)

☆U.S. GOVERNMENT PRINTING OFFICE: 1991- 548-187/ 40541

Made in the USA
San Bernardino, CA
06 March 2016